Ergebnisse der Mathematik
und ihrer Grenzgebiete

3. Folge

Volume 33

A Series of Modern Surveys
in Mathematics

Springer

Berlin
Heidelberg
New York
Barcelona
Budapest
Hong Kong
London
Milan
Paris
Santa Clara
Singapore
Tokyo

Carlos Andradas
Ludwig Bröcker
Jesús M. Ruiz

Constructible Sets in Real Geometry

Springer

Carlos Andradas
Departamento de Algebra
Universidad Complutense de Madrid
E-28040 Madrid
Spain
e-mail: carlos@sunal1.mat.ucm.es

Ludwig Bröcker
Mathematisches Institut
Universität Münster
Einsteinstraße 62
D-48149 Münster
Germany
e-mail: broe@escher.uni-muenster.de

Jesús M. Ruiz
Departamento de Geometría y Topología
Universidad Complutense de Madrid
E-28040 Madrid
Spain
e-mail: jesusr@eucmax.sim.ucm.es

```
Library of Congress Cataloging-in-Publication Data
Andradas, Carlos.
   Constructible sets in real geometry / Carlos Andradas, Ludwig
Bröcker, Jesús M. Ruiz.
      p.   cm.  -- (Ergebnisse der Mathematik und ihrer Grenzgebiete ;
3. Folge, Bd. 33)
   Includes bibliographical references and index.
   ISBN 3-540-60451-0 (alk. paper)
   1. Constructibility (Set theory)  2. Geometry, Algebraic.
I. Bröcker, Ludwig, 1940-   . II. Ruiz, Jesús M.  III. Title.
IV. Series.
QA248.A67  1996
516'.13--dc20                                      95-51853
                                                       CIP
```

Mathematics Subject Classification (1991): 12Jxx, 13Fxx, 13Jxx, 14Pxx

ISBN-13: 978-3-642-80026-9   e-ISBN-13: 978-3-642-80024-5
DOI: 10.1007/978-3-642-80024-5

SPIN 10518835   41/3143 - 5 4 3 2 1 0 - Printed on acid-free paper

Preface

The plan to write this book was laid out in April 1987 at Oberwolfach, during the Conference "Reelle Algebraische Geometrie". Afterwards we met at various conferences and seminars in Luminy, Madrid, Münster, Oberwolfach, Segovia, Soesterberg, Trento and La Turballe. We would like to thank the organizers and the institutions which supported these meetings. With pleasure we remember the special year on Real Algebraic Geometry and Quadratic Forms (Ragsquad) in Berkeley 1990/91 where an essential part of this book was written. Thanks to T.Y. Lam and R. Robson.

We are indepted to our Departments: Universidad Complutense de Madrid and Westfälische Wilhelms-Universität Münster, as well as the D.A.A.D. and the D.G.I.C.Y.T..

It is not possible to mention here all colleagues and friends who showed permanent interest in the project. They encouraged us to continue and complete this work. In particular, we are obliged to Jacek Bochnak, Mike Buchner, Michel Coste and Claus Scheiderer for proofreading and helpful suggestions. While the work was still in progress, Manfred Knebusch used parts of our manuscript for a course on Real Algebraic Geometry. His experience convinced us that it was worth pursuing a fully abstract approach. We were also in permanent contact with Murray Marshall, and the reader will recognize the mutual influence of ideas.

We are also indebted to Professor Reinhold Remmert and to the Springer-Verlag for publishing the book in this series.

Finally, we thank Erika Becker for her care in converting many pages of handwritten mathematics into beautiful TeX.

Madrid, Münster, Majadahonda   *C. Andradas, L. Bröcker, J. M. Ruiz*
March 1996

Contents

Introduction

Let A be the ring of polynomials in n indeterminates over $\mathbb{R}$. Then any subset S of $\mathbb{R}^n$ which is the solution set of a polynomial system $f_1(x) > 0, \ldots, f_k(x) > 0$ is also the solution set of a system of n inequalities $g_1(x) > 0, \ldots, g_n(x) > 0$, no matter how big k is. This observation, made about twelve years ago for $n \leq 3$ and proved in full generality five years later is the starting point of the present book.

Many similar problems can be raised: Firstly one may generalize the type of sets S which appear as solution of polynomial systems, and ask questions of the following type. What happens if the strict inequalities are replaced by non-strict ones? What if we consider arbitrary semialgebraic sets S, that is, sets described by finitely many polynomial inequalities where unions are also involved? Is it still possible to bound by a function depending only on n the number of polynomials needed for the description? How can one recognize those semialgebraic sets which are given by pure intersections of strict or non-strict inequalities, called basic open and basic closed respectively? If S is basic open, is its closure basic closed?

Secondly, one may change the underlying space $\mathbb{R}^n$ and the ring A, and ask corresponding questions in the new situation. For instance, take instead of $\mathbb{R}^n$ an algebraic subset V of $\mathbb{R}^n$, or replace $\mathbb{R}$ by an arbitrary real closed field R or even by an ordered field K with real closure R. The corresponding rings are then $\mathbb{R}[V]$, $R[V]$ and $K[V]$ respectively. One can also take a real analytic set together with its ring of analytic functions, as well as an analytic set germ with its ring of analytic function germs. Or, more generally, one may take for A any commutative ring with unit and consider as underlying space the real spectrum $\operatorname{Spec}_r(A)$ of A.

As the reader may realize, all these questions can be posed at a very basic level, although it is not apparent how to handle them. Their solution seems to require, at least until the present moment, advanced methods from analytic and algebraic geometry, model theory, commutative algebra, quadratic forms and real algebra, including Marshall's spaces of orderings and Coste-Roy's real spectra, both highly important in this book.

Our purpose is to introduce a way of attacking all the above questions in a systematic manner. To that end, we exhibit a theory whose objects include all the different kinds of *constructible sets* which appear in real geometry. For instance, if the underlying space is $\mathbb{R}^n$ and A is the ring of polynomials, they are the semialgebraic sets, which, by Tarski's principle, form the smallest class of sets containing all real algebraic varieties and being closed under boolean operations and projections. So they coincide with the *definable sets*, that is, sets that can be defined by *formulae in the language of ordered fields*. This allows the use of methods from model theory to prove in a very efficient way many properties of semialgebraic sets. In other situations Tarski's principle is no longer available. That is the case for semianalytic sets and germs, which are the constructible sets in real analytic geometry. Furthermore, the notion of constructible set is well established in abstract settings as real spectra and spaces of orderings.

Now, what is the essence of constructible sets? Firstly, their definition involves the relation $>$ which appears naturally in real geometry. Secondly, more than the concrete values of the defining functions, what matters are their signs. This leads us to consider a space X together with a monoid G of functions taking the values -1, 0 and 1, which stand for being < 0, $= 0$ and > 0, respectively. Under minor extra conditions we call (X, G) a *real space*, and there are natural notions of *basic open*, *constructible* and *Zariski-closed* sets, as well as the corresponding topologies. In addition, there is also an abstract notion of quadratic *form*. Then, to every constructible set there is attached a unique form, and the complexity of the former can be studied in terms of the latter. This opens the door to apply the theory of quadratic forms to constructible sets.

Clearly, not much can be done at this very general level. Thus, we define *spaces of signs* by imposing four axioms on real spaces. They are inspired by geometric properties of semialgebraic sets and algebraic properties of reduced quadratic forms. The theory of spaces of signs, which we develop in Chapters III, IV and V, is surprisingly rich: it combines the power of real spectra and spaces of orderings. In fact, any commutative ring with unit A gives rise to a space of signs (X, G), where $X = \mathrm{Spec}_r(A)$ and $G = \{\mathrm{sign}[f] \mid f \in A\}$. If the monoid G is a group, the field like situation, then (X, G) is just a space of orderings in the sense of Marshall.

Thus we need a good knowledge of both theories. A general outline of real spectra is given at the beginning of Chapter II and Marshall's spaces of orderings are studied in full detail in Chapter IV. One key point of the whole theory is that questions on spaces of signs can be reduced, via *local-global principles*, to finite *subspaces*, and then to combinatorics. Of high interest are certain subspaces called *fans*, which on the one hand admit quite a simple structure, and on the other govern the whole theory.

Even after we have developed spaces of signs, only part of the way is done. Coming back to the situation $X = \mathrm{Spec}_r(A)$, in order to get quantitative results one needs an algebraic description of the fans, or at least an estimation of their size. This information is hidden in the residue fields of A, and to read it one needs a good understanding of the real algebra of fields. The basic ideas of this topic are presented in Chapter VI. They grew during the 70's out of the study of quadratic forms and Witt rings over formally real fields, without any geometric application in mind. They are sufficient to give estimates for the fans occuring in real algebraic geometry, where the residue fields are algebraic function fields. The situation is more complicated in analytic geometry, where very little is known about the residue fields.

A natural class of rings which is still accessible is that of *excellent rings*, to which Chapter VII is devoted. It contains the algebras finitely generated over fields and also the rings of analytic functions and analytic function germs. Moving back and forth through henselizations and completions, where power series methods can be applied, a deep understanding of the real algebra of excellent rings is achieved. Along the way, we show constructibility of closures and characterize local constructibility of connected components.

Closing the circle, how do we come back to the geometric problems posed at the beginning? Let X be a real algebraic, a compact analytic set or an analytic set germ, and A the suitable ring of functions on X. Since A is excellent, we already know a lot about $\mathrm{Spec}_r(A)$, although in the analytic case some extra work is still required to estimate the size of fans in the space of signs associated to A. Finally, we need a transfer from $\mathrm{Spec}_r(A)$ to X, or, more precisely, a one-to-one correspondence between constructible sets in $\mathrm{Spec}_r(A)$ and X. In the algebraic case this transfer comes from the Artin-Lang homomorphism theorem. In the analytic case, this originated about ten years ago in the solution of Hilbert's 17th problem for meromorphic functions on X. All of this is done in Chapter VIII.

In the end, we can harvest the fruits of the theory: we answer the questions posed at the beginning, not only in the algebraic case but also in the analytic.

Look at the long path we have drawn from geometry to real spectra, spaces of signs, combinatorics and back to geometry. We have tried to display this path in Chapter I by examples and through heuristic considerations, before going into technicalities. This first chapter also contains a few remarks on how this work fits into the general stream of real geometry. Therefore it should be seen as an extended introduction.

Chapter I. A First Look at Semialgebraic Geometry

Summary. This chapter can be viewed as an introduction to the book and as motivation for the problems considered. It contains almost no proofs. In Section 1 we introduce the Tarski-Seidenberg theorem in several forms which are practical and sufficiently general, without entering too far into the terminology of model theory. In the next section we discuss some typical problems of semialgebraic geometry, trying to show how the topic of this book –the description of semialgebraic and more general sets by few generators– fits into the theory that we develop. For this kind of complexity problem we introduce in Section 3 the unifying terminology of real spaces, the spaces which occur in various contexts like semialgebraic geometry. semianalytic geometry, real spectra of rings and spaces of orderings of fields. This book deals with the relations between these. However, in Section 4 we first look at typical examples and illustrations in the semialgebraic situation. So this section is mostly recommended for motivation.

1. Real Closed Fields and Transfer Principles

The objects we have in mind at our first look at semialgebraic geometry are certain subsets of $\mathbb{R}^n$, which are defined by real polynomials. Here, one might first think of solutions of systems of polynomial equations, that is, affine algebraic varieties over the reals. But then, of course, one considers not merely the solutions in $\mathbb{R}^n$ but also in $\mathbb{C}^n$ or in $\mathbb{P}^n(\mathbb{C})$, where the variety shows its full beauty.

However, the situation changes if we are concerned with polynomial relations instead of polynomial equations: For instance, the set

$$S = \{x \in \mathbb{R} \mid \exists\, y \in \mathbb{R} : x^2 + y^2 = 1\}$$

is not algebraic and not even a boolean combination of algebraic subsets of $\mathbb{R}$. On the other hand, S can be described by inequalities:

$$S = \{x \in \mathbb{R} \mid -1 \leq x \leq 1\}.$$

Here we have an example of a fundamental result in real algebraic geometry: the Tarski-Seidenberg theorem, which we are going to explain. Before we do this, we recall the most important notions from the theory of formally real fields. More generally, consider a commutative ring A with unit. We set

$$\sum A^2 = \{a \in A \mid \exists r \in \mathbb{N}, a_1 \ldots, a_r \in A : a = a_1^2 + \cdots + a_r^2\},$$

that is, $\sum A^2$ is the set of all sums of squares in A.

By a total ordering of a field we mean always an ordering which is compatible with the field structure in the usual way.

Proposition and Definition 1.1   *For a commutative ring A with unit the following conditions are equivalent:*

   a) $-1 \notin \sum A^2$.
   b) There exists at least one prime ideal $\mathfrak{p}$ of A whose residue field $\kappa(\mathfrak{p})$ admits a total ordering.

If these conditions hold, the ring A is called formally real.

Proof.   [B-C-R 4.3], [Kn-Sch III.3], [Pr2 §1]. See also Proposition II.1.3.   □

In particular, a field K is formally real, if and only if it can be ordered. In this case

$$\sum K^2 = \{a \in K \mid a \geq 0 \text{ in all orderings of } K\},$$

see [B-C-R 1.1], [Kn-Sch I.1], [Pr2 §1] (for the corresponding formula for rings see Proposition II.1.15).

Proposition and Definition 1.2   *For a field K the following conditions are equivalent:*

   a) K is formally real, but no algebraic extension is formally real.
   b) K admits an ordering which cannot be extended to any algebraic extension of K.
   c) K is ordered such that each positive element is a square in K, and any polynomial $p(\mathrm{t}) \in K[\mathrm{t}]$ of odd degree admits a root in K.
   d) K is not algebraically closed, but $K(\sqrt{-1})$ is algebraically closed.

If these conditions hold, K is called real closed.

Proof.   [B-C-R 1.2], [Kn-Sch I.5], [Pr2 §3].   □

From *c)* we see that the behaviour of a real closed field R is very similar to that of the field $\mathbb{R}$ of the ordinary reals. Therefore the semialgebraic geometry which we are going to consider can be developed over a real closed field R in the same pleasant way as over $\mathbb{R}$. This is done not just for fruitless generalizations: Even if one is only interested in results over the reals, the more general concept is needed for intermediate steps. We will soon have occasion to see this.

Proposition and Definition 1.3   *Let $(K, <)$ be an ordered field and $\overline{K}$ its algebraic closure. Then there exists a real closed field R with $K \subset R \subset \overline{K}$ such that the unique ordering of R extends $<$, and any two real closed fields with that property are conjugate in $\overline{K}$ over K.*

The (essentially unique) field R is called the real closure of $(K, <)$.

Proof.   [B-C-R 1.2], [Kn-Sch I.11], [Pr2 §3]. $\square$

We need also some notions from model theory. However, we do not enter into this seriously and introduce them in a most elementary way:

(1.4) Formulae. Let A be a commutative ring with unit. A *formula $\Phi(\mathbf{x})$ with parameters* in A (more precisely, a *first order formula in the language of ordered fields with parameters in A*) is an expression which is built up by a finite number of conjunctions, disjunctions, negations, universal and existential quantifiers on the variables starting from atomic formulae, which are the expressions $f(\mathbf{x}) > 0$ where $f \in A[\mathbf{x}]$, $\mathbf{x} = (\mathbf{x}_1, \ldots, \mathbf{x}_n)$.

A formula $\Phi = \Phi(\mathbf{x})$ is called a *sentence*, if it does not have free variables, i.e. all occurences of $\mathbf{x}_1, \ldots, \mathbf{x}_n$ are bounded by some universal or existential quantifier. Now let $\alpha : A \to R_\alpha$ be a non-trivial homomorphism of A into a real closed field R_α, and let Φ be a sentence. Then we write

$$R_\alpha \models \Phi$$

if the sentence which one gets from Φ by applying α to the parameters holds in R_α.

For instance the expression $\exists \mathbf{y}(a\mathbf{x} + \mathbf{y} < 0)$ where $a \in A$, is a formula and the expression $\forall \mathbf{x} \, \exists \mathbf{y}(a\mathbf{x} + \mathbf{y} < 0)$ is a sentence.

Now we are able to state the results which are known under the names of Tarski principle or transfer principle or Tarski-Seidenberg theorem, in a general form which will be convenient for us.

Theorem 1.5 (Elimination of Quantifiers)   *Let $\Phi(\mathbf{x})$ be a formula with parameters in the ring A and free variables $\mathbf{x} = (\mathbf{x}_1, \ldots, \mathbf{x}_n)$. Then there exists a formula $\Psi(\mathbf{x})$ with parameters in A and the same free variables $\mathbf{x} = (\mathbf{x}_1, \ldots, \mathbf{x}_n)$ but without quantifiers, such that for all non trivial homomorphisms $\alpha : A \to R_\alpha$ of A into some real closed field R_α one has*

$$R_\alpha \models \forall \mathbf{x}(\Phi(\mathbf{x}) \leftrightarrow \Psi(\mathbf{x})).$$

Proof.   Let $\mathbf{y} = (\mathbf{y}_1, \ldots, \mathbf{y}_m)$ be the bounded variables in Φ. We replace also the parameters of Φ by new variables $\mathbf{z} = (\mathbf{z}_1, \ldots, \mathbf{z}_r)$, thus getting a formula $\Phi(\mathbf{x}, \mathbf{y}, \mathbf{z})$ with parameters in $\mathbb{Z}$ and free variables $\mathbf{x}, \mathbf{z}$.

Now by [B-C-R 1.4.4] the result holds for $A = \mathbb{Z}$, so we find $\Psi(\mathbf{x}, \mathbf{z})$ with parameters in $\mathbb{Z}$ and without quantifiers, such that for all real closed fields R :

$$R \models \forall \mathbf{x}, \mathbf{z}(\Phi(\mathbf{x}, \mathbf{y}, \mathbf{z}) \leftrightarrow \Psi(\mathbf{x}, \mathbf{z})).$$

In particular, for $R = R_\alpha$ we may insert for $\mathbf{z}$ the parameters of the original formula $\Phi(\mathbf{x})$. $\square$

Theorem 1.6 (Model Completeness)   *Let A be a commutative ring with unit and let $\alpha : A \to R_\alpha$ and $\beta : A \to R_\beta$ be homomorphisms into real closed fields such that we have a commutative diagram*

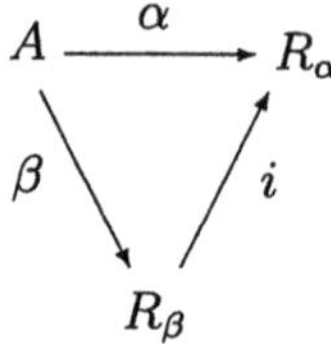

where i is an embedding. Then for any sentence Φ with parameters in A one has $R_\alpha \models \Phi$ if and only if $R_\beta \models \Phi$.

Proof.   [B-C-R 5.2.3]. $\square$

Let us demonstrate the power of this result by proving a classical theorem:

Theorem 1.7 (Artin-Lang Homomorphism Theorem)   *Let $(K, <)$ be an ordered field and let A be a commutative integral finitely generated K-algebra with quotient field $\mathrm{qf}(A) = L$. Let $g_1, \ldots, g_r \in A$, and assume that there is an extension of $<$ to L (which we denote again by $<$) such that $g_i > 0$, $i = 1, \ldots, r$. Let R be the real closure of $(K, <)$. Then there exists a homomorphism $\beta : A \to R$ such that $\beta(g_i) > 0$, $i = 1, \ldots, r$.*

Proof.   Take a representation: $A = K[\mathbf{t}]/(f_1, \ldots, f_m)$, $f_1, \ldots, f_m \in K[\mathbf{t}]$, $\mathbf{t} = (\mathbf{t}_1, \ldots, \mathbf{t}_n)$, and let R_α be the real closure of $(L, <)$, so that we have

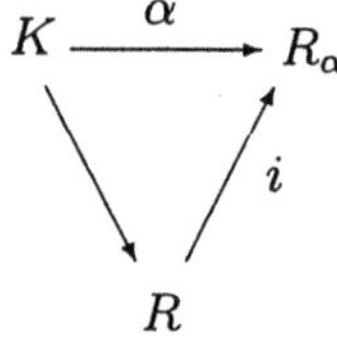

Now consider the following sentence Φ

$$\exists \mathbf{x}\ f_i(\mathbf{x}) = 0,\ g_j(\mathbf{x}) > 0;\ i = 1, \ldots, m;\ j = 1, \ldots, r$$

where $\mathbf{x} = (\mathbf{x}_1, \ldots, \mathbf{x}_n)$. Then $R_\alpha \models \Phi$, since we may insert $\mathbf{t}_i + (f_1, \ldots, f_m)$ for $\mathbf{x}_i$. Hence $R \models \Phi$ which shows that $K \to R$ can be extended to $\beta : A \to R$ and proves the theorem. $\square$

2. What is Semialgebraic Geometry?

We fix a real closed field R and n a positive integer. Let $\mathbf{t} = (\mathbf{t}_1, \ldots, \mathbf{t}_n)$. For $f_1, \ldots, f_m \in R[\mathbf{t}]$ we write

$$\{f_1\alpha_1, \ldots, f_m\alpha_m\} = \{x \in R^n \mid f_1(x)\alpha_1, \ldots, f_m(x)\alpha_m\}$$

where α_i stands for > 0, ≥ 0 or $= 0$. Note that

$$\{f_1 = 0, \ldots, f_m = 0\} = \{f = 0\}$$

for $f = f_1^2 + \cdots + f_m^2$.

(2.1) Semialgebraic Sets. A *semialgebraic set* $S \subset R^n$ is a finite boolean combination of sets of the form $\{f_1\alpha_1, \ldots, f_m\alpha_m\}$.

The semialgebraic sets $S \subset R^n$ form a boolean algebra. Moreover, any such S can be written in a *normal form*, say

$$S = \{f_0 = 0\} \cap (\{f_{11}\alpha_{11}, \ldots, f_{1s_1}\alpha_{1s_1}\} \cup \cdots \cup \{f_{t1}\alpha_{t1}, \ldots, f_{ts_t}\alpha_{ts_t}\})$$

where α_{ij} stands for > 0 or ≥ 0 or $= 0$, but such a description is not at all unique.

Roughly speaking, one main concern of semialgebraic geometry is the study of the connections between the geometric properties of semialgebraic sets and the formulae that describe them. This geometry is, in our opinion, very attractive, since it is close to the geometry of natural figures. It comes up also in practical disciplines like robotics or computational geometry. Before we get more specific let us present an alternative description of semialgebraic sets in terms of formulae.

Proposition and Definition 2.2   *A subset $S \subset R^n$ is semialgebraic if and only if it is* definable, *i.e. there exists a formula $\Phi(\mathbf{x})$ with parameters in R and free variables $\mathbf{x} = (\mathbf{x}_1, \ldots, \mathbf{x}_n)$ such that $S = \{x \in R^n \mid \Phi(x)\}$.*

Proof.   Let S be definable and $\Phi(\mathbf{x})$ a formula for S. Replacing $\mathbf{x}_i$ by $\mathbf{t}_i$ for $i = 1, \ldots, n$ we get a sentence $\Phi(\mathbf{t})$ with parameters in $R[\mathbf{t}]$. Now, according to Theorem 1.6, choose a sentence $\Psi(\mathbf{t})$ without quantifiers such that for all $\alpha : R[\mathbf{t}] \to R_\alpha$

$$R_\alpha \models \Phi(\mathbf{t}) \leftrightarrow \Psi(\mathbf{t}) \, .$$

This holds in particular for $\alpha : R[\mathbf{t}] \to R$; $\mathbf{t}_i \mapsto x_i$, that is, $\Phi(x) \leftrightarrow \Psi(x)$ for all $x \in R^n$. $\square$

To illustrate how practical this is, consider some consequences which are hard to obtain if one uses the definition of a semialgebraic set directly.

Corollary 2.3   *Let $S \subset R^n$ be semialgebraic. Then the closure $\mathrm{Adh}(S)$ of S, the interior $\mathrm{Int}(S)$ of S, the boundary $\mathrm{Bd}(S)$ of S and the image $f(S)$ by any polynomial map $f : R^n \to R^m$ are again semialgebraic.*

Here R^n is provided with the product topology, which comes from the order topology on R.

Proof.   For instance, $x \in \mathrm{Adh}(S)$ if and only if $\forall \varepsilon > 0 \, \exists x' \in S \, (\|x - x'\|^2 < \varepsilon)$, and so $\mathrm{Adh}(S)$ is definable. $\square$

Remarks 2.4   A semialgebraic set S is of finite nature with respect to many aspects. So let us mention a couple of facts which might illustrate this remark, although we do not touch upon them in the book.

   a) *S is homeomorphic to a union of simplices of a finite polyhedron* (see [B-C-R 9.2], [Df-Kn3 §2]).
   b) *S admits a natural d-dimensional measure, where $d = \dim(S)$. For this measure $S \cap B_r$ has a finite volume if $0 \le r < \infty$* (see [Yo]).

Some explanations are needed here. First of all, it is not clear whether these statements make sense for real closed fields other than $\mathbb{R}$. However, at least *a)* can be interpreted and proved for arbitrary real closed fields, working in the category of semialgebraic sets and maps. Then, one could try the model theoretic approach, which consists of: first, proving the result for $\mathbb{R}$; second, reformulating it in the form $\mathbb{R} \models \Phi$, where Φ is a suitable sentence; third, applying model completeness to get $R \models \Phi$, and, fourth, come back from this formulation to the standard mathematical statement. However this strategy does not work for *a)*. Concerning *b)* the situation is even worse, the statement is transcendental by nature. Thus we see two types of properties that are far from being *first order properties* and escape the direct approach by model theoretic means.

(2.5) Diagrams.   In order to attack quantitative questions in semialgebraic geometry one may replace in the normal forms introduced in 2.1 the polynomials f_0, f_{ij} by their degrees d_0, d_{ij} and adjoin the embedding dimension n. Thus we get a formal expression

$$\triangle = (n; d_0; \{d_{11}, \alpha_{11}; \ldots; d_{1s_1}, \alpha_{1s_1}\} \vee \cdots \vee \{d_{t1}, \alpha_{11}; \ldots; d_{ts_t}, \alpha_{ts_t}\}) \, .$$

Such a $\triangle$ is called a *diagram for S*; we shall also say that *S is of type $\triangle$*. The diagram is called *closed* (resp. *relatively open*) if all α_{ij} mean ≥ 0 (resp. > 0). It is called *algebraic* if all α_{ij} mean $= 0$. One may replace in $\triangle$ the degrees d_0, d_{ij} by other complexity functions. This is important, for instance, in the study of sets which are defined by Pfaffian functions ([Kh]).

As we pointed out before, one characteristic concern of semialgebraic geometry is the study of the connections between the properties of a set S and the form of its diagrams $\triangle$. There are two directions:

a) Given a diagram $\triangle$, which geometrical properties of a semialgebraic set S of type $\triangle$ depend only on $\triangle$?

b) Given some restriction on the semialgebraic set S, can we choose a special diagram $\triangle$ for S that reflects that restriction?

Typical examples for *a)* are the following:

Proposition 2.6 (Milnor, Thom) *Let $\triangle$ be algebraic (resp. closed). There is a bound $b(\triangle)$, only depending on $\triangle$, such that for all S of type $\triangle$ one has $\sum b_i(S) \leq b(\triangle)$ (resp. $b_0 \leq b(\triangle)$), where $b_i = i^{th}$ Betti number.*

Proof. [Mi], [B-C-R 11.5.4]. □

Proposition 2.7 (Yomdin) *There is a bound $v(\triangle)$, only depending on $\triangle$, such that for all S of type $\triangle$ with $\dim(S) = d$ one has* volume $(S \cap B_r) \leq r^d v(\triangle)$.

Proof. [Yo]. □

This book is concerned with question *b)*. A first result in this direction is:

Proposition 2.8 (Finiteness Theorem) *Let S be closed. Then S admits a closed diagram. Let S be open in its Zariski closure $\mathrm{Adh}_Z(S)$. Then S admits a relatively open diagram.*

Proof. [B-C-R 2.7] (compare also 3.5 below and Corollary II.1.13). □

We shall consider mainly sets S which are either open or closed. We shall show that even the *length* of a diagram $\triangle$ for S, that is, the number of polynomials for a description of S, can be bounded by a function of the dimension of S. Of particular significance are the sets which admit relatively open diagrams without unions. These are the so-called *basic open sets* $\{f_1 > 0, \ldots, f_s > 0\}$ in a real algebraic affine variety $X = V(R) \subset R^n$. It will be shown that s can be bounded by $\dim(X)$, and that this bound is sharp. Corresponding questions are considered not only for semialgebraic sets but also for germs of semianalytic sets, global semianalytic sets, constructible sets in real spectra of rings and open and closed sets in spaces of orderings. We shall introduce a unifying terminology for this in the next section.

3. Real Spaces

We start with very general concepts:

(3.1) Boolean Algebras and Stone Representation. Recall that a topological space B is called a *Boolean space* if it is compact Hausdorff and has a basis of open and closed sets. Recall also that a boolean space is *totally disconnected*, that is, its connected components are its points. We shall denote by

$\mathcal{L}(B)$ the boolean algebra of all open and closed subsets of B.

Now let $(\mathcal{C}, \cup, \cap)$ be an arbitrary boolean algebra. Then a *representation* of $\mathcal{C}$ is a boolean space B together with an algebra isomorphism $\mathcal{C} \to \mathcal{L}(B)$. The space B is called the *Stone space of* $\mathcal{C}$. This Stone space B is constructed as follows. Take for B the set of all ultrafilters ϕ of $\mathcal{C}$ and endow this set B with the *constructible topology* which has as a basis the sets $\widetilde{C} = \{\phi \mid C \in \phi\}$ for $C \in \mathcal{C}$.

Recall that a filter ϕ of a boolean algebra $\mathcal{L}$ (with a minimum) is called a *prime filter* if for all $L, M \in \mathcal{L}$ with $L \cup M \in \phi$ one has $L \in \phi$ or $M \in \phi$. So in a boolean algebra $\mathcal{C}$ the prime filters and the ultrafilters coincide. More generally, let $\mathcal{U} \subset \mathcal{C}$ be a subalgebra that generates $\mathcal{C}$. Then for any ultrafilter ϕ of $\mathcal{C}$ the restriction $\phi \cap \mathcal{U}$ is a prime filter of $\mathcal{U}$ and

$$\phi \mapsto \phi \cap \mathcal{U}$$

defines a one to one correspondence between the ultrafilters of $\mathcal{C}$ and the prime filters of $\mathcal{U}$.

For all of this we refer to [Bl-Sl].

(3.2) Real Spaces. A group of signs Σ is a compact topological group with an additional element 0. Here we will restrict to the case $\Sigma = \mathbb{F}_3 = \{0, -1, +1\}$ (with its multiplicative structure). The general case comes up in p-adic geometry and in the theory of signatures of higher level.

Now, a *real space* is a pair (X, G) consisting of a set X and a multiplicative monoid G of maps $g : X \to \mathbb{F}_3$, such that the following conditions hold:

$\mathbf{S_1}$: G contains the unit map $1 : X \to \mathbb{F}_3$; $x \mapsto +1$.

$\mathbf{S_2}$: G contains the map $-1 : X \to \mathbb{F}_3$; $x \mapsto -1$.

$\mathbf{S_3}$: G *separates points*, that is, for any two points $x, y \in X$ there is an $f \in G$ with $f(x) \neq f(y)$.

Note that an element $f \in G$ with $f(x) \neq 0$ for all $x \in X$ is a unit of G, since $f^2 = 1$. We will denote by G^* the group of units of G. Note also that the map $0 : X \to \mathbb{F}_3$; $x \mapsto 0$ need not belong to G. It is allowed that $X = \emptyset$, in which case (X, G) is called *trivial*. The next simplest case is $\#(X) = 1$ and $G = \{+1, -1\}$; this space is called *atomic*.

An *isomorphism* from a real space (X, G) onto another (X', G') is a bijection $\theta : X \to X'$ such that the map $\theta^* : G' \to G$; $g' \mapsto g' \circ \theta$ is a well-defined bijection. Then θ^* is an isomorphism of multiplicative monoids.

For $f \in G$ and $x \in X$, we will write

$$f(x) > 0, \ f(x) \geq 0, \ f(x) = 0, \ f(x) \neq 0, \ \text{etc.}$$

instead of

$$f(x) = +1, \ f(x) \in \{0, +1\}, \ f(x) = 0, \ f(x) \in \{-1, +1\}, \ \text{etc..}$$

For $f_i, g_j, h_k \in G$, we denote by

$$\{f_1 > 0, \ldots, f_r > 0, g_1 \geq 0, \ldots, g_s \geq 0, h_1 = 0, \ldots, h_t = 0\}$$

the set

$$\{x \in X \mid f_1(x) > 0, \ldots, f_r(x) > 0, g_1(x) \geq 0, \ldots, g_s(x) \geq 0, h_1(x) = 0, \ldots, h_t(x) = 0\}.$$

The sets of the form

$$\{f_1 > 0, \ldots, f_s > 0\}, \quad f_i \in G,$$

are called *basic open*, and the sets of the form

$$\{f_1 \geq 0, \ldots, f_s \geq 0\}, \quad f_i \in G,$$

are called *basic closed*. If $s = 1$ these sets are called *principal open* and *principal closed*, respectively. Also, the subsets of the form

$$\{f_1 \neq 0, \ldots, f_s \neq 0\}, \quad f_i \in G,$$

are called *Zariski basic open* or *Z-basic open*, and the sets of the form

$$\{f_1 = 0, \ldots, f_s = 0\}, \quad f_i \in G,$$

are called *Zariski basic closed* or *Z-basic closed*.

A finite union of basic open (resp. closed) sets is called *strictly open* (resp. *closed*). We denote by $\mathcal{C}(X)$ the boolean algebra generated by the basic open sets, and by $\mathcal{U}(X) \subset \mathcal{C}(X)$ the subalgebra (not closed under complementation) consisting of all strictly open sets. The sets $C \in \mathcal{C}(X)$ are called *constructible*. Clearly, any constructible set is a union of sets of the form

$$\{f_1 = 0, \ldots, f_r = 0, g_1 > 0, \ldots, g_s > 0\}.$$

We have similar notions in the Zariski case, *Z-strictly open* and *closed* sets, and *Z-constructible* sets, but this will be less important.

(3.3) Topologies on Real Spaces. There are two important topologies in a real space (X, G), namely:

   a) The *constructible topology*. This has as a basis the collection of all constructible sets.
   b) The *Harrison topology*. This has as a basis the collection of all basic open sets.

Correspondingly, we have the *constructible Zariski topology* and the *Zariski topology*. For these topologies we have the following inclusions.

$$
\begin{array}{ccc}
\text{constructible} & \supset & \text{Harrison} \\
\cup & & \cup \\
\text{constructible Zariski} & \supset & \text{Zariski}
\end{array}
$$

The constructible Zariski topology will not appear in our context. If we use topological notations without specification, we mean the Harrison topology. We use an index Z for the Zariski topology; for instance, $\mathrm{Adh}_Z(Y)$ stands for the Zariski closure of $Y \subset X$. Later on we shall see all these topologies working together, but here there is already a simple example: *for every basic open set* $C \subset X$ *it holds* $C \cap \mathrm{Adh}_Z(\mathrm{Bd}(C)) = \emptyset$.

Proof.   Suppose that $C = \{f_1 > 0, \dots, f_k > 0\}$, and let $x \in \mathrm{Bd}(C)$. Then $f_i(x) = 0$ for at least one $i = 1, \dots, k$. Therefore $x \in \{f_1 = 0\} \cup \cdots \cup \{f_k = 0\}$ and thus $\mathrm{Adh}_Z(\mathrm{Bd}(C)) \subset \{f_1 = 0\} \cup \cdots \cup \{f_k = 0\}$. On the other hand $\{f_i = 0\} \cap C = \emptyset$ for $i = 1, \dots, k$. $\square$ In the particular (but important) case when G is a group, the constructible and the Harrison topologies coincide, and the Zariski one is trivial.

(3.4) The Stone Space of a Real Space. Let (X, G) be a real space and let $\widetilde{X}$ be the Stone space of the boolean algebra $\mathcal{C}(X)$ of all constructible sets of X. We get a canonical real space $(\widetilde{X}, G)$ as follows. Let $\phi \in \widetilde{X}$ and $g \in G$. If $g = 1$, (resp. -1, 0), we put $g(\phi) = +1$, (resp. -1, 0). Otherwise, exactly one of the sets

$$\{g > 0\}, \ \{-g > 0\}, \ \{g = 0\}$$

belongs to ϕ, and according to this, we put

$$g(\phi) = +1, \ g(\phi) = -1, \ g(\phi) = 0.$$

To summarize this situation, we say that $\widetilde{X}$ is the Stone space of X. Write as in 3.1 $\widetilde{C} = \{\phi \mid C \in \phi\}$ for $C \subset X$. Then the map $C \mapsto \widetilde{C}$ is injective, preserves finite boolean operations, and induces a bijection between principal open (resp. closed) sets of X and $\widetilde{X}$. Thus we have two isomorphisms

$$\mathcal{C}(X) \to \mathcal{C}(\widetilde{X}) \quad \text{and} \quad \mathcal{U}(X) \to \mathcal{U}(\widetilde{X}).$$

Also, we see that the constructible topologies of $\widetilde{X}$ as Stone space and as a real space coincide: both are generated by the sets $\widetilde{C}$, $C \in \mathcal{C}(X)$. On the other hand, the *Harrison topology* of $\widetilde{X}$ is generated by the sets $\widetilde{U}$, $U \in \mathcal{U}(X)$. Again, we always consider the latter topology, if not otherwise stated. It follows that every constructible set $\widetilde{C}$ of $\widetilde{X}$ is compact with the constructible topology and quasicompact with the Harrison topology. An immediate but useful consequence is that a subset of $\widetilde{X}$ is constructible if and only if it is open and closed in the constructible topology.

We have the canonical map

$$\kappa : X \to \widetilde{X}; x \mapsto \phi_x = \{C \in \mathcal{C}(X) \mid x \in C\}$$

(in other words, ϕ_x is *the principal filter generated by* x). This map is injective and $\widetilde{C} \cap X = C$ for every constructible set $C \subset X$. Hence, κ is a dense embedding with respect to the constructible (resp. Harrison) topologies in both spaces. In fact, via this embedding, every constructible set $C \subset X$ is dense in $\widetilde{C}$; we see that $\widetilde{C}$ is the Stone space of C.

Later we shall see that a main step to solve many questions concerning constructible sets of X is to translate them to $\widetilde{X}$ using the above machinery. This of course will require additional properties of X, but we shall develop an abstract theory general enough to cover most geometric cases: algebraic varieties over formally real fields, real analytic germs, compact analytic spaces over $\mathbb{R}$.

(3.5) Extra Conditions for Real Spaces. Let (X, G) be a real space. If $C \subset X$ is constructible and open, it is not clear that C should be strictly open. In fact, this is the content of a further axiom:

FT: *Every open constructible set of X is strictly open.*

Note that this axiom **FT** is equivalent to: *Every closed constructible set of X is strictly closed.* Another important supplementary axiom is:

AC: *The closure of every constructible set of X is also constructible.*

Again an equivalent formulation is: *The interior of every constructible set of X is constructible.* There is still a third axiom:

CC: *The connected components of every constructible set of X are also constructible.*

As was already said, one main concern of the abstract theory is the comparison of X and the Stone space $\widetilde{X}$ of $\mathcal{C}(X)$. Here we have some easy connections:

a) **FT** *holds for* $\widetilde{X}$.

b) **FT** *holds for X if and only if $\widetilde{C}$ is open in $\widetilde{X}$ for any open constructible set $C \subset X$, if and only if $\widetilde{C}$ is closed in $\widetilde{X}$ for any closed constructible set $C \subset X$.*

c) **AC** *holds for $\widetilde{X}$ if and only if* **AC** *and* **FT** *hold for X. If that is the case, the map $C \mapsto \widetilde{C}$ preserves closures and interiors.*

d) **CC** *holds for $\widetilde{X}$ if and only if every constructible set of $\widetilde{X}$ has finitely many connected components. This is the case if every constructible set of X has finitely many connected components.*

e) *If every constructible set of X has finitely many connected components, which are constructible, then the map $C \mapsto \widetilde{C}$ preserves connected components.*

Proof. *a)* Let $C \subset X$ be constructible and suppose that $\widetilde{C}$ is open in $\widetilde{X}$. Then $\widetilde{C}$ is a union of basic open sets, and since it is compact, a finite union. Hence it is strictly open.

b) First note that $\widetilde{C}$ is strictly open, and in particular open, for every strictly open set $C \subset X$, which implies the only if part. Secondly, suppose that $C \subset X$

is constructible and $\tilde{C}$ is open in $\tilde{X}$. Since **FT** holds for $\tilde{X}$, $\tilde{C}$ is strictly open and so C is strictly open too. From this we get immediately the if part.

   c) Suppose that C is constructible in X and $\mathrm{Adh}(\tilde{C})$ is constructible in $\tilde{X}$. Since **FT** holds for $\tilde{X}$, $\mathrm{Adh}(\tilde{C})$ is strictly closed. Thus $\mathrm{Adh}(\tilde{C}) = \tilde{D}$ with $D \subset X$ strictly closed and it follows easily that $\mathrm{Adh}(C) = D$. This shows both **AC** and **FT** for X. Now it suffices to show that if **FT** and **AC** hold for X, $\mathrm{Adh}(\tilde{C}) = \widetilde{\mathrm{Adh}(C)}$ for every constructible set $C \subset X$. But if $\mathrm{Adh}(C)$ is strictly closed, then $\widetilde{\mathrm{Adh}(C)}$ is closed. Thus $\widetilde{\mathrm{Adh}(C)}$ contains $\mathrm{Adh}(\tilde{C})$. Now, if $\widetilde{\mathrm{Adh}(C)} \setminus \mathrm{Adh}(\tilde{C}) \neq \emptyset$, there would be an open constructible set $U \subset X$ such that $\tilde{U} \cap \widetilde{\mathrm{Adh}(C)} \neq \emptyset$, hence $\tilde{U} \cap \tilde{C} = \emptyset$. Then $U \cap C = \emptyset$, which implies $U \cap \mathrm{Adh}(C) = \emptyset$, and $\tilde{U} \cap \widetilde{\mathrm{Adh}(C)} = \emptyset$, contradiction.

   d) Any constructible set $\tilde{C} \subset \tilde{X}$ is compact in the constructible topology. On the other hand $\tilde{C}$ is the disjoint union of its connected components in the Harrison topology, which are closed in that topology and so closed in the finer constructible topology. Hence we see that the components are open in the constructible topology if and only if they are finitely many. As a set is constructible if and only if it is open and closed in the constructible topology, we are done. Finally, let $C \subset X$ be a constructible set with finitely many connected components $C_1, \ldots, C_m$. We claim that $\tilde{C}$ has also finitely many connected components. Indeed, otherwise we can construct a partition of $\tilde{C}$ into $n > m$ open and closed disjoint subsets. But then, since **FT** holds for $\tilde{X}$, these subsets are strictly open and strictly closed, say $\tilde{D_1}, \ldots, \tilde{D_n}$, where the D_j's are open and closed in C. We see that C must have at least n connected components, contradiction.

   e) Let $C \subset X$ be a constructible set, and $C_1, \ldots, C_m$ its connected components. By *b)*, $\tilde{C_1}, \ldots, \tilde{C_m}$ is a partition of $\tilde{C}$ by open and closed disjoint subsets. Now, by *d)*, **CC** holds for $\tilde{X}$, and $\tilde{C}$ has finitely many connected components, which are constructible, say $\tilde{D_1}, \ldots, \tilde{D_n}$. Thus, each $\tilde{C_i}$ is a union of $\tilde{D_j}$'s. Similarly, each D_j is a union of C_i's. We conclude that $m = n$ and $\tilde{D_{j(i)}} = \tilde{C_i}$ for some permutation $i \mapsto j(i)$. $\qquad\square$

(3.6) Forms over Real Spaces. Let (X, G) be a real space. For an n-tuple $\langle g_1, \ldots, g_n \rangle$, $g_i \in G$, and a point $x \in X$ we write

$$\langle g_1, \ldots, g_n \rangle(x) = g_1(x) + \cdots + g_n(x) \in \mathbb{Z}.$$

Two n-tuples $\langle g_1, \ldots, g_n \rangle$ and $\langle h_1, \ldots, h_n \rangle$ are called *equivalent*, and we write

$$\langle g_1, \ldots, g_n \rangle \simeq \langle h_1, \ldots, h_n \rangle,$$

if

$$\langle g_1, \ldots, g_n \rangle(x) = \langle h_1, \ldots, h_n \rangle(x)$$

for all $x \in X$. An equivalence class for this relation is called a *form over X*; if ρ is the equivalence class of $\langle g_1, \ldots, g_n \rangle$, we simply write $\rho = \langle g_1, \ldots, g_n \rangle$. The number n is called the *dimension of ρ*, and denoted by $\dim(\rho)$, and $\rho(x) = \langle g_1, \ldots, g_n \rangle(x)$ is called the *signature of ρ at x*. For two forms $\rho = \langle g_1, \ldots, g_n \rangle$ and $\tau = \langle h_1, \ldots, h_m \rangle$ we define the sum

$$\rho + \tau = \langle g_1, \ldots, g_n, h_1, \ldots, h_m \rangle$$

and the product

$$\rho \otimes \tau = \langle g_1 h_1, \ldots, g_i h_j, \ldots, g_n h_m \rangle.$$

With these operations, the collection of all forms over X has the structure of a semiring, in which cancellation holds with respect to addition. Clearly, we have:

$$(\rho + \tau)(x) = \rho(x) + \tau(x) \qquad \text{and} \qquad (\rho \otimes \tau)(x) = \rho(x)\tau(x),$$

for all $x \in X$.

We can also build a ring, the so-called *reduced Witt ring $W(X)$ of degenerate forms*. To that end, we call two forms ρ and τ *similar*, and write $\rho \sim \tau$, if $\rho(x) = \tau(x)$ for all $x \in X$. Hence, $\rho = \tau$ if and only if $\rho \sim \tau$ and $\dim(\rho) = \dim(\tau)$. Then, the reduced Witt ring consists of all similarity classes of forms. Of course, if ρ is the class of $\langle g_1, \ldots, g_n \rangle$, then $-\rho$ is the class of $\langle -g_1, \ldots, -g_n \rangle$.

The form ρ is called *isotropic* if there is a form τ with $\rho \sim \tau$ and $\dim(\rho) > \dim(\tau)$. Otherwise, ρ is called *anisotropic*. In case $0 \in G$, the form $\langle 0 \rangle$ is considered isotropic. A form φ of the type $\varphi = \langle a_1^2, a_1 \rangle \otimes \cdots \otimes \langle a_n^2, a_n \rangle$ is called an *n-fold Pfister form*, and denoted by $\varphi = \langle\langle a_1, \ldots, a_n \rangle\rangle$.

Let (X, G) be a real space and ρ a form over X. Then the *signature map*

$$\hat{\rho} : X \to \mathbb{Z}; \; x \mapsto \rho(x)$$

is continuous, if X is endowed with the constructible topology and $\mathbb{Z}$ with the discrete topology. On the other hand, we can use forms to describe any constructible set:

Proposition and Definition 3.7   *Let $C \subset X$ be constructible, $C \neq \emptyset, X$.*

  a) *There exists an integer $k > 0$ and a form ρ over X such that $\rho(x) = k$ for $x \in C$ and $\rho(x) = 0$ for $x \notin C$.*
  b) *For such a k, the form ρ can always be chosen anisotropic, and then it is unique.*
  c) *The minimal k for which a) holds is a power of two, say $k = 2^w$, and we denote $w = w(C)$. We denote by $\rho(C)$ the corresponding anisotropic form, and by $l(C)$ the dimension of $\rho(C)$.*

The number $w(C)$ is called the width *of C, the number $l(C)$ is called the* length *of C, and the form $\rho(C)$ is called the* defining form of C.

Proof.   First note that if *a)* holds for $k_1 > k_2$, then it also holds for the remainder of the division of k_1 by k_2. Thus, the minimal k for which *a)* holds divides all the others. Consequently, it is enough to show that *a)* holds for some power of 2. Let I be a finite set of indices and $C = \bigcup_{i \in I} C_i$, where for each $i \in I$ we have

$$C_i = \{f_{i1} = 0, \ldots, f_{ir_i} = 0, g_{i1} > 0, \ldots, g_{is_i} > 0\}.$$

Now, for every $J \subset I$, set

$$C_J = \bigcap_{j \in J} C_j.$$

Then, there exists $t(J)$ and w such that all the C_J's can be written as

$$C_J = \{a_{J1} = 0, \ldots, a_{Jt(J)} = 0, b_{J1} > 0, \ldots, b_{Jw} > 0\}.$$

(Note that since $1 \in G$, we can add trivial inequalities $1 > 0$ to get the same number w of b's for all the C_J's, but we cannot do the same with the equalities, since 0 need not belong to G.) Then, we put

$$\varphi_J = \langle 1, -a_{J1}^2 \rangle \otimes \langle 1, -a_{Jt(J)}^2 \rangle \otimes \langle\!\langle b_{J1}, \ldots, b_{Jw} \rangle\!\rangle,$$

and notice that $\varphi_J(x) = 2^w$ for $x \in C_J$, $\varphi_J(x) = 0$ otherwise. Finally, set

$$\rho = \sum_{J \subset I} (-1)^{\#(J)+1} \varphi_J.$$

$\square$

(3.8) The Invariants s, $\bar{s}$, t, $\bar{t}$, w and l. Let (X, G) be a real space.

a) We denote by $s(X)$ (resp. $\bar{s}(X)$) the smallest number $\leq \infty$ such that any basic open (resp. closed) set $C \neq \emptyset, X$, is an intersection of at most $s(X)$ (resp. $\bar{s}(X)$) principal open (resp. closed) sets. The number $s(X)$ is called the *stability index of X*.

b) We denote by $t(X)$ (resp. $\bar{t}(X)$) the smallest number $\leq \infty$ such that any strictly open (resp. closed) set $C \neq \emptyset, X$, is a union of at most $t(X)$ (resp. $\bar{t}(X)$) basic open (resp. closed) sets.

c) Finally, we denote by $w(X)$ (resp. $l(X)$) the smallest number $\leq \infty$ such that $w(C) \leq w(X)$ (resp. $l(C) \leq l(X)$) for any constructible set $C \neq \emptyset, X$.

In case $\#(X) \leq 1$ we define all these invariants to be 0. These invariants are related by several inequalities. For instance, it is clear from the proof of Proposition and Definition 3.7 that $w \leq s$. Also, it follows easily by complementation that $\bar{t} \leq s^t$ and $t \leq s^{\bar{t}}$. We shall see that in most important cases, all the invariants can be bounded effectively by a function of $s(X)$, and so all are finite if $s(X)$ is finite.

Equivalence and similarity of forms over X and over $\widetilde{X}$ are the same thing, as follows easily from the fact that X is dense in $\widetilde{X}$. It also follows that for every constructible set $C \subset X$ the defining forms of C and $\tilde{C}$ coincide. Hence, $w(C) = w(\tilde{C})$ and $l(C) = l(\tilde{C})$. In particular, the invariants w and l are equal for X and $\tilde{X}$. Clearly, this is also true for all the other invariants.

We use the preceding notations in the following example, which is of central interest in this work.

Examples 3.9 Let R be a real closed field, and let V be an affine irreducible algebraic R-variety. Then the following conditions for V are equivalent:

a) V is the Zariski closure of $V(R)$.
b) The function field $R(V)$ of V is formally real.

Proof. Indeed, let $A = R[V]$ be the ring of regular functions of V. The ring A is a finitely generated R-algebra and $R(V)$ is the quotient field of A. Now V is the Zariski closure of $V(R)$ if and only if for any nonzero element $f \in A$ there is some point $x \in V(R)$ with $f(x) \neq 0$. But applying the Artin-Lang theorem (Theorem 1.7) to f^2, the latter assertion is equivalent to the existence of some ordering in the field $R(V)$, because in our situation the homomorphisms $A \to R$ correspond to the points $x \in V(R)$. $\square$

Now a (not necessarily irreducible) R-variety is called *real* if $a)$ holds for V. So let V be a real affine algebraic R-variety and set $X = V(R)$. For $f \in R[V]$ define

$$\mathrm{sign}[f] : X \to \mathbb{F}_3 \,;\; x \mapsto \mathrm{sign}[f(x)] = \begin{cases} +1 & \text{if } f(x) > 0 \\ 0 & \text{if } f(x) = 0 \\ -1 & \text{if } f(x) < 0 \end{cases}$$

and let

$$G = \{\mathrm{sign}[f] \,:\, f \in R[V]\}.$$

Then (X, G) is a real space, and its constructible sets are just the semialgebraic subsets of X. The Harrison topology on X is called the *strong* topology. This is also induced from the product topology of R^n through any presentation $X \subset R^n$. The finiteness theorem (Proposition 2.8) says that **FT** holds for this space, that is, a semialgebraic set $S \subset X$ is open (resp. closed) in the strong topology if and only if it is strictly open (resp. strictly closed).

In Chapter VI we shall show that

$$\begin{aligned} s(X) &= d, \\ \overline{s}(X) &= \frac{1}{2}d(d+1), \end{aligned}$$

where $d = \dim(V)$, and these bounds are sharp. Also $t(X)$ and $\overline{t}(X)$ turn out to be bounded by a function of d, but our bounds are very high and probably not the best ones except for $d \leq 2$. Here $t = \overline{t} = 1$ if $d = 1$ and $t = 3, \overline{t} = 2$ if $d = 2$ (Proposition VI.5.6).

For this example the corresponding Stone space is of fundamental importance. It is the real spectrum of the ring $R[V]$, endowed with its canonical structure of real space. (see the remarks following Definition 4.3).

In the next section, we shall consider some examples of principal, non-principal, basic and non-basic semialgebraic sets. Before we do this, let us look at certain discrete real spaces, which surprisingly play an essential role for the investigation of semialgebraic sets.

Examples 3.10   *a)* Let X be a finite set and let $G = \{+1, -1\}^X$. So (X, G) is the simplest example of a real space. If (X, G) is trivial or atomic, $s(X) = \bar{s}(X) = t(X) = \bar{t}(X) = w(X) = 0$ and $l(X) = \#(X)$. So let $\#(X) > 1$. Then $s(X) = \bar{s}(X) = t(X) = \bar{t}(X) = 1$. In some sense, which will be clarified, this is a 1-dimensional situation.

b) Let us turn to a 2-dimensional situation. For this let Y be finite, $\#(Y) \geq 2$ and consider another 2-element set, say $\mathbb{Z}_2 = \{+1, -1\}$. Let $X = Y \times \mathbb{Z}_2$ and depict this set as

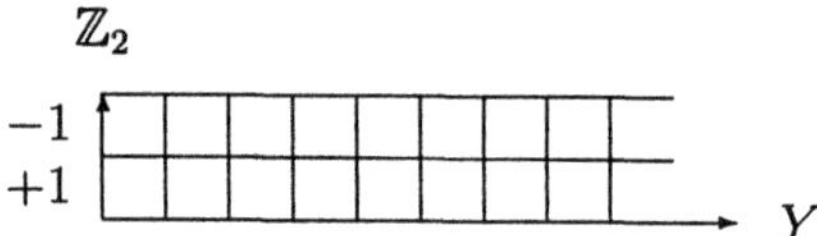

where every box stands for an element of X. A 4-element subset of X which is a union of two columns is called a *fan*. Now let $H = \{+1, -1\}^Y = \mathbb{Z}_2^Y$ and $G = H \times \widehat{\mathbb{Z}_2}$ (=dual of $\mathbb{Z}_2$), where for $x = (y, \alpha) \in X$, $g = (h, \beta) \in G$, we set $g(x) = h(y)\beta(\alpha)$. Now, it turns out that $P \subset X$ is of the type $\{g > 0\}$ (principal open and principal closed) if $\#(P \cap F) = 0 \bmod 2$ for all fans $F \subset X$. So a principal set would be

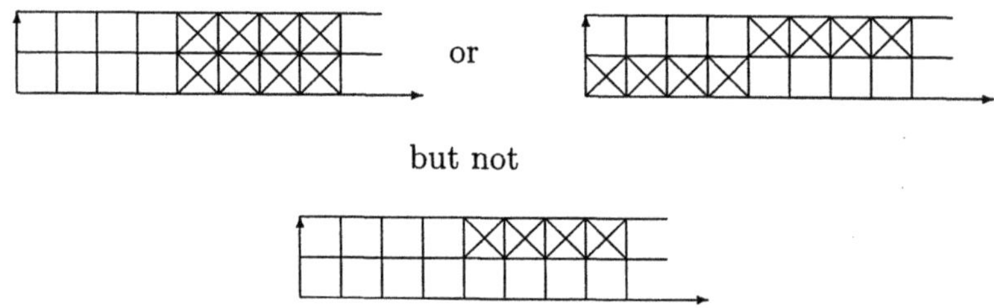

but not

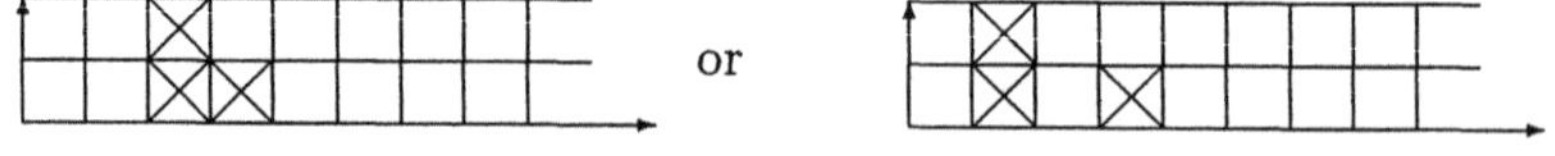

However, this one is still basic (because it is the intersection of the first two). After some reflection we conclude $s(X) = 2$ and see that $C \subset X$ is basic if and only if $\#(C \cap F) \neq 3$ for all fans $F \subset X$. Thus the following sets

or

are not basic. We also conclude $t(X) = 2$ and $w(X) = 2$. The computation showing that $l(X) = 4$ is a little more complicated (see Section IV.7).

We shall encounter these figures in the next section as pictures of semialgebraic sets in R^2, a correspondence which is not accidental. In fact, we shall show that the problems we investigate in this book can be reduced, via real algebra, to the combinatorial theory of spaces of orderings, which we shall develop in Chapter IV. Indeed, Example *a)* corresponds to a space $n \times E$ and Example *b)* to a space $(n \times E)[\mathbb{Z}_2]$, which we shall introduce in Section IV.2.

4. Examples

Again we fix an algebraic affine real R-variety V and set as above $X = V(R)$. First let $\dim(V) = 1$. Then X is a real curve in some R^n. As stated before, we shall see that $s = t = \overline{s} = \overline{t} = 1$, or, in other words, that every open (resp. closed) semialgebraic set $S \subset X$ can be described by a single inequality $f > 0$ (resp. $f \geq 0$) with $f \in R[V]$. This seems to be rather obvious (see the figure below), but we have not found references for this fact in the case of singular curves.

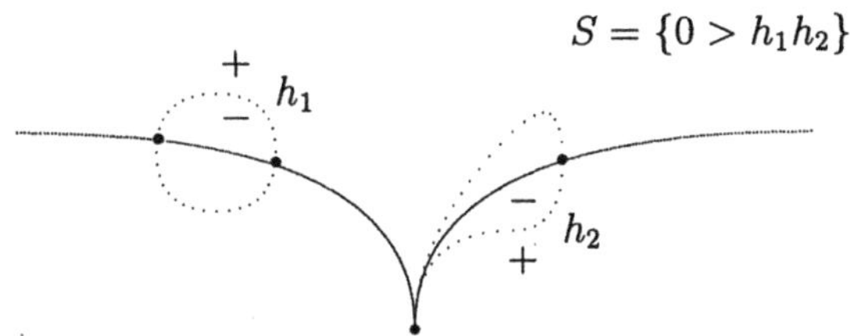

In the case that V is smooth this was essentially shown by Witt [Wt1] for $R = \mathbb{R}$; for arbitrary R see [Kn3]. Another proof, in which the methods we use here already appear, was given by Elman-Lam-Prestel [El-Lm-Pr].

However, the typical phenomena will not appear until $\dim(V) \geq 2$. In order to get a feeling for these phenomena it will be sufficient to consider $V = \mathbb{A}^2$, so $X = R^2$:

Examples 4.1   Consider $f = \mathrm{t}_2^2 - \mathrm{t}_1^2(\mathrm{t}_1 - 1)$. The interior $\mathrm{Int}(\{f \geq 0\})$ of the basic closed set $\{f \geq 0\}$ is not basic open. The reason is that $\mathrm{Int}(\{f \geq 0\})$ contains the origin and this belongs to the Zariski closure of the boundary of $\{f \geq 0\} \subset R^2$, so that we can apply the following result, whose proof in the abstract setting of real spaces (I.3.3) can be repeated here:

Proposition 4.2   *If S is basic open in X, then $\mathrm{Adh}_Z(\mathrm{Bd}(S)) \cap S = \emptyset$.*

In Section V.4 we shall show that $S \setminus \mathrm{Adh}_Z(\mathrm{Bd}(S))$ is basic open in X if S is basic closed in X. Conversely, one may ask whether $\mathrm{Adh}(S)$ is basic closed if S is basic open in X. This does not hold in general, but it is true if $\dim(V) \leq 3$

and $\dim(\mathrm{Sing}(V)) \leq 1$, where $\mathrm{Sing}(V)$ is the singular locus of V. The proof that will be given in Section VI.7 requires already the combinatorial techniques discussed in Section 3. In the other cases, $\dim(V) \geq 4$ or $\dim(\mathrm{Sing}(V)) \geq 2$, there are counterexamples (see Section VI.7).

The set $\mathrm{Int}(\{f \geq 0\})$ of Example 4.1, however, is not far from being basic open in the sense that it differs from a basic open set in a piece of codimension ≥ 1. This kind of information is often useful, as it makes possible to argue by induction on the dimension. To be precise we introduce a new definition:

Definition 4.3   *Let S, $S' \subset X$ be semialgebraic. We call S and S' generically equal and write $S \equiv S'$ if for all irreducible components V_i of V one has*

$$\dim(V_i(R) \cap (S \cup S') \setminus (S \cap S')) < \dim(V_i) \, .$$

In other words, S and S' are equal up to thin sets in X. We call S generically basic, if $S \equiv S'$ for a basic set S', and generically principal, if $S \equiv S'$ where S' is principal.

Note that generically we cannot distinguish between open and closed, since for all semialgebraic sets $S \subset X$ one has $\dim(\mathrm{Adh}(S) \setminus S) < \dim(S)$ ([B-C-R 2.8.12]). Obviously generic equality is an equivalence relation and its classes are stable under birational isomorphisms of irreducible varieties. This should enable us to consider semialgebraic sets (up to generic equality) over function fields. But what is a real point of a function field or a field in general? Here one has to generalize the notion of a real point by introducing certain ideal points. Let us give a preview of what will be dealt with in detail in the later chapters. We have

$$X = \mathrm{Hom}_R(R[V], R).$$

Now we generalize the right hand side to

$$\{\alpha : R[V] \to R_\alpha\}/ \sim,$$

where α ranges over all (non trivial) homomorphisms into real closed fields R_α and "$\sim$" is the equivalence relation generated by all commutative triangles

$$
\begin{array}{ccc}
R[V] & \xrightarrow{\ \ \alpha\ \ } & R_\alpha \\
 {}_\beta \searrow & & \nearrow {}_i \\
 & R_\beta &
\end{array}
$$

where i is an embedding. This space is called the *real spectrum of $R[V]$* and denoted by $\mathrm{Spec}_r(R[V])$. Now we define on $\mathrm{Spec}_r(R[V])$ the structure of a real space as follows. For $f \in R[V]$ define

$$\mathrm{sign}[f] : \mathrm{Spec}_r(R[V]) \to \mathbb{F}_3 \, ; \ \alpha \mapsto \mathrm{sign}[\alpha(f)] = \begin{cases} +1 & \text{if } \alpha(f) > 0 \\ 0 & \text{if } \alpha(f) = 0 \\ -1 & \text{if } \alpha(f) < 0 \end{cases}$$

and let

$$G = \{\operatorname{sign}[f] \; : \; f \in R[V]\}.$$

Clearly, $(\operatorname{Spec}_r(R[V]), G)$ is a real space, and we have basic, principal, constructible sets, and the constructible and the Harrison topologies.

In the same way, replacing $R[V]$ by an arbitrary commutative ring A, one gets $\operatorname{Spec}_r(A)$; this will be studied in Chapter II. The first observation is that we can identify

$$\operatorname{Spec}_r(A) \equiv \bigcup_{\mathfrak{p} \in \operatorname{Spec}(A)} \operatorname{Spec}_r(\kappa(\mathfrak{p}))$$

but, unlike the Zariski spectrum, the real spectrum of a field K is not at all trivial. As a matter of fact, $\operatorname{Spec}_r(K)$ has a very rich structure: it is a *space of orderings*. So here is the place where spaces of orderings enter the scene. Furthermore, we shall see that these $\operatorname{Spec}_r(\kappa(\mathfrak{p}))$'s glue together to give $\operatorname{Spec}_r(A)$ the structure of a *space of signs* (Section III.5).

Now returning to our situation $A = R[V]$, we have the canonical embedding

$$X \to \operatorname{Spec}_r(R[V]) \,;\, x \mapsto \text{evaluation at } x$$

and $\operatorname{Spec}_r(R[V])$ turns out to be the Stone space of the boolean algebra $\mathcal{C}(X)$ of all semialgebraic subsets of $X = V(R)$. So its points are the ultrafilters of semialgebraic subsets of X. Namely, for $f \in R[V]$ the set $\{x \in X \mid f(x) > 0\}$ (resp. $\{x \in X \mid f(x) = 0\}$) belongs to the ultrafilter $\alpha \in \operatorname{Spec}_r(R[V])$ if and only if $\alpha(f) > 0$ (resp. $\alpha(f) = 0$), [B-C-R 7.2], [Kn-Sch III.5]. It follows that $(\widetilde{X}, G) = (\operatorname{Spec}_r(R[V]), G)$ is the Stone space of the real space $(X, G) = (V(R), G)$, in the sense of 3.4.

For generic properties, that is, properties compatible with generic equality, we can suppose that V is irreducible, and it turns out that one has to look at $\operatorname{Spec}_r(R(V)) \subset \operatorname{Spec}_r(R[V])$, that is, the ideal points we asked for are just the orderings of the function field of V. In this case we also have a natural real space, namely $(\operatorname{Spec}_r(R(V)), G^*)$, where G^* is the group of units of G, which consists of all $\operatorname{sign}[f]$ for $f \in R(V)$, $f \neq 0$.

The following examples stand for their generic classes.

Examples 4.4   We work again in $V = \mathbb{A}^2$ so that $X = R^2$.

a) $S = \{t_1 \geq 0\}$ or $S = \{t_1 t_2 \geq 0\}$ are principal by definition:

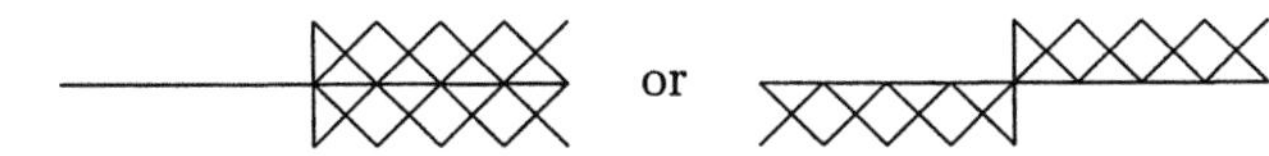

b) $S = \{t_1 \geq 0, t_2 \geq 0\}$ is basic by definition:

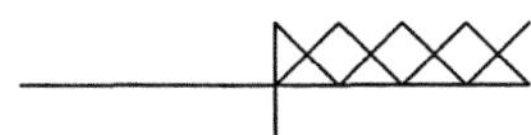

but not generically principal, and we invite the reader to show it directly.

c) (Lojasiewicz's example) The sets

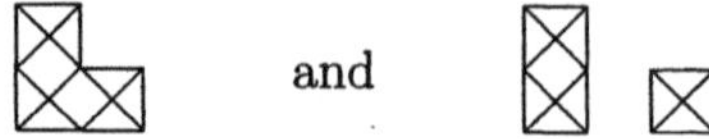

are not generically basic.

d) Examples as above show that, surprisingly, small movements of the figures can make the difference between being

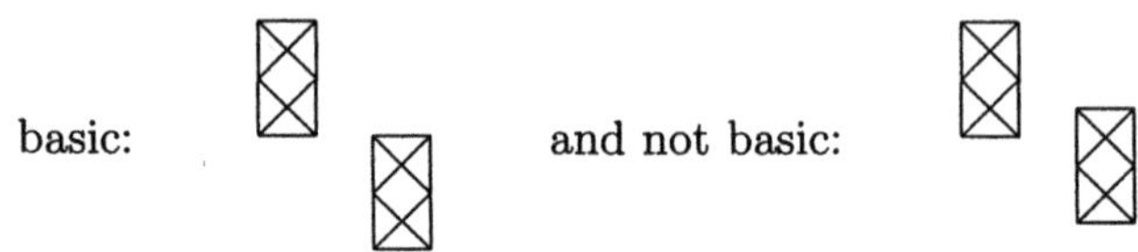

So in Examples *a)*, *b)* and *c)* we recognize the figures of Section 3.

Let us look at a unified method, which shows that the set in *b)* is not generically principal and the sets in *c)* are not generically basic. For this we consider certain elements of $\mathrm{Spec}_r(R[t_1, t_2])$ which we realize as prime filters. Let $\gamma \subset R^2$ be (the real part of) an irreducible curve. Choose two points (possibly equal) on γ and at each point one *half branch* ϕ_1 and ϕ_2, as indicated in the figure.

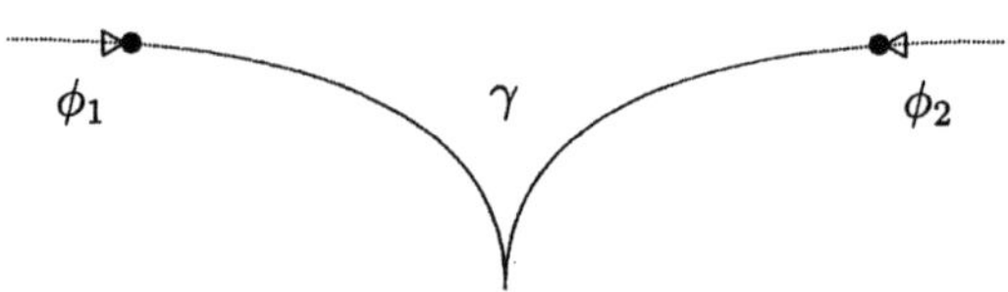

To each ϕ_i corresponds a unique ultrafilter, consisting of all semialgebraic sets which contain that half branch. So we may denote this ultrafilter again by ϕ_i. Now, according to the two sides of the half branch, ϕ_i is contained in exactly two other ultrafilters ϕ_{i1} and ϕ_{i2}. We draw this symbolically like

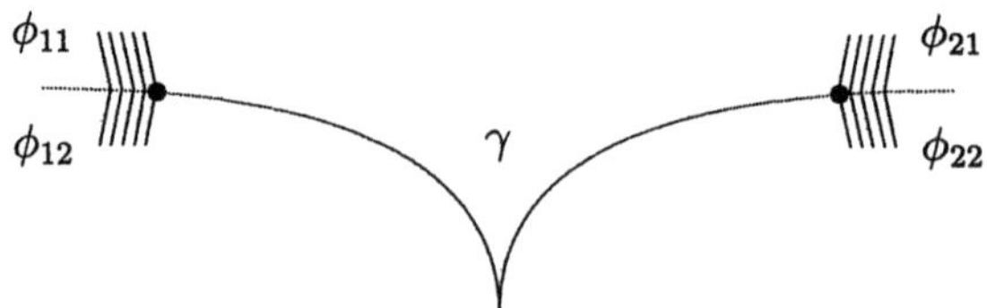

Let $F = \{\phi_{11}, \phi_{12}, \phi_{21}, \phi_{22}\}$ and call this again a *fan*. Then:

Proposition 4.5   *For all $f \in R[t_1, t_2]$ one has*

$$\#\{\phi_{ij} \in F \,|\, \{f > 0\} \in \phi_{ij}\} \equiv 0 \bmod 2 \;.$$

Proof.   Let h be an equation of our curve γ. Then $g_1 f = g_2 h^m$, where $m \geq 0$ and g_1, g_2 are polynomials that do not vanish on γ. In particular they have the same sign on both sides of every half branch ϕ_i; furthermore h changes sign along γ ([B-C-R 4.5.1]). Using all this information one easily checks the statement case by case. $\qquad\qquad\square$

It follows immediately that if a semialgebraic set $S \subset R^2$ is generically principal (resp. basic), then $\#\{\phi_{ij} \in F \mid S \in \phi_{ij}\}$ is even (resp. $\neq 3$). This is exactly the same as in the discrete example of the preceding section. Regarding example *b)* we choose F like

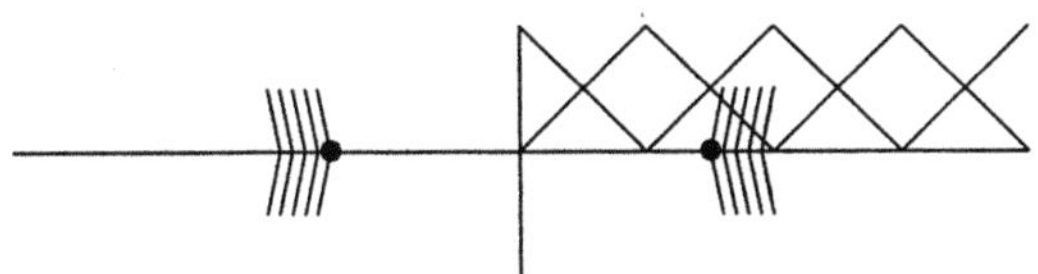

so here $\#\{\phi_{ij} \in F \mid S \in \phi_{ij}\} = 1$, hence S is not generically principal. For *c)* we choose F like

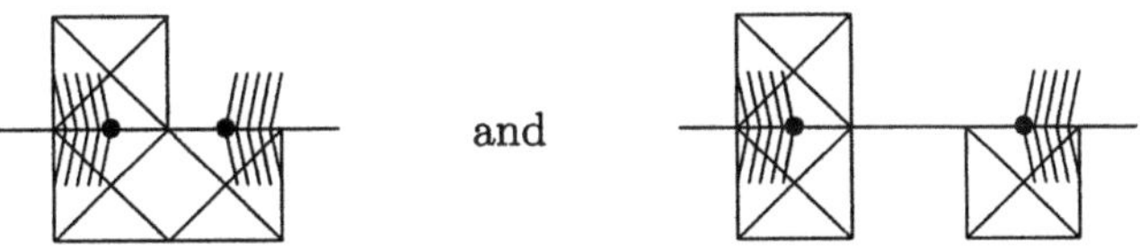

and

so here $\#\{\phi_{ij} \in F \mid S \in \phi_{ij}\} = 3$, and these S's are not generically basic.

This method is the right one to disprove the generic principal or basic property, since the fans, which we shall introduce in full generality in Chapter III, are the only obstructions to these properties. Fans appeared for the first time in the study of quadratic forms and Witt rings over fields. In geometry, the relevant fans needed to decide whether or not a set is principal or basic do not appear before blowing up, as in Example 4.8 below. The reason is that fans are birational rather than biregular objects, and so they cannot by depicted in any model of a given function field, but only in some suitable ones.

Examples 4.6   A basic set described by linear polynomials is convex. Also other familiar basic sets like balls, elipses, etc. are convex. However the converse is not true. For instance a sports field is convex but not generically basic:

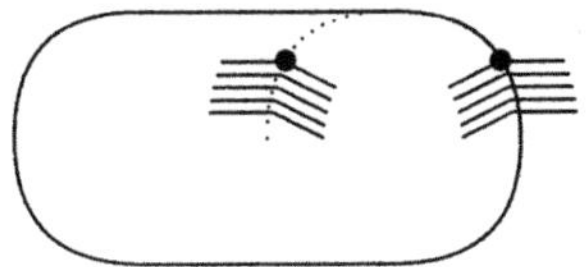

Example 4.7 Let $S \subset R^2$ be the interior of a regular n-gon where $n \in \mathbb{N}$ might be very large. So $S = \{l_1 > 0, \ldots, l_n > 0\}$ where the l_i's are linear forms.

As stated in 3.8, $s(R^2) = 2$, so that we also have a description of the form $S = \{g > 0, h > 0\}$ with $g, h \in R[t_1, t_2]$. Here is one proposal to do that: set $g = l_1 \cdots l_n$ and let h be the equation of the circle through the vertices of the n-gon. However, this is the solution for a very special situation, which gives us no idea how to attack the general case.

This example also shows that one cannot bound simultaneously the number of the polynomials and their degrees for a description of S, since in any description all linear forms l_i must occur as divisors of some polynomial. Besides, by Proposition 2.6, for a description of a semialgebraic set S with large Betti numbers either the number of the describing equations or their degrees must increase, because Betti numbers are bounded in terms of that number and those degrees.

Example 4.8 Consider a curve $\gamma \subset R^2$ which has four connected components and asymptotes as indicated in the figure.

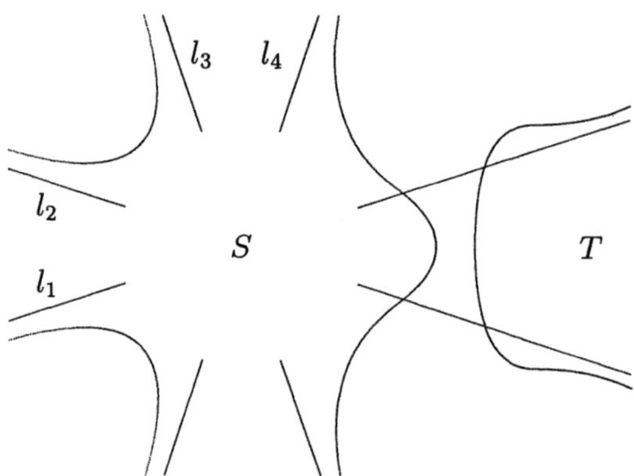

Such a curve, which has been considered by Mostowski [Mw], can be given by an equation of the form

$$0 = f = l_1 \cdot l_2 \cdot l_3 \cdot l_4 - c$$

where the l_i's are linear forms for the asymptotes and c is the equation of some circle of a certain radius centered on the positive part of the x-axis. We claim that the component S of $R^2 \setminus \gamma$ is not generically basic. However, the trick we used above does not apply directly. So let us look at the image of S under the stereographic map: $R^2 \to \mathbb{S}^2 \subset R^3$, which defines a birational isomorphism. On $\mathbb{S}^2$ our situation looks like

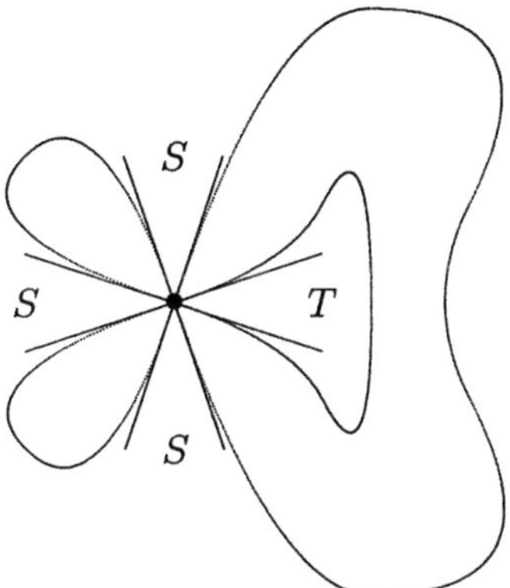

where the dot $\bullet$ stands for the north pole. Here we easily find a fan F such that S belongs to a unique filter of F, and conclude that S is not generically principal. However it is not clear yet why it is not generically basic. To see this, we blow-up the dot, thus getting a situation like

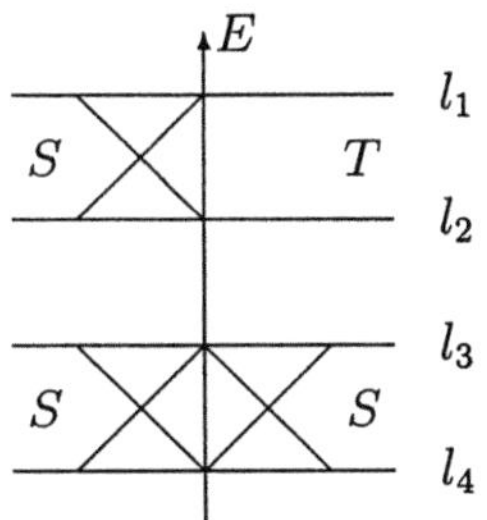

in a neighbourhood of the exceptional divisor E, and our former argument with fans becomes evident here. This shows also that one cannot find a non-zero polynomial $g \in R[\mathsf{t}_1, \mathsf{t}_2]$ which separates S and T generically, that is $g \geq 0$ on S, $g \leq 0$ on T.

Now consider $H(\mathsf{t}_1, \mathsf{t}_2, \mathsf{t}_3) = \mathsf{t}_3^2 - f(\mathsf{t}_1, \mathsf{t}_2)$. The set $Z = \{H = 0\} \subset R^3$ has two connected components S' and T' which project onto S and T respectively. We claim that S' and T' cannot be separated generically by an element $G \in R[Z] = R[\mathsf{t}_1, \mathsf{t}_2, \mathsf{t}_3]/(H)$. Indeed, otherwise $g = \mathrm{tr}(G)$ would separate S and T generically, where "tr" is the trace of $R(Z) = R(\mathsf{t}_1, \mathsf{t}_2)[\sqrt{f}]$ over $R(\mathsf{t}_1, \mathsf{t}_2)$: on Z we have an equation $G^2 - gG + h = 0$, and for every point $(x, y) \in \{f > 0\}$, $g(x, y)$ is the sum of the two roots $G(x, y, +\sqrt{f(x, y)})$ and $G(x, y, -\sqrt{f(x, y)})$ of the polynomial $\mathsf{z}^2 - g(x, y)\mathsf{z} + h(x, y)$, from which the claim follows.

This is what Mostowski pointed out by his example, but he used a different argument. It turns out also that S' is not generically basic, for otherwise S' and

T' would be both basic (for T' this can be checked directly), and then they could be separated by a polynomial, an assertion that requires a proof (see Section V.3).

Chapter II. Real Algebra

Summary. We consider general properties of the real spectrum of a commutative ring with unit. In Sections 1 and 2 we collect the basic facts, and for more information we refer to [B-C-R] and [Kn-Schd]. In Section 3 valuation theory enters the scene. It is of fundamental importance for the whole work. As an application we obtain in Section 4 the first results on going-up and going-down in the real spectrum. In Section 5 we present the notion and basic properties of abstract semialgebraic functions on constructible sets of the real spectrum; this is done in two equivalent ways which have both their advantages. These functions are used in Section 6 to construct cylindrical decompositions with respect to systems of polynomials. Finally, in Section 7 we introduce real strict localizations, which are the real analogues of the strict localizations used in etale cohomology.

1. The Real Spectrum of a Ring

Let A be a commutative ring with unit.

Definition 1.1 *The* real spectrum of A, *denoted by* $\mathrm{Spec}_r(A)$, *is the set*

$$\{\alpha : A \to R_\alpha\}/\sim,$$

where α runs over all non-trivial homomorphisms from A into some real closed field R_α and "$\sim$" is the equivalence relation generated by all commutative triangles

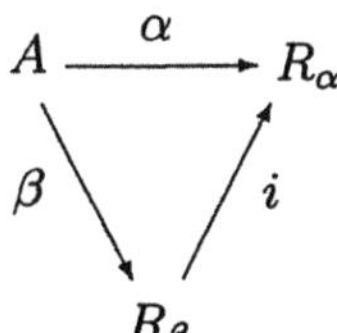

Let us present two alternative descriptions of $\mathrm{Spec}_r(A)$. First we need:

Proposition and Definition 1.2 *A subset $P \subset A$ is called a* precone *of A if the following conditions hold:*

  a) $P + P \subset P$; $P \cdot P \subset P$,
  b) $A^2 \subset P$,
  c) $-1 \notin P$.

For a precone P the following conditions are equivalent:

  d) *If $f, g \in A$ and $fg \in P$, then $f \in P$ or $-g \in P$.*
  d') *The* support *of P, $\mathrm{supp}(P) = P \cap -P$, is a prime ideal of A, and P induces a unique ordering $<_P$ in its residue field $\kappa(\mathrm{supp}(P))$.*

If they hold, P is called a prime cone.

Proposition 1.3   *The following sets are in a canonical one to one correspondence:*

  a) $\mathrm{Spec}_r(A)$.
  b) *The set of all prime cones $P \subset A$.*
  c) *The set of all pairs $(\mathfrak{p}, <)$, where $\mathfrak{p} \subset A$ is a prime ideal and $<$ is a total ordering in the residue field $\kappa(\mathfrak{p})$.*

Proof.   The essential fact for the proof is the existence and uniqueness of real closures. We just indicate how the correspondence looks like. Let $\alpha : A \to R_\alpha$ represent an element of $\mathrm{Spec}_r(A)$. Then the associated prime cone P is the set of all $f \in A$ such that $\alpha(f) \geq 0$. From a prime cone P we get the pair $(\mathrm{supp}(P), <_P)$ according to d'), and from a pair $(\mathfrak{p}, <)$ we find in turn the real closure $R = R_\alpha$ of $\kappa(\mathfrak{p})$ with respect to $<$ and thus $\alpha : A \to R_\alpha$. $\square$

In the sequel we shall use the same letter, say α, for the homomorphism $A \to R_\alpha$, for the element of $\mathrm{Spec}_r(A)$ that it represents and for the corresponding prime cone. Thus we feel free to jump among the different interpretations of $\alpha \in \mathrm{Spec}_r(A)$. Concerning the real closed fields R_α one can restrict to the case that the extension $R_\alpha \supset \kappa(\mathrm{supp}(\alpha))$ is algebraic, but this is not always convenient. We write $\kappa(\alpha)$ instead of R_α if this is chosen algebraic over $\kappa(\mathrm{supp}(\alpha))$. Clearly we obtain the identification

$$\mathrm{Spec}_r(A) \equiv \bigcup_{\mathfrak{p}} \mathrm{Spec}_r(\kappa(\mathfrak{p})).$$

In particular, it follows from Proposition and Definition I.1.1 that $\mathrm{Spec}_r(A) \neq \emptyset$ if and only if A is formally real.

(1.4) Functions. We shall consider the elements of A as functions on $\mathrm{Spec}_r(A)$ in the following sense: A *function on* $\mathrm{Spec}_r(A)$ is an assignment

$$\alpha \mapsto f(\alpha) \in R_\alpha, \quad \text{for all homomorphisms } \alpha : A \to R_\alpha,$$

which is compatible with the equivalence relation of Definition 1.1. Then $f(\alpha)$ is algebraic over $\kappa(\mathrm{supp}(\alpha))$. Moreover, sign conditions like $f(\alpha) > 0, f(\alpha) = 0, f(\alpha) \geq 0$ have a precise meaning. In particular, each $f \in A$ defines a function

on $\mathrm{Spec}_r(A)$, namely
$$\alpha \mapsto f(\alpha) = \alpha(f).$$
We have also the function $|f|$ given by
$$|f|(\alpha) = |f(\alpha)| = \begin{cases} f(\alpha) & \text{if } f(\alpha) \geq 0, \\ -f(\alpha) & \text{if } f(\alpha) < 0. \end{cases}$$
and $\mathrm{sign}[f]$, given by
$$\mathrm{sign}[f](\alpha) = \begin{cases} +1 & \text{if } f(\alpha) > 0, \\ 0 & \text{if } f(\alpha) = 0, \\ -1 & \text{if } f(\alpha) < 0. \end{cases}$$
The concept of a function on $\mathrm{Spec}_r(A)$ will be further developed in Section 4.

(1.5) Real Space Associated to A. Let $X = \mathrm{Spec}_r(A)$ and $G = \{\mathrm{sign}[f] \mid f \in A\}$. Then (X, G) is a real space, and we may adopt the notations and terminology of Section I.3. Here, by simplicity, for $f \in A$, we shall write $f > 0, f \geq 0, f = 0, f \neq 0$, etc. instead of $\mathrm{sign}[f] > 0, \mathrm{sign}[f] \geq 0, \mathrm{sign}[f] = 0, \mathrm{sign}[f] \neq 0$, etc. Once we have given $\mathrm{Spec}_r(A)$ a structure of real space, we have basic open and basic closed sets, principal open and principal closed sets, Z-open and Z-closed sets, strictly open and strictly closed sets, strictly Z-open and strictly Z-closed sets, constructible sets and Z-constructible sets. We also have the four topologies:

$$
\begin{array}{ccc}
\text{constructible} & \supset & \text{Harrison} \\
\cup & & \cup \\
\text{constructible Zariski} & \supset & \text{Zariski}
\end{array}
$$

The constructible topologies are compact ([B-C-R 7.1]) and, in general, the others are only quasicompact. As was said for real spaces, if we use topological notations without specification, we mean the Harrison topology; for the Zariski topology we use the index Z. If A is a field, then $G^* = G \setminus \{0\}$, and the Harrison and the constructible topologies coincide.

We write $s(A)$ instead of $s(\mathrm{Spec}_r(A))$, and similarly for the other invariants $\bar{s}$, t, $\bar{t}$, w and l. Of course, $s(A)$ is called the *stability index of A*.

The assignment $A \mapsto \mathrm{Spec}_r(A)$ is a contravariant functor
$$\left\{ \text{commutative rings with unit} \right\} \longrightarrow \left\{ \text{topological spaces} \right\}$$
with respect to all these topologies, that is, for any homomorphism $\varphi : A \to B$ of commutative rings with unit (here we always assume that $\varphi(1) = 1$) the induced map
$$\varphi^* : \mathrm{Spec}_r(B) \to \mathrm{Spec}_r(A)\,;\ (\beta : B \to R_\beta) \mapsto (\beta \circ \varphi : A \to R_\beta)$$
is continuous. Note that if φ is an epimorphism, then φ^* is an embedding; if φ is a localization, φ^* is an embedding too.

Theorem 1.6 (Ultrafilter Theorem)   *Let $\widetilde{X}$ denote the Stone space of the boolean algebra of all constructible subsets of $X = \mathrm{Spec}_r(A)$. For every $\alpha \in \widetilde{X}$ we denote by ϕ_α the* principal filter *of α, which is the filter consisting of all constructible sets containing α. Then the map*

$$\alpha \mapsto \phi_\alpha$$

is a homeomorphism with respect to the constructible and to the Harrison topologies.

Proof.   [B-C-R 7.1.12, 7.1.15]. $\square$

(1.7) Constructible Sets. It is clear from the definitions that constructible sets of $\mathrm{Spec}_r(A)$ are described by quantifier free sentences with parameters in A (I.1.4). Now by elimination of quantifiers (Theorem I.1.5) we have the following key result:

A subset $C \subset \mathrm{Spec}_r(A)$ is constructible if and only if there exists a first order sentence Φ with parameters in A such that

$$C = \{\alpha \in \mathrm{Spec}_r(A) \mid R_\alpha \models \Phi\}.$$

Remember also that $C \subset \mathrm{Spec}_r(A)$ is constructible if and only if it is open and closed in the constructible topology. In particular, a constructible set is quasicompact in all the topologies considered above.

Example 1.8   Let K be a field and $A = K[\mathbf{x}]$, $\mathbf{x} = (\mathbf{x}_1, \ldots, \mathbf{x}_n)$. By the Artin-Lang homomorphism theorem (Theorem I.1.7), the constructible points of $\mathrm{Spec}_r(A)$ are the homomorphisms $\alpha : A \to R$, where R is a real closure of K. Equivalently, they are the prime cones with maximal support. If K is real closed, they are just the ordinary points α for which $\mathrm{supp}(\alpha) = (\mathbf{x}_1 - a_1, \ldots, \mathbf{x}_n - a_n)$ with $(a_1, \ldots, a_n) \in K^n$. $\square$

A first application of the first order logic description of constructible sets is the following:

Proposition 1.9   *Let B be a finitely presented A-algebra, say $\varphi : A \to B$. Then $\varphi^* : \mathrm{Spec}_r(B) \to \mathrm{Spec}_r(A)$ sends constructible sets onto constructible sets.*

Proof.   By assumption there are indeterminates $\mathbf{t} = (\mathbf{t}_1, \ldots, \mathbf{t}_n)$ and elements $g_1, \ldots, g_s \in A[\mathbf{t}]$ such that B is isomorphic to $A[\mathbf{t}]/(g_1, \ldots, g_s)$. Then $\mathrm{Spec}_r(B)$ can be identified with the constructible set $\{g_1 = \cdots = g_s = 0\} \subset \mathrm{Spec}_r(A[\mathbf{t}])$. This reduces the problem to the case $B = A[\mathbf{t}]$, and by induction to the case of one single variable $\mathbf{t} = \mathbf{t}_1$.

Thus, let $C \subset \mathrm{Spec}_r(A[\mathbf{t}])$ be constructible and defined by a sentence Φ in $A[\mathbf{t}]$. Substituting in Φ each occurence of $\mathbf{t}$ by a new variable $\mathbf{x}$ we obtain a new formula $\Phi(\mathbf{x})$, which now is a formula in A. Then $\exists \mathbf{x}\Phi(\mathbf{x})$ is a sentence Ψ in A, and clearly

$$\varphi^*(C) = \{\alpha \in \mathrm{Spec}_r(A) \mid R_\alpha \models \Psi\},$$

which shows that $\varphi^*(C)$ is constructible. $\qquad\square$

(1.10) Proconstructible Sets. A set $Y \subset \mathrm{Spec}_r(A)$ is called *proconstructible* if it is an (arbitrary) intersection of constructible sets. Thus "proconstructible" means closed in the constructible topology. Hence, finite unions of proconstructible sets are again proconstructible.

Here we have an example. Let $\alpha \in \mathrm{Spec}_r(A)$ and denote by U_α the intersection of all neighbourhoods of α. This set is proconstructible because according to the definitions $U_\alpha = \bigcap_{f(\alpha)>0}\{f > 0\}$.

We shall also consider the following example. Let $\mathfrak{p} \in \mathrm{Spec}(A)$ and see $\mathrm{Spec}_r(\kappa(\mathfrak{p}))$ as a subset of $\mathrm{Spec}_r(A)$. Then

$$\mathrm{Spec}_r(\kappa(\mathfrak{p})) = \bigcap_{f \notin \mathfrak{p}}\{f \neq 0\} \cap \bigcap_{f \in \mathfrak{p}}\{f = 0\}$$

is proconstructible in $\mathrm{Spec}_r(A)$.

A proconstructible set, as a subset of $\mathrm{Spec}_r(A)$, can be endowed with the induced Harrison, constructible or Zariski topology. Being closed with respect to the finest, which is compact, the given proconstructible set is quasicompact with respect to all of them. Like constructible sets, proconstructible sets will always carry the Harrison topology, unless something different is explicitely said.

Proposition 1.11   *Let $Y \subset \mathrm{Spec}_r(A)$ be proconstructible. Then for $C \subset Y$ the following conditions are equivalent:*

a) C is open and closed with respect to the induced constructible topology.
b) $C = C' \cap Y$ for some constructible set C' of $\mathrm{Spec}_r(A)$.

Proof.   *a)$\Rightarrow$ b)* Since C is open in the constructible topology, for every $\alpha \in C$ there is a neighbourhood $U(\alpha) = U'(\alpha) \cap Y \subset C$, where $U'(\alpha)$ is constructible in $\mathrm{Spec}_r(A)$. Since C is also closed, it is compact, and the assertion follows immediately.

b)$\Rightarrow$ a) is obvious. $\qquad\square$

Correspondingly one shows:

Proposition 1.12   *Let $Y \subset \mathrm{Spec}_r(A)$ be proconstructible. Then for $C \subset Y$ the following conditions are equivalent:*

a) C is open with respect to the induced Harrison topology and closed with respect to the induced constructible topology.
b) $C = C' \cap Y$ for some strictly open set C' of $\mathrm{Spec}_r(A)$.

Corollary 1.13 (Finiteness Theorem)   *Let $C \subset \mathrm{Spec}_r(A)$ be constructible. If C is open (resp. closed), then C is strictly open (resp. strictly closed).*

Proof.   Setting $Y = \mathrm{Spec}_r(A)$ in Proposition 1.12 we get the case that C is open. The case that C is closed follows by complementation. $\square$

In view of the ultrafilter theorem (Proposition 1.6) this latter corollary is a particular case of the general discussion of I.3.3. We conclude this section by quoting the fundamental abstract Positivstellensatz and deducing from it the Hörmander-Lojasiewicz inequality in a very general form.

Theorem 1.14 (Positivstellensatz)   *Let* $f_1, \ldots, f_r, g_1, \ldots, g_s, h_1, \ldots, h_t \in A$. *Then the following assertions are equivalent:*

a) $\{f_1 = 0, \ldots, f_r = 0, g_1 \neq 0, \ldots, g_s \neq 0, h_1 \geq 0, \ldots, h_t \geq 0\} = \emptyset$.
b) There exists an equation of the form

$$g_1^{2n_1} \cdots g_s^{2n_r} + \sum_{m_1 + \cdots + m_t \leq d} a_m h_1^{m_1} \cdots h_t^{m_t} = b_1 f_1 + \cdots + b_r f_r,$$

for suitable $n_i, m_j, d \in \mathbb{N}$, $b_i \in A$ *and* $a_m \in \sum A^2$.

Proof.   [B-C-R 4.4.1], [Kn-Schd III.9, p.140-143]. $\square$

Corollary 1.15   *For* $t \in A$ *the following properties are equivalent:*

a) $t(\alpha) \geq 0$ *for all* $\alpha \in \mathrm{Spec}_r(A)$.
b) There exist $q_1, q_2 \in \sum A^2$ *and* $n \in \mathbb{N}$ *such that* $t^{2n+1} + t q_1 = q_2$.

Proof. $b) \Rightarrow a)$ is obvious.
    $a) \Rightarrow b)$ We have $\{t \neq 0, -t \geq 0\} = \emptyset$, thus by the Positivstellensatz $t^{2n} + q_1 - t q_3 = 0$ with $q_1, q_3 \in \sum A^2$. Hence $t^{2n+1} + t q_1 = q_2$, where $q_2 = t^2 q_3$. $\square$

Proposition 1.16 (Hörmander-Lojasiewicz Inequality)   *Let* $C \subset \mathrm{Spec}_r(A)$ *be closed and constructible, and let* $f, g \in A$ *verify* $\{f = 0\} \cap C \subset \{g = 0\}$. *Then there are an odd positive integer* l *and* $h \in A$ *such that:*

a) $|g|^l \leq (1 + h^2)|f|$, *and*
b) $\mathrm{sign}[(1 + h^2)f + g^l] = \mathrm{sign}[f]$

over the set C.

Proof.   First we prove *a)* in case C is basic closed, that is, $C = \{h_1 \geq 0, \ldots, h_t \geq 0\}$. Then the hypothesis reads

$$\{g \neq 0, f = 0, h_1 \geq 0, \ldots, h_t \geq 0\} = \emptyset,$$

and by the Positivstellensatz we find:

$$g^{2d} + h = bf, \quad h = \sum a_m h_1^{m_1} \cdots h_t^{m_t}, \ a_m \in \sum A^2, \ b \in A.$$

Consequently over C we have $g^{2d} \leq g^{2d} + |h| = g^{2d} + h = |b||f|$, and so

$$|g|^{2d+1} \leq |bg||f| \leq \left(1 + (bg)^2\right)|f|,$$

that is, *a)*.

For unions, let $a)$ hold with $l_i = 2d_i + 1, h_i$ over C_i, $i = 1, 2$. Suppose, for instance, $n = d_1 - d_2 \geq 0$. Then

$$|g|^{l_1} \leq h|f| \leq (1 + h^2)|f|, \quad h = 1 + h_1^2 + g^{2n}(1 + h_2^2),$$

which is $a)$, over $C_1 \cup C_2$. For $b)$ note that the latter inequality is strict on $\{f \neq 0\}$.

Now by Corollary 1.13 any closed constructible set is of the form $C_1 \cup \cdots \cup C_s$ for suitable basic closed sets C_i, and we are done. $\square$

2. Specializations, Zero Sets and Real Ideals

We continue considering the real spectrum of a commutative ring A with unit. First we study closures of points.

Proposition and Definition 2.1   *For two prime cones $\alpha, \beta \in \mathrm{Spec}_r(A)$ the following assertions are equivalent:*

a) $\beta \subset \alpha$ *(here α and β are seen as prime cones of A).*
b) $\alpha \in \mathrm{Adh}(\beta)$.
c) *For all $f \in A$, if $f(\beta) \geq 0$, then $f(\alpha) \geq 0$.*
d) *For all $f \in A$, if $f(\alpha) > 0$, then $f(\beta) > 0$.*

If these conditions hold, we call α a specialization *of β, and β a* generization *of α, and write $\beta \to \alpha$.*

Consequently, the proconstructible set $U_\alpha = \bigcap_{f(\alpha)>0}\{f > 0\}$ which we considered in 1.10 is exactly the set of all generizations of α.

Proposition and Definition 2.2   *We have:*

a) *The specializations of each $\alpha \in \mathrm{Spec}_r(A)$ form a chain.*
b) *Each $\alpha \in \mathrm{Spec}_r(A)$ has a unique specialization α' which is closed.*
c) *The subspace of all closed points of $\mathrm{Spec}_r(A)$ is compact. This subspace is called the* maximal real spectrum *of A and denoted by $\mathrm{Max}_r(A)$.*
d) *The map $\mathrm{Spec}_r(A) \to \mathrm{Max}_r(A); \alpha \mapsto \alpha'$ defined by b) is a closed continuous retraction.*

Proof.   [B-C-R 7.1.22-24], [Kn-Schd III.6, p.126-128], See also Corollary III.1.19. $\square$

Proposition 2.3   *Let $Y \subset \mathrm{Spec}_r(A)$ be proconstructible. Then*

$$\mathrm{Adh}(Y) = \{\alpha \in \mathrm{Spec}_r(A) \,|\, \beta \to \alpha \text{ for some } \beta \in Y\}.$$

Proof. Apparently, one has "$\subset$". Conversely, let $\alpha \in \mathrm{Adh}(Y)$. Then for any $f_1, \ldots, f_r \in A$ with $f_1(\alpha) > 0, \ldots, f_r(\alpha) > 0$ we have $U = \{f_1 > 0, \ldots, f_r > 0\} \cap Y \neq \emptyset$. Now, such a U is closed in the constructible topology, and Y is compact in that topology, so that the intersection of all those U's contains some element $\beta \in Y$, and for this we get $\beta \to \alpha$. $\qquad\square$

The next result presents a typical example of specializations:

Proposition 2.4   *Let A be a local henselian ring with residue field k (see Definition 7.1). Then for any $\beta \in \mathrm{Spec}_r(A)$ there is $\alpha \in \mathrm{Spec}_r(k)$ with $\beta \to \alpha$.*

Proof. For $a \in A$ let $\bar{a} \in k$ be the residue class. Now assume that $\bar{a} = 1$. Then by Hensel's lemma, $P(\mathsf{t}) = \mathsf{t}^2 + \mathsf{t} + 1 - a$ has a root in A, which shows that $a(\beta) > 0$. Thus $u(\beta)v(\beta) > 0$ for any two units $u, v \in A$ whose residue classes in k coincide. Therefore $\alpha = \{\bar{b} \mid b \in \beta\} \subset k$ is a prime cone of $\mathrm{Spec}_r(k) \subset \mathrm{Spec}_r(A)$ and $\beta \to \alpha$. $\qquad\square$

(2.5) Dimension. For $\alpha \in \mathrm{Spec}_r(A)$ we set

$$\dim(\alpha) = \textit{dimension of } \alpha = \dim(A/\mathrm{supp}(\alpha)),$$

and

$$\mathrm{ht}(\alpha) = \textit{height of } \alpha = \mathrm{ht}(\mathrm{supp}(\alpha)).$$

Moreover, for a subset $Y \subset \mathrm{Spec}_r(A)$, we set

$$\dim(Y) = \textit{dimension of } Y = \sup\{-1, \dim(\alpha) \mid \alpha \in Y\}.$$

In particular, we set $\dim_r(A) = \dim(\mathrm{Spec}_r(A))$. This notion of dimension depends clearly on the ring A under consideration. In order to have a more intrinsic notion we set, for a specialization $\beta \to \alpha$,

$$\dim(\beta \to \alpha) = \textit{dimension of } \beta \to \alpha = \dim(A_{\mathrm{supp}(\alpha)}/\mathrm{supp}(\beta));$$

if this dimension is d, we say that β is a *d-dimensional generization of α.* For $Y \subset \mathrm{Spec}_r(A)$ and $\alpha \in \mathrm{Spec}_r(A)$,

$$\dim_\alpha(Y) = \textit{dimension of } Y \textit{ at } \alpha = \sup\{-1, \dim(\beta \to \alpha) \mid \beta \in Y, \beta \to \alpha\}.$$

One has the following facts:

  a) If $\beta \to \alpha$, $\dim(\beta) \geq \dim(\alpha)$.
  b) If $Y \subset Y'$, $\dim(Y) \leq \dim(Y')$ and $\dim_\alpha(Y) \leq \dim_\alpha(Y')$ for all α.
  c) If Y is proconstructible, $\dim(Y) = \dim(\mathrm{Adh}(Y))$ and $\dim_\alpha(Y) = \dim_\alpha(\mathrm{Adh}(Y))$ for all α.

(For *c)* use Proposition 2.3.)

Next, we explain how the correspondence between ideals of A and zero sets of $\mathrm{Spec}_r(A)$ looks like. To that end we need another notion:

(2.6) Real Ideals. An ideal I is called *real* if for any $a_1, \ldots, a_r \in A$ with $a_1^2 + \cdots + a_r^2 \in I$ it follows $a_1, \ldots, a_r \in I$. A real ideal is clearly radical. For an arbitrary ideal $I \subset A$ the intersection of all real prime ideals $\mathfrak{p} \supset I$ is called the *real radical of* I and denoted by $\sqrt[r]{I}$. As a matter of fact, $\sqrt[r]{I}$ is the smallest real ideal containing I. For the following facts we refer to [B-C-R 4.1] or [Kn-Sch III.2, p.103-105]:

a) Let $\mathfrak{p} \in \mathrm{Spec}(A)$. Then $\mathfrak{p}$ is real if and only if the residue field $\kappa(\mathfrak{p})$ is formally real.

b) Let A be noetherian and I a radical ideal of A. Then I is real if and only if the minimal prime ideals containing I are real.

c) For any ideal I of A,

$$\sqrt[r]{I} = \{f \in A \mid f^{2m} + a_1^2 + \cdots + a_r^2 \in I \text{ for some } m \geq 1 \text{ and } a_i \in A\}$$

Note that $\mathrm{Spec}_r(A) = \mathrm{Spec}_r(A/\sqrt[r]{(0)})$. Also that $\dim_r(A) = \dim(A/\sqrt[r]{(0)})$, if all minimal primes of A are real (for instance, if A is noetherian).

(2.7) Zero Sets and Ideals. For a subset $I \subset A$ we define its *zero set* $\mathcal{Z}(I) \subset \mathrm{Spec}_r(A)$ by

$$\mathcal{Z}(I) = \{\alpha \in \mathrm{Spec}_r(A) \mid f(\alpha) = 0 \text{ for all } f \in I\} = \bigcap_{f \in I} \{f = 0\}.$$

Note that $\mathcal{Z}(I)$ is Zariski closed and that any Zariski closed subset of $\mathrm{Spec}_r(A)$ can be written in this way.

Conversely, for any subset $Y \subset \mathrm{Spec}_r(A)$ we define its *zero ideal* $\mathcal{J}(Y) \subset A$ by

$$\mathcal{J}(Y) = \{f \in A \mid f(\alpha) = 0 \text{ for all } \alpha \in Y\} = \bigcap_{\alpha \in Y} \mathrm{supp}(\alpha),$$

which is obviously a real ideal.

These operators $\mathcal{Z}$ and $\mathcal{J}$ verify the same elementary properties as their analogues in $\mathrm{Spec}(A)$. In particular,

$$\mathrm{Adh}_Z(Y) = \mathcal{Z}\mathcal{J}(Y)$$

for any subset $Y \subset \mathrm{Spec}_r(A)$. However the composition in the reverse order is determined by the less trivial abstract real Nullstellensatz:

Theorem 2.8 (Real Nullstellensatz) $\mathcal{J}\mathcal{Z}(I) = \sqrt[r]{I}$ *for any ideal $I \subset A$.*

Proof. As $\mathcal{J}\mathcal{Z}(I)$ is a real ideal that contains I, it must contain also $\sqrt[r]{I}$. Conversely, let $g \in \mathcal{J}\mathcal{Z}(I)$. This means that $\{g = 0\} \supset \bigcap_{f \in I}\{f = 0\}$ and by compactness of the constructible topology we find $f_1, \ldots, f_r \in I$ such that $\{g = 0\} \supset \{f_1 = \cdots = f_r = 0\}$, that is, $\{f_1 = \cdots = f_r = 0, g \neq 0\} = \emptyset$. The conclusion follows now from the Positivstellensatz (Theorem 1.14) and the description of the real radical given in 2.6 c). $\square$ We conclude this section by showing a result which will be needed in the fundamental Section V.2.

Proposition 2.9   *Let* $Y \subset \mathrm{Spec}_r(A)$ *be a proconstructible set whose zero ideal* $\mathfrak{p} = \mathcal{J}(Y)$ *is prime. Then there exists an element* $\alpha \in Y$ *such that* $\mathfrak{p} = \mathrm{supp}(\alpha)$. *Consequently,* $\mathrm{Adh}_Z(Y) = \mathcal{Z}(\mathrm{supp}(\alpha))$ *and* $\dim(Y) = \dim(\mathrm{Adh}_Z(Y))$.

Proof.   For any $f \notin \mathfrak{p}$ the set $Y_f = Y \setminus \{f = 0\}$ is non-empty and proconstructible. Now consider the family $\{Y_f \mid f \notin \mathfrak{p}\}$. Since $\mathfrak{p}$ is prime and $Y_{f_1 \cdot f_2} = Y_{f_1} \cap Y_{f_2}$, this family is stable under finite intersections. Furthermore, all the Y_f's are closed in the constructible topology, which is compact, so that $\bigcap_{f \notin \mathfrak{p}} Y_f \neq \emptyset$. Thus there exists some point α in that intersection, that is, $\alpha \in Y$ and $f(\alpha) \neq 0$ for $f \notin \mathfrak{p}$. Since obviously $\mathrm{supp}(\alpha) \supset \mathfrak{p}$, we conclude $\mathrm{supp}(\alpha) = \mathfrak{p}$.

$$\square$$

3. Real Valuations

Let K be a field. A *valuation of* K is given by a *valuation ring* $V \subset K$ whose field of fractions is K. This defines uniquely a group of values $\Gamma_v = \{\text{cyclic } V\text{−submodules of } K\}$ and a valuation map

$$v : K \to \Gamma_v \, ; \, a \mapsto v(a).$$

However we always write Γ_v additively with the reversed ordering. Thus $v(ab) = v(a) + v(b)$ for $a, b \in K$ and $v(0) = \infty$. We denote by $\mathfrak{m}_v$ the maximal ideal of V, by k_v its residue field $V/\mathfrak{m}_v$, and by $\lambda_v : K \to k_v \cup \infty$ the corresponding place; sometimes for $a \in V$ we may merely write $\overline{a}$ instead of $\lambda_v(a)$. If there is no risk of confusion we omit the subscript v. We denote the valuation of K both by its ring V and by its map v. For generalities on valuations we refer to [Bk CA, Ch.VI]. Apart from that, we omit the proofs which can be found in [B-C-R 10.1-2], [Kn-Sch II.1-8, p.50-78] or [Pr §§7-8].

Definition 3.1   *A valuation* v *of* K *is called* real *if its residue field* k_v *is formally real.*

Proposition and Definition 3.2   *For a valuation ring* V *of* K *and an element* $\beta \in \mathrm{Spec}_r(K)$ *(i.e. a total ordering* $< \, = <_\beta$ *of* K*) the following assertions are equivalent:*

    a) V is convex, i.e. if $x \in K, y \in V$ and $0 < x < y$, then $x \in V$.
    b) Any ideal $I \subset V$ is convex, i.e. if $x \in V, y \in I$ and $0 < x < y$, then $x \in I$.
    c) The maximal ideal $\mathfrak{m}_v$ of V is convex.
    d) If $x \in \mathfrak{m}_v$, then $1 + x > 0$.
    e) If $x \in \mathfrak{m}_v$ and $u \in V \setminus \mathfrak{m}_v$, then $\mathrm{sign}[u] = \mathrm{sign}[u + x]$.
    f) There exists $\alpha \in \mathrm{Spec}_r(k)$ such that $\beta \to \alpha$ in $\mathrm{Spec}_r(V)$. In particular, the valuation is real

If these conditions hold, V and β are called compatible.

Proof.   *a)$\Rightarrow$b)* Assume $0 < x < y$, $x \in V$, $y \in I$. Then $0 < x/y < 1$ and $z = x/y \in V$ by *a)*. Hence $x = zy \in I$.

b)$\Rightarrow$c) is obvious.

c)$\Rightarrow$d) If $1 + x < 0$ with $x \in \mathfrak{m}_v$, then $0 < 1 < -x$ and $1 \in \mathfrak{m}_v$ by *c)*.

d)$\Rightarrow$e) Since u is a unit, $v = u^{-1} \in V$ and $x/u = vx \in \mathfrak{m}_v$. Hence from $u(u + x) < 0$ we get $1 + x/u < 0$, against *d)*.

e)$\Rightarrow$f) The condition *e)* guarantees that we can define a total ordering in k as follows: any non-zero element of k is of the form $\overline{u}$ for some unit u of V, and we put $\overline{u} > 0$ in case $u > 0$. From this we get *f)*.

f)$\Rightarrow$a) Assume $0 < x < y$, $x \in K \setminus V$, $y \in V$. Then $1/x \in \mathfrak{m}_v$ and $z = y/x \in \mathfrak{m}_v$. But then $1 < z$, and since $\beta \to \alpha$, we deduce $1 \leq 0$ in k. Contradiction. $\quad\square$

The following result, due to Baer and Krull, is a kind of converse of property *f)* above:

Proposition 3.3   *Let V be a valuation ring of K and $\alpha \in \mathrm{Spec}_r(k_v)$.*

a) There exists $\beta_0 \in \mathrm{Spec}_r(K) \subset \mathrm{Spec}_r(V)$ such that $\beta_0 \to \alpha$.

b) There is a bijective map

$$\mathrm{Hom}(\Gamma_v, \{+1, -1\}) \to \{\beta \in \mathrm{Spec}_r(K) \,|\, \beta \to \alpha\}\,;\, \chi \mapsto \beta = \chi \cdot \beta_0,$$

defined by

$$\mathrm{sign}[f](\beta) = \chi(v(f)) \cdot \mathrm{sign}[f](\beta_0) \quad \text{(for } f \neq 0).$$

The set $\mathrm{Hom}(\Gamma_v, \{+1, -1\})$ can be seen as the group $\widehat{\Gamma_v}$ of the real characters of Γ_v; we shall study this in more detail in Chapter IV. As a trivial application of this proposition we get: *If $\Gamma_v \simeq \mathbb{Z}^d$, then α has 2^d generizations in $\mathrm{Spec}_r(K) \subset \mathrm{Spec}_r(V)$.* This situation appears very naturally in geometry connected to regular points of dimension d.

Lemma 3.4   *Let A be a regular local domain of dimension d with residue field k and quotient field K. Fix a regular system of parameters $x_1, \ldots, x_d$. Then there exists a unique valuation ring V of K with $\Gamma_v = \bigoplus_{1 \leq i \leq d} v(x_i)\mathbb{Z}$ (lexicographically ordered) and $k_v = k$.*

In particular, every ordering of k has at least 2^d generizations in $\mathrm{Spec}_r(A)$ which are orderings of K.

Proof.   Set $A_1 = A/(x_1)$. This is a regular local domain of dimension $d - 1$, and the classes $\overline{x}_2, \ldots, \overline{x}_d \in A_1$ form a regular system of parameters. On the other hand the quotient field K_1 of A_1 is the residue field of the rank 1 discrete valuation ring $A_{(x_1)} \subset K$. So the assertion follows by induction. $\quad\square$

Consider for instance a real closed field R and $A = R[\mathrm{x}_1, \mathrm{x}_2]_\mathfrak{m}$, where $\mathfrak{m} = (\mathrm{x}_1, \mathrm{x}_2)$, and let u_1, u_2 any other generators of $\mathfrak{m}$. This gives rise to a rank 2 discrete valuation ring V of K, with $k_v = R$ and $\mathrm{Spec}_r(k_v)$ consisting of a single point that we denote by α. Then the four generizations β_{ij} of α in $\mathrm{Spec}_r(V) \subset \mathrm{Spec}_r(A)$ can be visualized in the picture

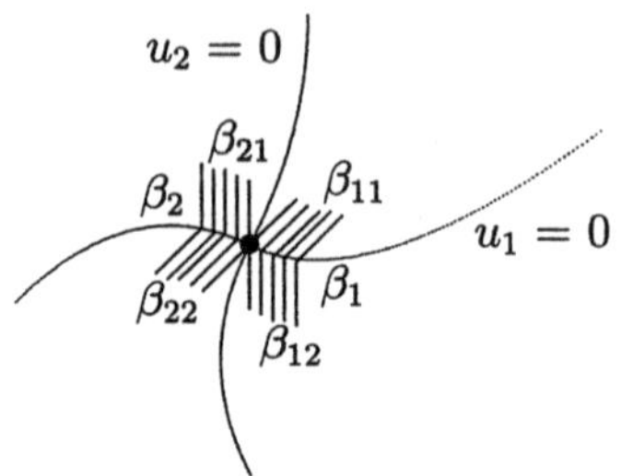

Here β_1 and β_2 stand for the generizations of α in $\mathrm{Spec}_r(A/(u_1)) = \{u_1 = 0\} \subset \mathrm{Spec}_r(A)$. By Theorem 1.6 we have actually the same situation as in Examples I.4.4 and ff. Note that we also visualize the four specialization chains $\beta_{ij} \to \beta_i \to \alpha$ in $\mathrm{Spec}_r(A)$. Of course one can draw a similar picture for any algebraic R-variety of dimension 2 at a non-singular real point, and one can imagine how the picture for higher dimensions would look like.

We recall that a *valued field* is a pair (K, v), where K is a field and v a valuation of K. An *extension of* (K, v) is another valued field (K', v') such that $K' \supset K$ and $v' \cap K = v$; we write then $(K', v') \supset (K, v)$. Such an extension is called *immediate* if the canonical embeddings $k_v \to k_{v'}$ and $\Gamma_v \to \Gamma_{v'}$ are isomorphisms.

Proposition 3.5   *Let $(K', v') \supset (K, v)$ be an immediate extension of valued fields and let $\beta \in \mathrm{Spec}_r(K)$ be compatible with v. Then there exists a unique $\beta' \in \mathrm{Spec}_r(K')$ which extends β and is compatible with v'.*

Proof.   This follows directly from Proposition 3.2 *f)* and Proposition 3.3.   □

Now let us fix a subring A of K. We are going to consider valuations of K in relation to A.

(3.6) Convex Hulls. Let $\beta \in \mathrm{Spec}_r(K)$. Then

$$V = \{x \in K \mid -a \leq_\beta x \leq_\beta a \ \text{ for some } a \in A\}$$

(the ring of elements *bounded by A*) is a valuation ring of K. In fact it is the smallest valuation ring that contains A and is compatible with β ([B-C-R 10.1.11]). This V is called the *convex hull of A in K with respect to β.*

In particular, if we take $A = \mathbb{Z}$ we get the smallest valuation ring $V = V(\beta)$ of K that is compatible with β. One sees from Proposition and Definition 3.2 that all bigger valuation rings are again compatible with β, and they are linearly ordered by inclusion.

Definition 3.7   *Let $\alpha \in \mathrm{Spec}_r(A)$. A valuation ring V of K is said to be centered at α if the following conditions hold:*

a) V dominates the local ring $A_{\mathrm{supp}(\alpha)}$, and

b) the total ordering α extends from $\kappa(\mathrm{supp}(\alpha))$ to k_v via the embedding $\kappa(\mathrm{supp}(\alpha)) \subset k_v$ induced by a).

Let us see how this situation occurs.

(3.8) Construction of Valuations Centered at a Prime Cone. Let $\alpha \in \mathrm{Spec}_r(A)$ and $\beta \in \mathrm{Spec}_r(K)$ be such that the restriction of β to A is a generization of α. Let V be the convex hull of $A_{\mathrm{supp}(\alpha)}$ in K with respect to β. Then V is centered at α.

Proof.   Indeed, suppose $x \in \mathrm{supp}(\alpha)$ but $x \notin \mathfrak{m}_v$. Then $1/x \in V$ and so there is $a \in A_{\mathrm{supp}(\alpha)}$ such that $-a \leq_\beta 1/x \leq_\beta a$. Taking $a + 1$ if necessary, we may assume a is a unit of $A_{\mathrm{supp}(\alpha)}$, which means that the residue class $\overline{a}$ in $\kappa(\mathrm{supp}(\alpha))$ is not zero. Moreover, we have $0 \leq_\beta 1/a \leq_\beta x$ and, since $\beta \to \alpha$ and $x \in \mathrm{supp}(\alpha)$, it follows $0 \leq_\alpha 1/\overline{a} \leq_\alpha 0$, a contradiction. This shows condition a), and b) is then immediate. $\square$

Now let W be another valuation of K centered at α and compatible with β. Then by definition $W \supset A_{\mathrm{supp}(\alpha)}$ and so $W \supset V$. Hence there is a prime ideal $\mathfrak{p}$ of V with $W = V_\mathfrak{p}$ and $\mathfrak{p} \cap A = \mathrm{supp}(\alpha)$. Conversely, by Proposition and Definition 3.2, any localization $V_\mathfrak{p}$ with $\mathfrak{p} \cap A = \mathrm{supp}(\alpha)$ is a valuation ring of K centered at α and compatible with β. Thus when β varies we obtain all valuations of K centered at α.

Let us look at a more general situation.

(3.9) Convex Hull Associated to a Specialization. Let A be an arbitrary commutative ring with unit, and $\beta \to \alpha$ a specialization in $\mathrm{Spec}_r(A)$. We set

$$A_{\beta\alpha} = A_{\mathrm{supp}(\alpha)}/\mathrm{supp}(\beta), \quad K = \mathrm{qf}(A_{\beta\alpha}) = \kappa(\mathrm{supp}(\beta)).$$

Now consider $\beta \to \alpha$ in $\mathrm{Spec}_r(A_{\beta\alpha})$, so that β is a total ordering of K. Then the convex hull of $A_{\beta\alpha}$ in K with respect to β is the smallest valuation ring of K centered at α and convex with respect to β. This convex hull will be denoted by $V_{\beta\alpha}$.

(3.10) Real Closed Valuation Rings Associated to a Specialization. We start with the same situation as above. Thus we have

$$A_{\beta\alpha} \subset \kappa(\mathrm{supp}(\beta)) \subset R_\beta,$$

where R_β is a real closed extension of $\kappa(\mathrm{supp}(\beta))$. We are going to construct a valuation ring $W_{\beta\alpha} \subset R_\beta$. This construction will be compatible with the equivalence relation of Definition 1.1, in the sense that for all commutative triangles

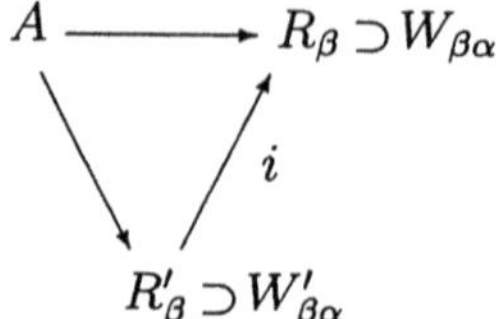

one has $W'_{\beta\alpha} = W_{\beta\alpha} \cap R'_\beta$. First, let $\mathcal{W}$ be the family of all convex valuation rings of R_β that dominate $A_{\beta\alpha}$. This family forms a chain and has not only a smallest member –the convex hull of $A_{\beta\alpha}$ in R_β– but also a biggest member: the union of all the elements of $\mathcal{W}$. This union is called the *real closed valuation ring associated to* $\beta \to \alpha$ and is denoted by $W_{\beta\alpha}$; its residue field will be denoted by $k_{\beta\alpha}$, its value group by $\Gamma_{\beta\alpha}$ and its place by $\lambda_{\beta\alpha}$. As was explained in 3.8, $W_{\beta\alpha} \subset R_\beta$ is centered at α. The valuation ring $W_{\beta\alpha}$ shares with the others in $\mathcal{W}$ three important properties:

a) They are henselian.
b) Their residue fields are real closed extensions of $\kappa(\mathrm{supp}(\alpha))$.
c) Their value groups are divisible.

These properties follow from the general result below, which can be deduced from the decomposition law for valuations (or [Pr §8]):

Proposition 3.11   *Let (R, v) be a real closed valued field, $V \neq R$. Then the value group Γ_v is divisible. Moreover the residue field k_v is algebraically closed or real closed. The valuation ring V is henselian if and only if k_v is real closed.*

The fact that makes the difference between $W_{\beta\alpha}$ and $V_{\beta\alpha}$ is the following:

Proposition 3.12   *Let A be a commutative ring with unit and consider a chain $\gamma \to \beta \to \alpha$ in $\mathrm{Spec}_r(A)$. Then we get a diagram*

$$
\begin{array}{ccc}
A_{\gamma\alpha} & \longrightarrow & W_{\gamma\alpha} \subset W_{\gamma\beta} \subset R_\gamma \\
\downarrow & & \Big\downarrow \lambda_{\gamma\beta} \Big| \quad \Big\downarrow \lambda_{\gamma\beta} \\
A_{\beta\alpha} & \longrightarrow & W_{\beta\alpha} \subset k_{\gamma\beta} = R_\beta
\end{array}
$$

Proof.   First of all we can always take $R_\beta = k_{\gamma\beta}$ by 3.10 a). Now, since $W_{\gamma\alpha}$ and $W_{\gamma\beta}$ are both convex valuation rings of R_γ, either $W_{\gamma\alpha} \subset W_{\gamma\beta}$ or $W_{\gamma\beta} \subset W_{\gamma\alpha}$. But $W_{\gamma\alpha}$ dominates $A_{\gamma\alpha}$, $W_{\gamma\beta}$ dominates $A_{\gamma\beta}$ and $\mathrm{supp}(\beta) \subset \mathrm{supp}(\alpha)$, so it must be $W_{\gamma\alpha} \subset W_{\gamma\beta}$. Furthermore, since $W_{\gamma\alpha}$ dominates $A_{\gamma\alpha}$, its image $\lambda_{\gamma\beta}(W_{\gamma\alpha})$ dominates $A_{\beta\alpha}$. By the maximality of $W_{\beta\alpha}$ we conclude $\lambda_{\gamma\beta}(W_{\gamma\alpha}) \subset W_{\beta\alpha}$. On the other hand, $\lambda_{\gamma\beta}^{-1}(W_{\beta\alpha})$ is a convex valuation ring of R_γ dominating $A_{\gamma\alpha}$, and now by the maximality of $W_{\gamma\alpha}$, we conclude $\lambda_{\gamma\beta}^{-1}(W_{\beta\alpha}) \subset W_{\gamma\alpha}$. Thus $\lambda_{\gamma\beta}(W_{\gamma\alpha}) \supset W_{\beta\alpha}$, as wanted. $\square$

Examples 3.13   Fix any field k and $\alpha \in \mathrm{Spec}_r(k)$. Let t be an indeterminate, $A = k[[t]]$ the ring of formal power series in t and $K = k((t))$ its quotient field. So $A = V$ is already a valuation ring of K, with $\Gamma_v = \mathbb{Z}$ and $k_v = k$. Furthermore, V is centered at α. In fact, by Proposition 3.2, α has two generizations α_+ and α_- in $\mathrm{Spec}_r(V)$, according to the conditions $t(\alpha_+) > 0$ and $t(\alpha_-) < 0$. In this situation we get:

  a) *The real closure* $\kappa(\alpha_+)$ *of* $\kappa(\mathrm{supp}(\alpha_+))$ *is the field of* formal Puiseux series *with coefficients in* $\kappa(\alpha)$, *i.e.*

  $$\kappa(\alpha_+) = \bigcup_{m \geq 1} \kappa(\alpha)((\mathrm{t}^{1/m}))$$

  with the obvious operations. The positive elements of $\kappa(\alpha_+)$ *are the series* $a_p \mathrm{t}^{\frac{p}{m}} + a_{p+1} \mathrm{t}^{\frac{p+1}{m}} + \cdots$ *with* $p \in \mathbb{Z}$ *and* $a_p > 0$

  b) *The real closed valuation ring* $W_{\alpha_+\alpha}$ *associated to* $\alpha_+ \to \alpha$ *is the* ring of formal Puiseux series *with coefficients in* $\kappa(\alpha)$, *i.e.*

  $$W_{\alpha_+\alpha} = \bigcup_{m \geq 1} \kappa(\alpha)[[\mathrm{t}^{1/m}]].$$

  The value group of $W_{\alpha_+\alpha}$ *is* $\mathbb{Q}$.

The explicit description of the ordering of $\kappa(\alpha_+)$ given in *a)* shows that the order topology of this field is in fact defined by the valuation w associated to $W_{\alpha_+\alpha}$: for every $n \geq 0$ the set $U^{(n)} = \{f \in W_{\alpha_+\alpha} \mid w(f) \geq n\}$ is a neighbourhood of 0, and the collection of all these $U^{(n)}$'s is a neighbourhood basis of 0. An easy but useful consequence is that every formal Puiseux series is the limit of its truncations.

$\square$

4. Real Going-Up and Real Going-Down

Let A be a commutative ring with unit and A' an A-algebra via a homomorphism $\varphi : A \to A'$.

Definition 4.1   *In the situation above we say that*

  a) *The point* $\gamma' \in \mathrm{Spec}_r(A')$ *lies over the point* $\gamma \in \mathrm{Spec}_r(A)$ *if* $\varphi^*(\gamma') = \gamma$.
  b) *The* real going-up *holds for* φ *if for any specialization* $\beta \to \alpha$ *in* $\mathrm{Spec}_r(A)$ *and any* $\beta' \in \mathrm{Spec}_r(A')$ *lying over* β, *there exists* $\alpha' \in \mathrm{Spec}_r(A')$ *lying over* α *such that* $\beta' \to \alpha'$.
  c) *The* real going-down *holds for* φ *if for any specialization* $\beta \to \alpha$ *in* $\mathrm{Spec}_r(A)$ *and any* $\alpha' \in \mathrm{Spec}_r(A')$ *lying over* α, *there exists* $\beta' \in \mathrm{Spec}_r(A')$ *lying over* β *such that* $\beta' \to \alpha'$.

Of course, these two notions are patterned on the standard going-up and going-down of commutative algebra, but they turn out to be quite a more involved matter, specially the going-down. Before proving they hold in some special situations, we deduce some useful properties they imply.

Proposition 4.2   *Let $\varphi : A \to A'$ be as above. Then:*

a) The real going-up holds for φ if and only if φ^* is a closed map.

b) If the real going-down holds for φ, then

$$(\varphi^*)^{-1}(\mathrm{Adh}(Y)) = \mathrm{Adh}((\varphi^*)^{-1}(Y))$$

for every proconstructible set $Y \subset \mathrm{Spec}_r(A)$.

c) If the real going-down holds for φ and A' is finitely presented, then φ^* is an open map.

d) If φ^* is an open map, then the going-down holds for φ.

Proof.   *a)* Suppose first that the going-up holds for φ, and let C be closed in $\mathrm{Spec}_r(A')$. Then $\varphi^*(C)$ is compact in the constructible topology, which is Hausdorff, and so it is closed in that topology. In other words, $\varphi^*(C)$ is proconstructible, and by Proposition 2.3 $\mathrm{Adh}(\varphi^*(C))$ consists of the specializations of $\varphi^*(C)$. Hence, given $\alpha \in \mathrm{Adh}(\varphi^*(C))$ there is $\beta' \in C$ such that $\varphi^*(\beta') \to \alpha$. Then, by the real going-up, $\beta' \to \alpha'$ for some $\alpha' \in \mathrm{Spec}_r(A')$. As C is closed and $\beta' \in C$, we conclude $\alpha' \in C$ and $\alpha = \varphi^*(\alpha') \in \varphi^*(C)$.

Now let φ^* be closed, and consider $\beta \to \alpha$ in $\mathrm{Spec}_r(A)$, $\beta' \in \mathrm{Spec}_r(A')$ lying over β. Then $\varphi^*(\mathrm{Adh}(\beta'))$ is a closed set that contains β, and since $\beta \to \alpha$ it must contain α too. Whence there is $\alpha' \in \mathrm{Adh}(\beta')$ lying over α, and we are done.

b) Since φ^* is continuous, the inclusion "$\supset$" is clear. Then pick $\alpha' \in (\varphi^*)^{-1}(\mathrm{Adh}(Y))$. We have $\varphi^*(\alpha') \in \mathrm{Adh}(Y)$ and by Proposition 2.3 there is $\beta \in Y$ such that $\beta \to \varphi^*(\alpha')$. Now by the real going-down there is a generization β' of α' lying over β. Thus $\beta' \in (\varphi^*)^{-1}(Y)$ and $\alpha' \in \mathrm{Adh}((\varphi^*)^{-1}(Y))$.

c) We have to show that for any open constructible set $U' \subset \mathrm{Spec}_r(A')$ its image $U = \varphi^*(U')$ is open in $\mathrm{Spec}_r(A)$. But, by Proposition 1.9, U is constructible, and according to Proposition 2.3 we shall show that if $\beta \to \alpha$ and $\alpha \in U$, then also $\beta \in U$. So pick $\alpha' \in U'$ lying over α. By the real going-down there is $\beta' \to \alpha'$, β' lying over β. Since U' is open, $\beta' \in U'$ and hence $\beta = \varphi^*(\beta') \in U$.

d) Consider $\beta \to \alpha$ in $\mathrm{Spec}_r(A)$, and $\alpha' \in \mathrm{Spec}_r(A')$ lying over α. Now let $\{U\}$ be the family of all open constructible neigbourhoods of α'. As φ^* is open, and $\beta \to \alpha$, we see that $\beta \in \varphi^*(U)$, or equivalently $(\varphi^*)^{-1}(\beta) \cap U \neq \emptyset$, for every U. As $(\varphi^*)^{-1}(\beta)$ is proconstructible, by compactness we deduce $(\varphi^*)^{-1}(\beta) \cap \bigcap U \neq \emptyset$, and any point β' in that intersection is a generization of α' lying over β. Thus the going-down holds for φ. $\square$

Now we prove:

Proposition 4.3 (Real Going-up for Integral Extensions)   *Let $\varphi : A \to A'$ be as above and suppose that A' is integral over $\varphi(A)$. Then the real going-up holds for φ.*

Proof. Since the real going-up is immediate in case φ is onto, we can suppose that φ is injective, and identify A with $\varphi(A)$, so that φ becomes an inclusion $A \subset A'$. Then, given $\beta \to \alpha$ in $\mathrm{Spec}_r(A)$ and $\beta' \in \mathrm{Spec}_r(A')$ lying over β, consider the diagram

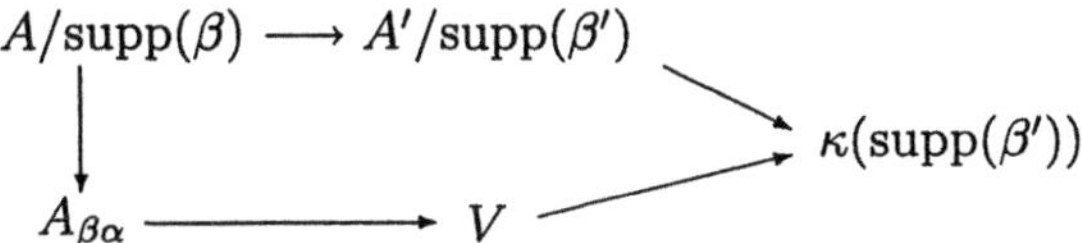

where V is the convex hull of $A_{\beta\alpha}$ in $\kappa(\mathrm{supp}(\beta'))$ with respect to β' and all the arrows are embeddings. Since $A'/\mathrm{supp}(\beta')$ is integral over $A/\mathrm{supp}(\beta)$ and V is integrally closed, we see that $A'/\mathrm{supp}(\beta') \subset V$. Then $\mathfrak{m}_v$ lies over $\mathfrak{p}/\mathrm{supp}(\beta')$ for some prime ideal $\mathfrak{p}$ of A', and V dominates the local ring $A'_{\mathfrak{p}}/\mathrm{supp}(\beta')$. Now V is convex with respect to β', and so $\beta' \to \alpha_1$ for some $\alpha_1 \in \mathrm{Spec}_r(k_v)$. Finally, let α' be the restriction of α_1 to the residue field of $A'_{\mathfrak{p}}/\mathrm{supp}(\beta')$, which is $\kappa(\mathfrak{p})$. Then $\beta' \to \alpha'$ and $\varphi^*(\alpha') = \alpha$. $\square$

Concerning the real going-down the situation is more difficult. First let us look at a typical example that illustrates the differences with the ordinary going-down.

Examples 4.4   Consider the homomorphism

$$\varphi : A = \mathbb{R}\{t\} \to A' = \mathbb{R}\{x_1, \ldots, x_n\} \,;\, x \mapsto x_1^2 + \cdots + x_n^2,$$

where the braces $\{\cdots\}$ mean convergent power series. The real going-down does not hold for φ: Take in A the specialization $\beta \to \alpha$ defined by:

$f(\beta) > 0$ if $f(t) > 0$ for all $t < 0$ close enough to 0,
$f(\alpha) > 0$ if $f(0) > 0$.

Then the prime cone α' of A' given by:

$f(\alpha') > 0$ if $f(0, \ldots, 0) > 0$

lies over α. However nothing lies over β, since $t(\beta) < 0$ and $\varphi(t) = x_1^2 + \cdots + x_n^2$ is always > 0.

Thus the going-down fails, despite the good properties of φ: its generic fiber is non-singular, and the special one is an isolated irreducible singularity, normal for $n \geq 3$ and factorial for $n \geq 5$. In Section VII.7 we shall look again at this and analyse more carefully the relationship between the real going-down and some regularity conditions on φ. $\square$

We end this section with:

Proposition 4.5 (Real Going-down for Polynomial Extensions) *The real going-down holds for the canonical inclusion* $\varphi : A \to A[t_1, \ldots, t_n]$, *where the* t_i's *are indeterminates.*

Proof. We consider only one indeterminate $t = t_1$, the general case following by induction. Thus let $\beta \to \alpha$ an specialization in $\mathrm{Spec}_r(A)$ and let $\alpha' \in \mathrm{Spec}_r(A[t])$ lie over α. We have $\kappa(\alpha) \subset k(\alpha')$ and $t(\alpha') \in k(\alpha')$. So we distinguish two cases:

Case 1: $t(\alpha') \in \kappa(\alpha)$. Then consider the real closed valuation $W_{\beta\alpha}$. Its residue field $k_{\beta\alpha}$ contains $\kappa(\alpha)$ and so there is some $\tau \in W_{\beta\alpha}$ whose residue class is $t(\alpha')$. Then we extend the canonical homomorphism $A \to W_{\beta\alpha}$ to $A[t]$ by sending $t \mapsto \tau$. The resulting map $A[t] \to W_{\beta\alpha}$ defines $\beta' \in \mathrm{Spec}_r(A[t])$, which does the job.

Case 2: $t(\alpha') \notin \kappa(\alpha)$. Then $t(\alpha')$ is transcendental over $\kappa(\alpha)$ and we may still denote $t = t(\alpha')$. Clearly, α' can be seen as a total ordering in $\kappa(\alpha)[t]$ (i.e. $\alpha' \in \mathrm{Spec}_r(\kappa(\alpha)[t])$ and $\mathrm{supp}(\alpha') = (0)$). Consider the place $W_{\beta\alpha} \to k_{\beta\alpha}$. Since $\kappa(\alpha) \subset k_{\beta\alpha}$ and $\kappa(\alpha)$ is real closed, there is a total ordering α^* in $k_{\beta\alpha}[t]$ which extends α'. Thus we have the diagram

$$
\begin{array}{ccccc}
A & \longrightarrow & A[t] & \longrightarrow W_{\beta\alpha}[t] & \subset k(\beta)[t] \\
\downarrow & & \downarrow & \downarrow & \\
\kappa(\alpha) & \longrightarrow & \kappa(\alpha)[t] & \longrightarrow k_{\beta\alpha}[t] &
\end{array}
$$

Now, in order to obtain β' we first lift α^* to some total ordering β^* of $W_{\beta\alpha}[t]$. For this consider the prime ideal $\mathfrak{m}_{\beta\alpha}[t]$ of $W_{\beta\alpha}[t]$, where $\mathfrak{m}_{\beta\alpha}$ is the maximal ideal of $W_{\beta\alpha}$. Then the localization $V = \left(W_{\beta\alpha}[t]\right)_{\mathfrak{m}_{\beta\alpha}[t]}$ is a valuation ring of $k(\beta)(t)$ with residue field $k_{\beta\alpha}(t)$.

In fact, let $x = P/Q$, where $P, Q \in W_{\beta\alpha}[t]$. Since $W_{\beta\alpha}$ is a valuation ring some coefficient c occuring in either P or Q divides all the others. Hence we can write

$$
x = c^{-1}P/c^{-1}Q, \quad c^{-1}P, c^{-1}Q \in W_{\beta\alpha}[t].
$$

Now some coefficient of either $c^{-1}P$ or $c^{-1}Q$ is 1 and consequently $c^{-1}P$, $c^{-1}Q$ are not both in $\mathfrak{m}_{\beta\alpha}[t]$. But this means that either $x \in V$ or $x^{-1} \in V$.

Now, by Proposition 3.3 *a)* we first lift α^* to a total ordering β_0 of $k(\beta)(t)$, then restrict this to a total ordering β^* of $W_{\beta\alpha}[t]$, and finally to a third one β' of $A[t]$. Clearly $\beta' \to \alpha'$ and $\varphi^*(\beta') = \beta$, $\square$

5. Abstract Semialgebraic Functions

Let A be a commutative ring with unit. In 1.4 we saw how the elements $f \in A$ can be considered as functions on $\mathrm{Spec}_r(A)$ with values in real closed fields. Of course, these are not ordinary functions, since the real closed fields involved are not unique. However, we used only statements like $f(\alpha) > 0$ for $\alpha \in \mathrm{Spec}_r(A)$, and these are well defined. So we also could define $|f|$ and $\mathrm{sign}[f]$. In this section we make the concept of a function more precise and introduce the so-called abstract semialgebraic functions. These functions will be used to study the topology of $\mathrm{Spec}_r(A)$.

(5.1) Functions. We recall that for homomorphisms $\alpha : A \to R_\alpha$ and $\beta : A \to R_\beta$ from A into real closed fields we defined the equivalence relation "$\sim$" which is generated by all commutative triangles

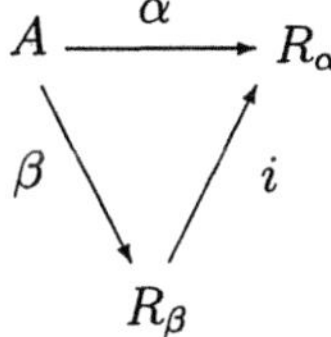

Now we fix for the rest of this section a proconstructible set $Y \subset \mathrm{Spec}_r(A)$. Consider an assignment

$$\alpha \mapsto f(\alpha) \in R_\alpha,$$

for all representatives $\alpha : A \to R_\alpha$ of points of Y. Such an assignment is called a *function on Y* if it is compatible with "$\sim$". Then $f(\alpha)$ is algebraic over $\kappa(\mathrm{supp}(\alpha))$. We denote by $\mathcal{F}(Y)$ the ring of all functions on Y. In particular, each $f \in A$ defines a function on Y, which we denote by $f|Y$. We denote by $A(Y)$ the set of all these "restrictions" $f|Y$, which apparently is a subring of $\mathcal{F}(Y)$.

Let $\Phi(\mathbf{x})$ be a first order formula with parameters in A and with one single free variable $\mathbf{x}$, and let f be a function on Y. Then the statement

$$R_\alpha \models \Phi(f(\alpha))$$

is well defined for $\alpha \in Y$. Indeed, this follows from model completeness (Theorem I.1.6). Now a function f on Y is called *definable* if there exists a first order formula $\Phi(\mathbf{x})$ with parameters in A and with one single free variable $\mathbf{x}$ such that

$$R_\alpha \models \forall \mathbf{x}(\Phi(\mathbf{x}) \leftrightarrow \mathbf{x} = f(\alpha)),$$

for all $\alpha \in Y$. We denote by $\mathcal{D}(Y)$ the set of all definable functions on Y.

Next we introduce a further condition which ensures continuity: A definable function f on Y is called *semialgebraic* if for all specializations $\beta \to \alpha$ in Y it holds $f(\beta) \in W_{\beta\alpha}$ and $\lambda_{\beta\alpha}(f(\beta)) = f(\alpha)$, where $W_{\beta\alpha}$ and $\lambda_{\beta\alpha}$ are the valuation ring and the place constructed in 3.10. We denote by $\mathcal{S}(Y)$ the set of all semialgebraic functions on Y.

Proposition 5.2   *Let $Y \subset \mathrm{Spec}_r(A)$ be proconstructible. Then $\mathcal{D}(Y)$ and $\mathcal{S}(Y)$ are subrings of $\mathcal{F}(Y)$ and we get the ring inclusions*

$$A(Y) \subset \mathcal{S}(Y) \subset \mathcal{D}(Y) \subset \mathcal{F}(Y).$$

Proof. First we show that $\mathcal{D}(Y)$ is a ring. So let $f, g \in \mathcal{D}(Y)$ have formulae $\Phi(\mathbf{x})$ and $\Psi(\mathbf{y})$ respectively. Then a formula for $f - g$ is

$$\exists \mathbf{x}, \mathbf{y}\left((\mathbf{z} = \mathbf{x} - \mathbf{y}) \wedge \Phi(\mathbf{x}) \wedge \Psi(\mathbf{y})\right),$$

and correspondingly we find a formula for fg. On the other hand, it is clear from the definition that $\mathcal{S}(Y)$ is a ring. Finally, a formula for a given $f \in A(Y)$ is $\mathbf{x} = f$, and f is semialgebraic by the construction of the place $\lambda_{\beta\alpha}$ associated to a specialization $\beta \to \alpha$. Hence $A(Y) \subset \mathcal{S}(Y)$. $\qquad\square$

We are going to present an alternative description of the ring $\mathcal{S}(Y)$.

(5.3) Sections. Let $\varphi : A \to A[\mathbf{t}]$ be the algebra of polynomials over A and $Y \subset \mathrm{Spec}_r(A)$ our proconstructible set. A section (on Y) $s : Y \to \mathrm{Spec}_r(A[\mathbf{t}])$ of the map $\varphi^* : \mathrm{Spec}_r(A[\mathbf{t}]) \to \mathrm{Spec}_r(A)$ is called *definable* if its image $s(Y) \subset \mathrm{Spec}_r(A[\mathbf{t}])$ is constructible in $(\varphi^*)^{-1}(Y)$ (see Proposition 1.11).

For a definable section s on Y the following assertions are equivalent:

  a) $s : Y \to \mathrm{Spec}_r(A[\mathbf{t}])$ *is continuous.*
  b) If $\beta \to \alpha$ *in* Y, *then* $s(\beta) \to s(\alpha)$ *in* $\mathrm{Spec}_r(A[\mathbf{t}])$.

(This follows easily from Proposition 2.3.) However, a third condition is still needed: A section s on Y is called *semialgebraic* if it is definable, continuous and, moreover, its image $s(Y)$ is closed in $(\varphi^*)^{-1}(Y)$. There is an example showing that the latter condition does not follow from the other two.

Examples 5.4 We consider $f \in A$ as a semialgebraic section on Y in the following way. Let $\alpha : A \to R_\alpha$ be a representative of a point $\alpha \in Y$. We extend α to $\gamma : A[\mathbf{t}] \to R_\alpha$ by putting $\gamma(\mathbf{t}) = f(\alpha)$. Thus we get a section, to be denoted by

$$s_f : Y \to \mathrm{Spec}_r(A[\mathbf{t}])\,; \ \alpha \mapsto \gamma.$$

Clearly the image of s_f is the constructible set

$$s_f(Y) = \{\gamma \in (\varphi^*)^{-1}(Y) \,|\, \mathbf{t} - f = 0\},$$

which is closed in $(\varphi^*)^{-1}(Y)$. Furthermore, s_f is continuous. In fact, suppose $\beta \to \alpha$ in Y. Then the prime cones $s_f(\beta)$ and $s_f(\alpha)$ are

$$s_f(\beta) = \{p(\mathbf{t}) \in A[\mathbf{t}] \,|\, p(f)(\beta) \geq 0\},$$
$$s_f(\alpha) = \{p(\mathbf{t}) \in A[\mathbf{t}] \,|\, p(f)(\alpha) \geq 0\}.$$

Since $p(f) \in A$ we conclude $s_f(\beta) \subset s_f(\alpha)$, that is, $s_f(\beta) \to s_f(\alpha)$.

By an analogous construction we get a correspondence between functions and sections more generally: Let f be a function on Y and $\alpha \in Y$, say $\alpha : A \to R_\alpha$. Then we extend α to $\gamma : A[\mathbf{t}] \to R_\alpha$ by putting $\gamma(\mathbf{t}) = f(\alpha)$. Then $\gamma \in \mathrm{Spec}_r(A[\mathbf{t}])$ is well defined and we set $s_f(\alpha) = \gamma$. Thus we obtain a section on Y, $s_f : Y \to \mathrm{Spec}_r(A[\mathbf{t}])$.

For the converse, we first look at the fibers of φ^*. Let $\alpha \in Y$ and $\gamma \in (\varphi^*)^{-1}(\alpha)$, say $\gamma : A[t] \to R_\gamma$. Then γ factorizes in the form

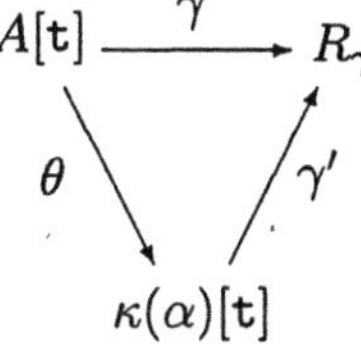

where θ is the canonical extension of $\alpha : A \to \kappa(\alpha)$. This shows that θ^* is a continuous bijection $\gamma' \mapsto \gamma$ from $\mathrm{Spec}_r(\kappa(\alpha)[t])$ onto the fiber $(\varphi^*)^{-1}(\alpha)$ (actually it is a homeomorphism, but we do not need this here). Now assume that s is a definable section on Y and set $\gamma = s(\alpha)$. Since s is definable, $\gamma = s(Y) \cap (\varphi^*)^{-1}(\alpha)$ is a constructible point of $(\varphi^*)^{-1}(\alpha)$, and consequently $\gamma' = (\theta^*)^{-1}(\gamma)$ is a constructible point of $\mathrm{Spec}_r(\kappa(\alpha)[t])$. Hence $\gamma'(t) = a$ for some element $a \in \kappa(\alpha) \subset R_\gamma$ (Examples 1.8) and we define $f_s(\alpha) = a$. Then f_s is a well defined function on Y.

Proposition 5.5   *With the notations introduced above:*

    a) The assignment $f \mapsto s_f$ defines a bijection from $\mathcal{D}(Y)$ onto the set of all definable sections on Y, whose inverse is $s \mapsto f_s$.

    b) Under this bijection a function f is semialgebraic if and only if the associated section s is semialgebraic.

Proof.   a) Let $f \in \mathcal{D}(Y)$ and let $\Phi(\mathbf{x})$ be a formula for f. Then $\Phi(\mathbf{t})$ is a sentence with parameters in $A[t]$ and this sentence defines a constructible set C in $(\varphi^*)^{-1}(Y) \subset \mathrm{Spec}_r(A[t])$. Clearly $s_f(Y) = C$ and so the section s_f is definable. Conversely, for a definable section s on Y the constructible set $s(Y) \subset (\varphi^*)^{-1}(Y)$ is defined by some sentence $\Phi(\mathbf{t})$ with parameters in $A[t]$. Then we get a formula $\Phi(\mathbf{x})$ for the function f_s by replacing each occurrence of t by the variable $\mathbf{x}$.

    b) Assume that s is a semialgebraic section on Y, and let $\beta \to \alpha$ in Y, say $\beta : A \to R_\beta$. We have to show that $f_s(\beta) \in W_{\beta\alpha}$ and $\lambda_{\beta\alpha}(f_s(\beta)) = f_s(\alpha)$ (see 3.10). Let $\delta = s(\beta), \gamma = s(\alpha)$. By assumption $\delta \to \gamma$ in $\mathrm{Spec}_r(A[t])$. Since $\gamma(t) = f_s(\alpha)$ is algebraic over $\kappa(\mathrm{supp}(\alpha))$, we find $g, h \in A$ such that

$$g(\alpha) \neq 0,\ |\mathbf{t}(\gamma)| < h(\alpha)/g(\alpha).$$

As $\varphi^*(\gamma) = \alpha$ we can also write $-h(\gamma) < \mathbf{t}(\gamma)g(\gamma) < h(\gamma)$, and from $\delta \to \gamma$ we obtain $-h(\delta) < \mathbf{t}(\delta)g(\delta) < h(\delta)$. Now $\varphi^*(\delta) = \beta$, so that

$$|\mathbf{t}(\delta)| < h(\beta)/g(\beta).$$

This means that $\mathbf{t}(\delta) = f_s(\beta) \in R_\beta$ is bounded by an element of $A_{\beta\alpha}$. Consequently $f_s(\beta) \in W_{\beta\alpha}$. We deduce also that $A[t]/\mathrm{supp}(\delta) \subset W_{\beta\alpha}$, so that

$W_{\beta\alpha}$ dominates a local ring $A[\mathbf{t}]_{\mathfrak{q}}/\mathrm{supp}(\delta)$ for some prime ideal $\mathfrak{q} \subset A[\mathbf{t}]$ with $\mathfrak{q} \supset \mathrm{supp}(\delta)$. Now consider the diagram

$$A/\mathrm{supp}(\beta) \subset A[\mathbf{t}]/\mathrm{supp}(\delta) \subset W_{\beta\alpha} \subset R_\beta = R_\delta$$

$$\downarrow \qquad\qquad\qquad \downarrow \qquad\qquad \downarrow \lambda_{\beta\alpha} \qquad\qquad\qquad\qquad (*)$$

$$A/\mathrm{supp}(\alpha) \subset \quad A[\mathbf{t}]/\mathfrak{q} \quad \subset k_{\beta\alpha}$$

The restriction of the total ordering of $k_{\beta\alpha}$ defines $\gamma_1 \in \mathrm{Spec}_r(A[\mathbf{t}])$ with support $\mathfrak{q}$ such that $\delta \to \gamma_1$. By construction $\varphi^*(\gamma_1) = \alpha$ and $\lambda_{\beta\alpha}(\mathbf{t}(\delta)) = \mathbf{t}(\gamma_1)$. Finally, $\delta \in s(Y)$, which is closed, so that $\gamma_1 \in s(Y)$ too. Thus $\gamma_1 = \gamma$ and

$$\lambda_{\beta\alpha}(f_s(\beta)) = \lambda_{\beta\alpha}(\mathbf{t}(\delta)) = \mathbf{t}(\gamma_1) = \mathbf{t}(\gamma) = f_s(\alpha).$$

Conversely, let $f \in \mathcal{S}(Y)$ and let $\beta \to \alpha$ in Y, say $\beta : A \to R_\beta$. We consider $s_f(\beta)$ and $s_f(\alpha)$ as prime cones in $A[\mathbf{t}]$. Then

$$s_f(\beta) = \{p(\mathbf{t}) \in A[\mathbf{t}] \mid p_\beta(f(\beta)) \geq 0\},$$
$$s_f(\alpha) = \{p(\mathbf{t}) \in A[\mathbf{t}] \mid p_\alpha(f(\alpha)) \geq 0\},$$

where $p_\alpha(\mathbf{t})$ is obtained by replacing the coefficients a_i of $p(\mathbf{t})$ by $\alpha(a_i)$, and $p_\beta(\mathbf{t})$ correspondingly. By assumption $p_\beta(f(\beta)) \in W_{\beta\alpha}$ and

$$\lambda_{\beta\alpha}\big(p_\beta(f(\beta))\big) = p_\alpha\big(\lambda_{\beta\alpha}(f(\beta))\big) = p_\alpha(f(\alpha)).$$

Since $\lambda_{\beta\alpha}$ preserves the ordering, we see that $s_f(\beta) \subset s_f(\alpha)$, i.e. $s_f(\beta) \to s_f(\alpha)$.

It remains to prove that $s_f(Y)$ is closed in $(\varphi^*)^{-1}(Y)$. So let $\gamma \in \mathrm{Adh}(s_f(Y))$, $\alpha = \varphi^*(\gamma) \in Y$. Then our claim is that $s_f(\alpha) = \gamma$. Since $s_f(Y)$ is constructible, there exists $\beta \in Y$ with $\delta = s_f(\beta) \to \gamma$. Then γ and $s_f(\alpha)$ are both specializations of $\delta = s_f(\beta)$ in $(\varphi^*)^{-1}(\alpha)$, and by Proposition 2.2 either $\gamma \to s_f(\alpha)$ or $s_f(\alpha) \to \gamma$. But since $s_f(\alpha)$ is closed in $(\varphi^*)^{-1}(\alpha)$ it must be $\gamma \to s_f(\alpha)$. Again by assumption, we obtain the diagram $(*)$ above, and $\gamma_1 \in \mathrm{Spec}_r(A[\mathbf{t}])$ such that $\delta \to \gamma_1$, $\varphi^*(\gamma_1) = \alpha$. Moreover $\mathbf{t}(\gamma_1) = \lambda_{\beta\alpha}(\mathbf{t}(\delta)) = \mathbf{t}(s_f(\alpha))$, so $s_f(\alpha) = \gamma_1$. On the other hand, $A[\mathbf{t}]_{\delta\gamma} \subset R_\delta = R_\beta$ dominates $A_{\beta\alpha}$, and so $W_{\delta\gamma} \subset W_{\beta\alpha} = W_{\delta\gamma_1}$. Thus $\gamma_1 \to \gamma$. Summing up we get $s_f(\alpha) = \gamma_1 \to \gamma \to s_f(\alpha)$, hence $s_f(\alpha) = \gamma$.
$\qquad\square$

Proposition 5.6   *Let f be a function on Y. Then:*

a) If $f \in \mathcal{D}(Y)$, the set $\{f > 0\} = \{\alpha \in Y \mid f(\alpha) > 0\}$ is constructible in Y.
b) If $f \in \mathcal{S}(Y)$ the set $\{f > 0\}$ is also open in Y.

Proof.   *a)* Let $\Phi(\mathbf{x})$ be a formula for f. Then

$$\{f > 0\} = \{\alpha \in Y \mid R_\alpha \models \exists \mathbf{x}(\mathbf{x} > 0 \wedge \Phi(\mathbf{x}))\}.$$

b) According to the preceding proposition we consider f as a semialgebraic section s_f. Then

$$\{f > 0\} = \varphi^*(s_f(Y) \cap \{\mathbf{t} > 0\}) = (s_f)^{-1}(\{\mathbf{t} > 0\}),$$

which is open in Y.
$\qquad\square$

6. Cylindrical Decomposition

Again we fix a commutative ring A with unit. We shall use the abstract semialgebraic functions of the previous section in order to decompose $\operatorname{Spec}_r(A)$ according to a system of polynomials. The construction is well known in semialgebraic geometry [B-C-R 2.3] and can be generalized to the present situation without difficulty.

Proposition 6.1   *Let $p_1, \ldots, p_s \in A[\mathbf{t}]$. Then there is a partition of $\operatorname{Spec}_r(A)$ into constructible sets $C_0, \ldots, C_m$, and for each $i = 1, \ldots, m$ a collection of semialgebraic functions $f_{i1}, \ldots, f_{ie_i}$ on C_i such that*

a) For $\alpha \in C_0$, all the polynomials $(p_1)_\alpha, \ldots, (p_s)_\alpha$ vanish.

b) For $\alpha \in C_i$, $f_{i1}(\alpha) < \cdots < f_{ie_i}(\alpha)$ are the different real roots of the non-zero polynomials among $(p_1)_\alpha, \ldots, (p_s)_\alpha \in \kappa(\alpha)[\mathbf{t}]$.

c) For $\alpha \in C_i$ and $a \in \kappa(\alpha)$ the sign of every $(p_l)_\alpha(a)$ depends only on the signs of $a - f_{i1}(\alpha), \ldots, a - f_{ie_i}(\alpha)$.

(Recall that for $p \in A[\mathbf{t}]$ we get $p_\alpha \in \kappa(\alpha)[\mathbf{t}]$ by applying α to the coefficients of p.)

Proof.   We may assume that the family $p_1, \ldots, p_s$ is stable under derivation, after adjoining the missing derivatives. In fact, at the end we shall have to drop those f_{ij} which do not correspond to the roots of the original p_i's. Let $C_0 \subset \operatorname{Spec}_r(A)$ be the set of the α's for which all $(p_l)_\alpha$ vanish; this set C_0 is constructible and we have condition *a)*. Next consider all systems of conditions on $\alpha \in \operatorname{Spec}_r(A)$ of the following form:

i) $(p_l)_\alpha$ vanishes for $l \in \Lambda$ and $(p_l)_\alpha$ does not vanish for $l \notin \Lambda$, where $\Lambda \neq \{1, \ldots, s\}$.

ii) The $(p_l)_\alpha, l \in \Lambda$, all together have exactly e different real roots, say $a_1 < a_2 < \cdots < a_e \in R_\alpha$, with $e \leq \sum_l \deg(p_l)$.

iii) Each $(p_l)_\alpha, l \in \Lambda$, has a given constant sign on everyone of the sets

$$(-\infty, a_1), \{a_1\}, (a_1, a_2), \{a_2\}, \ldots, (a_{e-1}, a_e), \{a_e\}, (a_e, \infty).$$

There are $m < \infty$ choices of such a system of conditions, and each one can be expressed by means of a sentence Φ_i. Then we put $C_i = \{\alpha \in \operatorname{Spec}_r(A) \mid R_\alpha \models \Phi_i\}$ for $i = 1, \ldots, m$. Moreover, for $\alpha \in C_i$ we set $f_{ij}(\alpha) = a_j$ according to condition *i)*. Then by definition the f_{ij}'s are definable functions on C_i for which conditions *b)* and *c)* hold true. It remains to show the validity of the continuity condition of 5.1. So let $\beta \to \alpha$ be a specialization in C_i. Consider the real closed valuation $W_{\beta\alpha}$ (see 3.10) and the corresponding diagram

$$A_{\beta\alpha} \longrightarrow W_{\beta\alpha} \subset k(\beta)$$

$$\downarrow \qquad\quad \downarrow \lambda_{\beta\alpha}$$

$$k(\mathrm{supp}(\alpha)) \subset k_{\beta\alpha}$$

Since the family $p_1, \ldots, p_s$ is stable under derivation, $f_{ij}(\alpha)$ is a simple root of some $(p_l)_\alpha$. But $(p_l)_\beta \in W_{\beta\alpha}[\mathrm{t}]$ and its residue class in $k_{\beta\alpha}[\mathrm{t}]$ is $(p_l)_\alpha$. Then, as $W_{\beta\alpha}$ is henselian (3.10 a)), there exists $t_j \in W_{\beta\alpha}$ with

$$(p_l)_\beta(t_j) = 0 \quad \text{and} \quad \lambda_{\beta\alpha}(t_j) = f_{ij}(\alpha).$$

Since $W_{\beta\alpha}$ is convex, we get $t_1 < \cdots < t_{e_i}$. Whence $t_j = f_{ij}(\beta)$ and we are done. $\qquad\qquad\qquad\qquad\qquad\qquad\qquad\qquad\qquad\qquad\qquad\qquad\qquad\qquad\square$

For the sequel we are interested in the extension $\varphi : A \to A[\mathrm{t}]$ and in $\mathrm{Spec}_r(A[\mathrm{t}])$. Therefore we consider semialgebraic functions on $C \subset \mathrm{Spec}_r(A)$ as semialgebraic sections $C \to \mathrm{Spec}_r(A[\mathrm{t}])$, according to Proposition 5.5. We shall use the above f_{ij} for the construction of cylindrical decompositions. First we need:

Proposition and Definition 6.2   *Let $C \subset \mathrm{Spec}_r(A)$ be constructible (and connected). Then also the following subsets of $\mathrm{Spec}_r(A[\mathrm{t}])$ are constructible (and connected):*

    a) The graph

$$C_f = \{f(\alpha) \in \mathrm{Spec}_r(A[\mathrm{t}]) \,|\, \alpha \in C\}$$

      of any semialgebraic function f on C.

    b) The slice

$$C_{fg} = \{\gamma \in \mathrm{Spec}_r(A[\mathrm{t}]) \,|\, \varphi^*(\gamma) \in C \text{ and } f(\varphi^*(\gamma)) < \mathrm{t}(\gamma) < g(\varphi^*(\gamma))\}$$

of any two semialgebraic functions $f < g$ on C (including the cases $f \equiv -\infty$, $g \equiv +\infty$).

Proof.   Part a) is obvious. Moreover, slices can be defined by first order sentences, so they are constructible. The fiber over $\alpha \in C$ of the map $\varphi^* : C_{fg} \to C$ is the set

$$\{\gamma \in (\varphi^*)^{-1}(\alpha) \,|\, f(\alpha) < \mathrm{t}(\gamma) < g(\alpha)\}$$

and this is the image of the interval

$$(f(\alpha), g(\alpha)) \subset \mathrm{Spec}_r(\kappa(\alpha)[\mathrm{t}])$$

by the continuous bijection $\theta^* : \mathrm{Spec}_r(\kappa(\alpha)[\mathrm{t}]) \to (\varphi^*)^{-1}(\alpha)$ described in 5.4 to interpret f as a section. Since the interval is connected, so is the fiber over α. If also C is connected, we conclude that C_{fg} is connected too. $\qquad\qquad\square$

Proposition 6.3 (Cylindrical Decomposition)   *Let $C \subset \operatorname{Spec}_r(A[\mathsf{t}])$ be a constructible set. Then there is a partition of $\operatorname{Spec}_r(A)$ into constructible sets $C_0, \ldots, C_m$ such that, for each $i = 0, \ldots, m$, $(C \cap \varphi^*)^{-1}(C_i)$ is a disjoint union of graphs and slices over C_i.*

Proof.   Consider the polynomials $p_1, \ldots, p_s \in A[\mathsf{t}]$ which occur in some quantifier free description of C. Then $\gamma \in C$ if and only if certain sign conditions hold for these $p_l(\gamma)$'s. But γ can be identified with some $a \in \kappa(\alpha)$, where $\alpha = \varphi^*(\gamma)$, such that $p_l(\gamma) = (p_l)_\alpha(a)$. Thus the result follows from Proposition 6.1. $\square$

7. Real Strict Localization

We have already seen (Proposition 2.4) that henselian rings have a good behaviour with respect to convexity and specialization in the real spectrum. This, together with other properties that we shall describe in the sequel, make henselian rings specially well suited for real algebra. In this section we shall study them in some detail, as well as the way to attach to a given local ring a natural henselian ring. The general reference for henselian rings is [Rd].

We shall work always with commutative local rings with unit and the homomorphisms are always assumed to be local. Following the standard usage, the maximal ideal of a local ring A will be denoted by $\mathfrak{m}_A$, the residue field by k_A and the canonical map $A \to k_A$ by $a \mapsto \overline{a}$.

Definition 7.1   *A local ring A is called* henselian *if every monic polynomial in $A[\mathsf{t}]$ whose class mod $\mathfrak{m}_A$ in $k_A[\mathsf{t}]$ has a simple root $x \in k_A$ has a root $a \in A$ such that $\overline{a} = x$.*

As an example we remember that any complete local ring is henselian [Nag 30.3, p. 104]. The key notion to deal with henselian rings in our context is the following:

Definition 7.2   *A local A-algebra B is called* local-etale A-algebra *if it is isomorphic to $A[\mathsf{t}]_{\mathfrak{p}}/(f)$, where $\mathfrak{p}$ is a prime ideal of $A[\mathsf{t}]$ lying over $\mathfrak{m}$ and $f \in \mathfrak{p}$ is a polynomial whose derivative f' with respect to the indeterminate t does not belong to $\mathfrak{p}$.*

Here there is a useful characterization of local-etale A-algebras:

Proposition 7.3 (Jacobian Criterion)   *Put $C = A[\mathsf{x}_1, \ldots, \mathsf{x}_n]$ and let $I, \mathfrak{p}$ be ideals of C with $I \subset \mathfrak{p}$ and $\mathfrak{p}$ prime. Then $B = C_{\mathfrak{p}}/I \cdot C_{\mathfrak{p}}$ is a local-etale A-algebra if and only if there are $f_1, \ldots, f_n \in I$ which generate the localization $I \cdot C_{\mathfrak{p}}$ and such that*

$$\det\big(\partial(f_1, \ldots, f_n)/\partial(\mathsf{x}_1, \ldots, \mathsf{x}_n)\big) \notin \mathfrak{p}.$$

Proof.   [Rd Ch.V, Th.5, p.60]. $\square$

The connection between henselian rings and local-etale algebras is described by:

Proposition 7.4   *The following assertions are equivalent:*

  a) A is henselian.
  b) For every local-etale A-algebra B such that the canonical homomorphism $k_A \to k_B$ is bijective, the structure homomorphism $A \to B$ is bijective.

Proof. [Rd Ch.VII, Prop.3, p.76].   □

This proposition gives the clue to produce henselian rings attached to a given local ring A: one has to consider direct limits of local-etale A-algebras. These limits have a distinguished name:

Definition 7.5   *An A-algebra B is called a* local-ind-etale *A-algebra if it is the direct limit of a filtrant family of local-etale A-algebras, the transition homomorphisms being local.*

The following properties are consequences of the definition and the properties of local-etale A-algebras:

Proposition 7.6   *Let B be a local-ind-etale A-algebra. Then:*

  a) For any integer q, $\mathfrak{m}_A^q B = \mathfrak{m}_B^q$.
  b) B is reduced (resp. normal, regular) if and only if A is so.
  c) B is noetherian if and only if A is so. In that case, for every prime ideal $\mathfrak{p}$ of A, the ring $B \otimes_A \kappa(\mathfrak{p})$ is a finite product of separable algebraic extensions of $\kappa(\mathfrak{p})$.
  d) For every local A-algebra C the canonical map

$$\mathrm{Hom}_{A-\mathrm{loc}}(B, C) \to \mathrm{Hom}_{k_A}(k_B, k_C)$$

  is injective. If moreover C is henselian, this map is bijective.

Proof.   *a)-c)* are [Rd Ch.VIII, Th.3, p.94] and *d)* is [Rd Ch.VIII, Prop.1, p.81].
                                                                   □

Proposition and Definition 7.7   *We have:*

  a) The family $\{A_i\}$ of local-etale A-algebras such that the canonical homomorphism $k_A \to k_{A_i}$ is bijective is filtrant with respect to the partial order relation: $i \le j$ if there is a local A-homomorphism $\varphi_{ij} : A_i \to A_j$.
  b) The direct limit B of the family $\{A_i\}$ of a) is a local-ind-etale A-algebra and the canonical homomorphism $k_A \to k_B$ is bijective. Moreover it is a henselian ring.
  c) The ring B verifies the following universal property: For every local homomorphism $\varphi : A \to C$ into a henselian ring C there exists a unique homomorphism $\varphi' : B \to C$ of local A-algebras making commutative the diagram

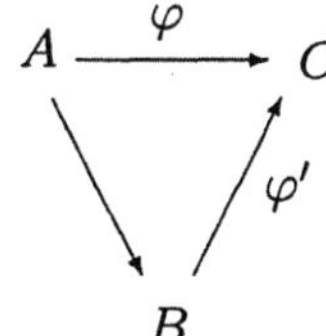

The ring B is called the henselization *of A and denoted by A^h; likewise, the homomorphism φ' of c) is called the* henselization *of φ and denoted by φ^h.*

For every ideal $I \subset A$ there is a canonical isomorphism $(A/I)^h = A^h/IA^h$.

Proof. It is completely analogous to the one of Proposition 7.10 below and we omit it.   □

(7.8) Henselization of Integrally Closed Domains. We describe here a more classical construction of the henselization ([Rd Ch.X, §2, p.110]).

Let A be a local domain which is integrally closed in its quotient field K and fix a separably algebraic closure L of K. The set $B^* \subset L$ of elements which are integral over A is called the *separable integral closure of A.* Then choose a maximal ideal $\mathfrak{n}^*$ of B^*. The group Σ of all K-automorphisms σ of L such that $\sigma(\mathfrak{n}^*) = \mathfrak{n}^*$ is called the *splitting group of $\mathfrak{n}^*$*, and the invariant subring $B \subset B^*$ of Σ is called the *splitting ring of $\mathfrak{n}^*$*. Put $\mathfrak{n} = \mathfrak{n}^* \cap B$ and consider the localization $B_{\mathfrak{n}}$. This ring $B_{\mathfrak{n}}$ is uniquely determined up to isomorphism over A, by the conjugacy properties of the Galois group of L over K, and it is isomorphic to the henselization of A: $A^h = B_{\mathfrak{n}}$.

Proposition 7.9   *The henselization V^h of a valuation ring V is a valuation ring. Furthermore $V^h \supset V$ is an immediate extension.*

Proof.   A valuation ring is integrally closed in its quotient field, and the henselization V^h can be constructed as shown in 7.8. But this is the standard definition of henselization of a valuation, and the result is well-known ([Rb Ch.F, Th.2, Th.3 and Cor.1, p. 177-184]).   □

Now consider a point $\alpha \in \mathrm{Spec}_r(A)$ with $\mathrm{supp}(\alpha) = \mathfrak{m}_A$. We fix in k_A the ordering induced by α.

Proposition and Definition 7.10   *We have:*

    a) *The family $\{A_i\}$ of local-etale A-algebras such that k_{A_i} is an algebraic ordered extension of k_A via the canonical homomorphism $k_A \to k_{A_i}$ is filtrant with respect to the partial order relation: $i \leq j$ if there is a local A-homomorphism $\varphi_{ij} : A_i \to A_j$ such that the induced homomorphism $k_{A_i} \to k_{A_j}$ is order preserving.*

    b) *The direct limit B of the family $\{A_i\}$ of a) is a local-ind-etale A-algebra and k_B is algebraic over k_A via the canonical homomorphism $k_A \to k_B$. Moreover it is a henselian ring with real closed residue field (and consequently $k_B = \kappa(\alpha)$).*

c) The ring B verifies the following universal property: For every local homomorphism $\varphi : A \to C$ into a henselian ring C whose residue field is real closed such that the induced homomorphism $k_A \to k_C$ is order preserving, there exists a unique homomorphism $\varphi' : B \to C$ of local A-algebras making commutative the diagram

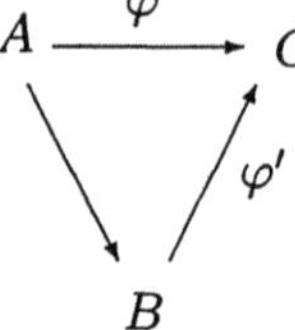

The ring B is called the real strict localization of A at α and denoted by A_α; *likewise, the homomorphism φ' of c) is called the* real strict localization of φ at α *and denoted by* φ_α.

Furthermore, there are unique isomorphisms $A_\alpha = (A^h)_\alpha$ and, for every ideal $I \subset A$, $(A/I)_\alpha = A_\alpha/IA_\alpha$.

Proof.   a) Let $A_i = A[\mathbf{x}_i]_{\mathfrak{p}_i}/(f_i), i = 1, 2$, be two elements of the family with residue fields $k_i, i = 1, 2$. Let K be a finite ordered extension of both residue fields, and let x_i stand for the image of $\mathbf{x}_i$ in K via the canonical homomorphism $A[\mathbf{x}_i] \to A_i \to k_{A_i} \to K$. We consider the homomorphism $\sigma : A[\mathbf{x}_1, \mathbf{x}_2] \to K$ defined by sending $\mathbf{x}_i$ to x_i for $i = 1, 2$, and set $C = A[\mathbf{x}_1, \mathbf{x}_2]_{\mathfrak{p}}/(f_1, f_2)$, where $\mathfrak{p}$ is the kernel of σ. By the jacobian criterion (Proposition 7.3), C is a local-etale A-algebra and we have a commutative diagram

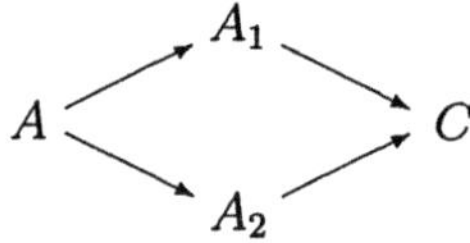

which shows *a).*

b) Obviously k_B is an ordered algebraic extension of k_A. Assume that L is an ordered algebraic extension of k_B and let $t \in L$. Let $f \in A[\mathbf{t}]$ be a monic polynomial whose class mod $\mathfrak{m}_A$, $\overline{f} \in k_A[\mathbf{t}]$, is the irreducible polynomial of t over k_A. Then the filtrant family $\{A_i\}$ contains the A-algebra $C = A[\mathbf{t}]_{\mathfrak{p}}/(f)$, where $\mathfrak{p}$ is the kernel of the map $A[\mathbf{t}] \to k_A[\mathbf{t}] \to L$ given by $\mathbf{t} \mapsto t$. Therefore there is a homomorphism $C \to B$, which induces an inclusion $k_C \to k_B$, and $t \in k_B$. This shows that k_B is real closed.

To see that B is henselian, consider a monic polynomial $f \in B[\mathbf{t}]$ whose class mod $\mathfrak{m}_B$, $\overline{f} \in k_B[\mathbf{t}]$, has a simple root $y \in k_B$. Choose some $C = A[\mathbf{x}_1]_{\mathfrak{p}}/(g)$ in the filtrant family with $f \in C[\mathbf{t}]$ via the homomorphism $C \to B$. Let us denote by x the image of $\mathbf{x}_1$ in k_B by the canonical homomorphism $C \to k_C \to k_B$, and define $\sigma : A[\mathbf{x}_1, \mathbf{x}_2] \to k_B$ by $\mathbf{x}_1 \mapsto x, \mathbf{x}_2 \mapsto y$. Then, by the jacobian criterion again, the filtrant family contains the A-algebra $C = A[\mathbf{x}_1, \mathbf{x}_2]_{\mathfrak{p}}/(f, g)$, where $\mathfrak{p}$ is the kernel of σ. Consequently, there is a homomorphism $C \to B$ and this homomorphism sends the class of x_2 in C to a root $z \in B$ of f such that $\bar{z} = y$.

The universal property follows readily from Proposition 7.6 $d)$, and the isomorphisms $A_\alpha = (A^h)_\alpha$ and $(A/I)_\alpha = A_\alpha/IA_\alpha$ are formal consequences of it. $\square$

We describe now the real spectrum of A_α. This description shows where the name of real strict localization comes from. Recall the notation introduced in 1.10:

$$U_\alpha = \{\beta \in \mathrm{Spec}_r(A) \mid \beta \to \alpha\}.$$

Theorem 7.11   *The canonical homomorphism $\varphi : A \to A_\alpha$ induces an injective continuous map $\varphi^* : \mathrm{Spec}_r(A_\alpha) \to \mathrm{Spec}_r(A)$ which is a homeomorphism onto U_α with respect to the Harrison topology.*

Proof.   Let $\beta' \in \mathrm{Spec}_r(A_\alpha)$. Since A_α is henselian and its residue field is $\kappa(\alpha)$, Proposition 2.4 implies $\beta' \to \alpha$. Thus $\varphi^*(\beta') \in U_\alpha$. Now let $\beta \in U_\alpha$, say $\beta : A \to R_\beta$, and consider the real closed valuation $W_{\beta\alpha}$ of R_β defined in 3.10, so that $\beta : A \to W_{\beta\alpha}$. Since $W_{\beta\alpha}$ is henselian (3.10 $a)$), by the universal property there is a unique local A-homomorphism $A_\alpha \to W_{\beta\alpha}$. This induces a prime cone β' of A_α with $\varphi^*(\beta') = \beta$. Therefore U_α is the image of φ^*.

Now we shall see that φ^* is injective. Let $\gamma : A_\alpha \to R_\gamma$ be a point in $\mathrm{Spec}_r(A_\alpha)$, $\beta = \varphi^*(\gamma) \in \mathrm{Spec}_r(A)$ and $\beta' \in \mathrm{Spec}_r(A_\alpha)$ the point lying over β constructed in the preceding paragraph. We must show that $\gamma = \beta'$. To that end we consider the following diagram:

$$
\begin{array}{ccc}
 & W_{\gamma\alpha} \subset R_\gamma & \\
{\scriptstyle\gamma}\nearrow & \uparrow{\scriptstyle u} & \cup \\
\beta : A \xrightarrow{\ \varphi\ } A_\alpha \xrightarrow{\ \beta'\ } & W_{\beta\alpha} \subset \kappa(\beta) &
\end{array}
$$

where u is a local homomorphism. Indeed, since k_{A_α} is algebraic over k_A, every element of the bigger field is bounded by one of the smaller. As $\gamma \to \alpha$, this implies that every element of A_α is bounded in γ by some element of A. It follows that A_α is contained in the convex hull V of $W_{\beta\alpha}$ in R_γ with respect to γ. On the other hand V dominates $W_{\beta\alpha}$ (3.9) and consequently A. This implies that V dominates A_α, and by the maximality of $W_{\gamma\alpha}$ we conclude $W_{\gamma\alpha} \supset V$ and consequently $W_{\gamma\alpha} \cap \kappa(\beta) \supset W_{\beta\alpha}$. The converse inclusion follows from the maximality of $W_{\beta\alpha}$, and so $W_{\gamma\alpha} \cap \kappa(\beta) = W_{\beta\alpha}$. This implies that u is local. Now $\varphi^*(\gamma) = \beta$ gives

$$\gamma \circ \varphi = u \circ \beta = u \circ (\beta' \circ \varphi) = (u \circ \beta') \circ \varphi.$$

Then from the universal property of A_α, which can be applied because $W_{\gamma\alpha}$ is henselian, we get $\gamma = u \circ \beta'$, that is, γ and β' are the same prime cone.

Therefore, φ^* is a continuous bijection from $\mathrm{Spec}_r(A_\alpha)$ onto U_α with respect to the Harrison and to the constructible topologies, and we want to show that it is in fact a homeomorphism respect to the first one.

To do that, we need some remarks concerning the constructible topology, and for the time being that is the only topology we consider in the various spaces involved. First of all, since $\mathrm{Spec}_r(A_\alpha)$ and U_α are compact and the map $\varphi^* : \mathrm{Spec}_r(A_\alpha) \to U_\alpha$ is a continuous bijection, it is a homeomorphism. This has the following consequence. Let $C \subset \mathrm{Spec}_r(A_\alpha)$ be constructible. Then C is open and closed and by the previous remark $\varphi^*(C)$ is also open and closed in U_α. Now U_α is closed in $\mathrm{Spec}_r(A)$, and we conclude that $\varphi^*(C)$ is closed in $\mathrm{Spec}_r(A)$. Since we are considering the constructible topology, this means that $\varphi^*(C)$ is proconstructible.

Now we return to the Harrison topology, to see that $\varphi^* : \mathrm{Spec}_r(A_\alpha) \to U_\alpha$ is closed. Since φ^* is bijective and every closed set of $\mathrm{Spec}_r(A_\alpha)$ is an intersection of closed constructible sets, we have to show that the image $\varphi^*(C)$ of every closed constructible set $C \subset \mathrm{Spec}_r(A_\alpha)$ is a closed set in U_α. But since $\varphi^*(C)$ is proconstructible, Proposition 2.3 says that its closure in U_α consists of the specializations of its elements. Consequently, it suffices to show that $\varphi^*(C)$ is closed under specializations in U_α.

So, let $\varphi^*(\gamma') = \gamma \to \beta = \varphi^*(\beta')$, with $\gamma' \in C$ and $\beta' \in \mathrm{Spec}_r(A_\alpha)$. By Proposition 3.12 and the description of $(\varphi^*)^{-1}$ we gave before, we have a diagram

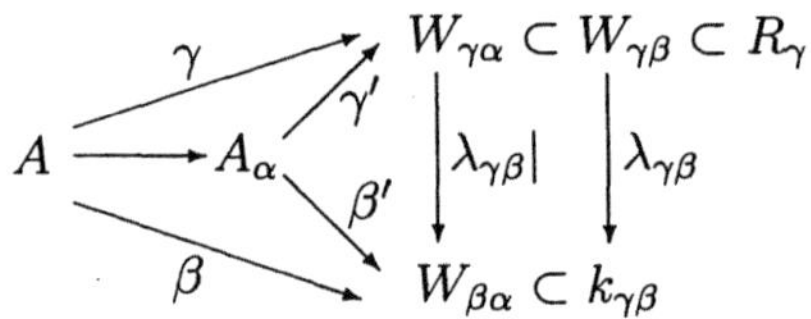

where one easily checks using the universal property of A_α that $\lambda_{\gamma\beta} \circ \gamma' = \beta'$. But this shows that $\gamma' \to \beta'$. Finally, as $\gamma' \in C$ and C is closed, we conclude $\beta' \in C$ and $\beta = \varphi^*(\beta') \in \varphi^*(C)$ as wanted. $\square$

Example 7.12 Let k be an ordered field and R its real closure. Let A be the localization of the polynomial ring $k[\mathbf{x}]$, $\mathbf{x} = (\mathbf{x}_1, \ldots, \mathbf{x}_n)$, at the maximal ideal $\mathfrak{m}$ generated by the $\mathbf{x}_i$'s and consider the prime cone $\alpha : A \to R$. Then the henselization of A is $A^h = k[[\mathbf{x}]]_{\mathrm{alg}}$, and the real strict localization of A at α is $A_\alpha = R[[\mathbf{x}]]_{\mathrm{alg}}$. The latter ring is also the real strict localization of A^h at α. Here $[[\cdots]]_{\mathrm{alg}}$ denotes *algebraic power series*, that is, formal power series which are algebraic over the polynomials. The theory of algebraic power series is throroughly similar to that of formal power series, including Weierstrass's preparation and division theorems and M. Artin's approximation theorem (see [B-C-R 8.2, 8.3]). Now, by the preceding results, we have canonical homeomorphisms:

$$\mathrm{Spec}_r(A) \equiv \{\beta \in \mathrm{Spec}_r(k[\mathbf{x}]) \mid \mathrm{supp}(\beta) \subset \mathfrak{m}\},$$

$$\mathrm{Spec}_r(A^h) \equiv \{\beta \in \mathrm{Spec}_r(k[\mathbf{x}]) \mid \mathfrak{m} \text{ is convex with respect to } \beta\},$$

$$\mathrm{Spec}_r(A_\alpha) \equiv \{\beta \in \mathrm{Spec}_r(k[\mathbf{x}]) \mid \beta \to \alpha\}.$$

$\square$

We close this section with a result concerning connected components which will be of use in Chapters VII and VIII. Given a topological space X, a subset $Y \subset X$ and a point $z \in X$ we shall denote by $cc(Y)$ the number $\leq \infty$ of connected components of Y and by $cc_z(Y)$ the number $\leq \infty$ of connected components of Y adherent to $z \in X$. Then:

Proposition 7.13   *Let C be a constructible set of* $\mathrm{Spec}_r(A)$ *and* $\alpha \in \mathrm{Adh}(C)$. *Consider the real strict localization* $\varphi : A \to A_\alpha$ *and let* $C_\alpha = (\varphi^*)^{-1}(C) \subset \mathrm{Spec}_r(A_\alpha)$. *Then:*

a) *For every integer $p \leq cc(C_\alpha)$ there are arbitrarily small open constructible neighbourhoods U of α such that $U \cap C = C_1 \cup \cdots \cup C_p$, where the C_i's are disjoint open subsets of $U \cap C$ adherent to α.*

b) $cc(C_\alpha) = cc(C \cap U_\alpha) \geq cc_\alpha(C)$.

Proof.   a) If $p \leq cc_\alpha(C)$, $C \cap U_\alpha$ must be the union of p disjoint non-empty open subsets $T_1, \ldots, T_p$. Then they are also closed and, as $C \cap U_\alpha$ is proconstructible, by Proposition 1.12 we have $T_i = U_i \cap C \cap U_\alpha$ for some open constructible set U_i of $\mathrm{Spec}_r(A)$. Furthermore, we have $U_\alpha = \bigcap U$, for any basis $\{U\}$ of open constructible neighbourhoods of α, and we can write

$$C \cap \bigcap U \setminus (U_1 \cup \cdots \cup U_p) = \emptyset \quad \text{and} \quad C \cap \bigcap U \cap U_i \cap U_j = \emptyset \text{ for } i \neq j.$$

By compactness of the constructible topology there is some U such that

$$C \cap U \setminus (U_1 \cup \cdots \cup U_p) = \emptyset \quad \text{and} \quad C \cap U \cap U_i \cap U_j = \emptyset \text{ for } i \neq j.$$

Finally we put $C_i = C \cap U \cap U_i$ for $i = 1, \ldots, p$, and it remains to see that each C_i is adherent to α. But $T_i \subset C_i$, and $\emptyset \neq T_i \subset U_\alpha$; hence there is $\beta \in C_i$ with $\beta \to \alpha$, which means that $\alpha \in \mathrm{Adh}(C_i)$.

b) Since φ^* is a homeomorphism from $\mathrm{Spec}_r(A_\alpha)$ onto U_α (Proposition 7.10), it induces a bijection between the connected components of C_α and the ones of $C \cap U_\alpha$, which gives the first equality. On the other hand, if D is a connected component of C, it is closed in C, hence proconstructible. Then by Proposition 2.3, $\alpha \in \mathrm{Adh}(D)$ if and only if $D \cap U_\alpha \neq \emptyset$. Now choose a connected component T of $C \cap U_\alpha$ meeting $D \cap U_\alpha$. Necessarily $T \subset D$, and $D \mapsto T$ defines an injection from the connected components of C adherent to α into the connected components of $C \cap U_\alpha$, and we obtain $cc(C \cap U_\alpha) \geq cc_\alpha(C)$. $\square$

Notes

The real spectrum was introduced by Coste-Roy at the end of the seventies ([Co-Ry1,2]) by model theoretic reasons. In particular, they showed the fundamental equivalence between constructibility and first order definability. There are now several very good presentations of the theory: [Lm3], [Be2], [Kn-Sch] and of course [B-C-R]. Here we refer to the latter and make some small modifications that fit our setting.

The first version of the finiteness theorem was given by Lojasiewicz [L1] without any reference to model theory; among the many other later proofs we quote here [Rc]. The real Nullstellensatz was proved independently and almost simultaneously by Dubois ([Du]) and Risler ([Rs]) at the end of the sixties, although a weak form of it had passed unnoticed in an earlier paper by Krivine ([Kv]). The first Positivstellensatz is due to Stengle ([St1]) and many others followed, as for instance several by Swan ([Sw]). The final unified treatment for all of them was given by Colliot-Thélène ([C-T]) in his contribution to the foundational 1981 real geometry conference at Rennes. The Hörmander-Lojasiewicz inequality in its abstract form was first published in [An-Br-Rz], but had already been proved by Coste (oral communication) using model theory. The notions of specialization and generization are classical in algebraic geometry and for real spectra were developed by Coste-Roy ([Co-Ry2]), who showed also their connection with real valuations; this was immediately pursued by Brumfiel in [Bf2]. However, the study of real valuations and their relationship with orderings is much older, appearing already in the papers of Baer ([Ba]) and Krull ([Kr]). Here we have not touched the classical theme of valuations and singularities, which has also a real version (see [An-Rz3]). The real going-up for integral extensions and the real going-down for polynomial extensions were proved in [Co-Ry2]. The systematic use of valuation rings associated to specializations was introduced by Schwartz ([Schw2]) and Delfs ([Df]) to study the sheaf of continuous abstract semialgebraic functions. The invention of this sheaf is largely due to Brumfiel ([Bf3,4]), with Schwartz's crucial finding that a definable continuous abstract function need not be semialgebraic. The progressive description given here tries to be closer to intuition. The cylindrical decomposition of Section 6 was introduced in [Al-An] to study constructibility in power series rings.

Finally, the notion of real strict localization was introduced by Roy ([Ry]) to be the real counterpart of Grothendieck's strict localization and the key ingredient in the construction of the sheaf of abstract Nash functions. This idea appeared first in [Ar-Mz]. All of this was later clarified in [Al-Ry]. Our Theorem 7.11 is an expanded version of the corresponding result of the latter paper, while Proposition 7.13 comes from [Rz7]. It is well worth mentioning here that the properties of real strict localizations are intimately tied to a very important open problem of the theory: the verification of a property called idempotency. This highly abstract property has been brought into new attention by Quarez [Qz]), who has shown that it is equivalent to Efroymson's extension and separation problems on Nash functions ([Ef]).

Chapter III. Spaces of Signs

Summary. Spaces of signs are defined by imposing four suplementary axioms on real spaces; then, spaces of orderings are a special class of these spaces of signs. This is done in Section 1, where we also define subspaces and draw the first consequences of our definitions. Section 2 contains the fundamental properties of forms, mainly in the case of spaces of orderings. The important notion of fan is introduced in Section 3, together with its elementary properties. In Section 4 we consider local spaces of orderings and localizations, which behave very much like in ring theory. Localizations are used in Section 5 to show that the real space associated to a commutative ring with unit is actually a space of signs, and also in Section 6, to prove that the subspaces of a space of signs are again spaces of signs.

1. The Axioms

Let (X, G) be a real space; we have a pairing

$$X \times G \to \mathbb{F}_3 \; ; \; (x, g) \mapsto g(x).$$

Since G separates points (Axiom $\mathbf{S}_3$) this pairing gives an inclusion $X \subset \widehat{G} = \mathrm{Hom}(G, \mathbb{F}_3)$. Now, $\widehat{G}$ inherits from $\mathbb{F}_3^G$ the product topology and $\widehat{G}$ is closed in $\mathbb{F}_3^G$. So $\widehat{G}$ is compact in this topology, which we call constructible. In fact, the constructible topology on X is just the induced one. Note also that a necessary condition for $\sigma \in \widehat{G}$ to be in X is that $\sigma(-1) = -1$. Now, for $x, y \in X$ we may form the product $\sigma = xy$ in $\widehat{G}$, which is no longer in X since $\sigma(-1) = 1$. However, products of odd order may again be in X, and it will be very important to study them. If G is a group, then $g^2 = 1$ for all $g \in G$ and $\widehat{G} = \mathrm{Hom}(G, \{+1, -1\}) = \mathrm{Hom}(G, \mathbb{C}^*)$ is the usual dual group when we provide G with the discrete topology (see [Mo]). This will be essential in Chapter IV.

Of course, not very much can be said about the structure of (X, G) in this generality. So we are going to impose further conditions on (X, G).

Proposition and Definition 1.1 *The following conditions for (X, G) are equivalent:*

a) X is compact with respect to the constructible topology.

b) The canonical embedding of X into its Stone space

$$\kappa : X \to \tilde{X} \; ; \; x \mapsto \text{ principal filter of } x,$$

is a homeomorphism with respect to the constructible topologies.

c) X is closed in $\hat{G}$.

If these conditions hold (X, G) is called a prespace of signs. If, moreover, G is a group, then (X, G) is called a prespace of orderings.

Example 1.2   Let A be a commutative ring with unit. Consider the real space (X, G) associated to A, that is, $X = \mathrm{Spec}_r(A)$ and $G = \{\mathrm{sign}[f] \mid f \in A\}$ (II.1.5). Then (X, G) is a prespace of signs (Theorem II.1.6). If $A = K$ is a field, then $(X, G \setminus \{0\})$ is a prespace of orderings. For instance, the prespace of orderings associated to the field $\mathbb{C}$ is trivial, and the prespace of orderings associated to the field $\mathbb{R}$ is atomic (I.3.2). $\square$

Now let (X, G) be an arbitrary prespace of signs. Note that $\kappa : X \to \tilde{X}$ is also a homeomorphism with respect to the Harrison topologies. In the sequel, if not otherwise mentioned, we always consider the Harrison topology. Recall that if (X, G) is a prespace of orderings, the constructible and the Harrison topologies coincide. The following facts follow from I.3.4-5:

Proposition 1.3

a) Let $C \subset X$ be constructible. Then C is quasicompact.

b) (Finiteness Theorem) Let $C \subset X$ be constructible. Then C is open (resp. closed) if and only if C is strictly open (resp. strictly closed).

c) A subset $C \subset X$ is constructible if and only if C is open and closed with respect to the constructible topology.

Definition 1.4   A Zariski closed set $V \subset X$ is called irreducible if it is not the union of two other Zariski closed sets. An irreducible Zariski closed set is called a subvariety of X.

Clearly, the Zariski closure $\mathrm{Adh}_Z(x)$ of a point $x \in X$ is a subvariety. Conversely, we have:

Proposition 1.5   Let $V \subset X$ be a non-empty subvariety and set

$$V^* = V \cap \bigcap_{\substack{g \in G \\ V \not\subset \{g=0\}}} \{g \neq 0\}.$$

Then

$$V^* = \{x \in V \mid V = \mathrm{Adh}_Z(x)\} \neq \emptyset.$$

Proof.   Since V is irreducible, the collection ϕ of all the sets $V \cap \{g \neq 0\}$ appearing in the intersection is a filter, and, by Definition III.1.1, ϕ is contained in some principal filter. On the other hand, a point x belongs to the intersection if and only if its principal filter ϕ_x contains ϕ, that is, if and only if $g(x) = 0$ for every g vanishing on V. The result follows immediately from these remarks. $\square$

Definition 1.6   *Let (X, G) be a real space. Let $Y \subset X$ be any subset of the form*

$$Y = Y_1 \cap Y_2 \, , \; Y_1 = \bigcap_{g \in G_1} \{g \geq 0\} \, , \; Y_2 = \bigcap_{g \in G_2} \{g > 0\}$$

for $G_1, G_2 \subset G$. Moreover, let $G|Y$ denote the set of all restrictions $g|Y$ for $g \in G$. Then $(Y, G|Y)$ is called a subspace *of (X, G). Usually, we just say that Y is a subspace of X.*

Note that the Y_1-part of a subspace Y can include *equalities*: the condition $g = 0$ is equivalent $-g^2 \geq 0$; also, that the Y_2-part can include *non-equalities*: $g \neq 0$ is $g^2 > 0$.

Example 1.7   Let $u : A \to B$ a homomorphim of commutative rings with unit, and $u^* : \mathrm{Spec}_r(B) \to \mathrm{Spec}_r(A)$ the corresponding map of real spectra. Then, for each $\alpha \in \mathrm{Spec}_r(A)$, the fiber $Y \subset \mathrm{Spec}_r(B)$ over α, that is, the set Y of all prime cones lying over α, is a subspace of $\mathrm{Spec}_r(B)$. On the other hand, it is known that Y is homeomorphic to the real spectrum of the ring $B \otimes_A \kappa(\alpha)$. Hence, Y can be equiped with two different monoids and we get two different real spaces. Namely, we can consider on Y the functions $\mathrm{sign}[f]$ with $f \in B$, or the functions $\mathrm{sign}[g]$ with $g \in B \otimes_A \kappa(\alpha)$. Clearly the second choice gives a bigger monoid. This will be important in Section VI.7.

Proposition 1.8   *Let (X, G) be a prespace of signs (resp. of orderings) and let $(Y, G|Y)$ be a subspace. Then $(Y, G|Y)$ is again a prespace of signs (resp. of orderings). Moreover, for $y_1, \ldots, y_m \in Y$ such that $y_1 \cdots y_m = z \in X$ (product formed in $\widehat{G}$) one has $z \in Y$.*

Proof.   The topology induced on Y by the constructible topology on X coincides with the constructible topology of $(Y, G|Y)$. But Y is compact with respect to the former. The second statement is clear. $\square$

Of course, we could consider more general subspaces. The reason why we restrict ourselves to the above ones will become clear in Section V.

(1.9) Canonical Decomposition. Let (X, G) be a prespace of signs and let $V \subset X$ be a subvariety. Consider the subset $V^* \subset X$ as in Proposition 1.5. Then, V^* is a subspace of X. Moreover $(G|V^*)^* = (G|V^*) \setminus \{0\}$ and we denote this group by $G(V^*)$. Thus, $(V^*, G(V^*))$ is a prespace of orderings. By Proposition III.1.5 we have a canonical decomposition

$$X = \bigcup_V V^*$$

where V runs over all subvarieties of X. Of course, if (X, G) is associated to a ring A, this decomposition corresponds to the decomposition

$$\operatorname{Spec}_r(A) = \bigcup_{\substack{\mathfrak{p} \in \operatorname{Spec}(A) \\ \mathfrak{p} \ \text{real}}} \operatorname{Spec}_r(k(\mathfrak{p})),$$

that is, the subvarieties V are the zero sets of the real prime ideals $\mathfrak{p}$ and the prespace of orderings $(V^*, G(V^*))$ are associated to the residue fields $\kappa(\mathfrak{p})$.

□

Let (X, G) be a prespace of signs and let $x, y \in X$. As for real spectra, we write $x \to y$ (y is a *specialization* of x and x is a *generization* of y) if $y \in \operatorname{Adh}(x)$. It is easily seen that $x \to y$ and $\operatorname{Adh}_Z(y) = \operatorname{Adh}_Z(x)$ imply $x = y$. Clearly, a point $x \in X$ is closed if and only if it has no specialization $y \neq x$; in view of this "maximality" characterization, we shall denote by $X_{\max}$ the set of all closed points of X. Furthermore, by the same arguments used for real spectra, we have:

Proposition 1.10   *Let $C \subset X$ be constructible. Then the following conditions are equivalent:*

> *a) C is closed (resp. open),*
> *b) For all $x, y \in X$ with $x \in C$ and $x \to y$ (resp. $y \to x$) one has $y \in C$.*

Proposition 1.11   *Let $C \subset X$ be constructible and Z-closed. Then there are $g_1, \ldots, g_m \in G$ such that $C = \{g_1 = 0, \ldots, g_m = 0\}$.*

Proof.   Let $U = X \setminus C$ and consider the constructible topology on X. Now, $U \subset \bigcup_{C \subset \{f = 0\}} \{f \neq 0\}$ and U is compact. □

(1.12) Proconstructible Sets. As usual, a subset $D \subset X$ is called *proconstructible*, if it is an arbitrary intersection of constructible sets. Thus proconstructible means closed in the constructible topology. For example, all subspaces are proconstructible.

A proconstructible set D can be endowed with the induced constructible, Harrison or Zariski topology. Being closed with respect to the finest, D is compact in the constructible topology and quasicompact in the Harrison and the Zariski topologies.

The following two result can be easily proved:

> *a) $E \subset D$ is open and closed in the constructible topology if and only if $E = E' \cap D$ for some constructible set $E' \subset X$.*
> *b) $E \subset D$ is open in the Harrison topology and closed in constructible topology if and only if $E = E' \cap D$ for some strictly open set $E' \subset X$.*

The latter statement generalizes the finiteness theorem. □

However, more specific features cannot be shown without further assumptions. So, let us introduce the following two axioms for a prespace of signs (X, G).

PE: For any two $a, b \in G$ there exists $c \in G$ such that $\{a = 0,\ b = 0\} = \{c = 0\}$.

By **PE** we get that any Z-closed and constructible set $W \subset X$ can be written in the form $W = \{f = 0\}$ for $f \in G$ and replacing f by f^2, if needed, we can achieve that $f(x) = +1$ for $x \notin W$. This f is called a *positive equation* for W.

HL: Let $C \subset X$ be constructible and closed. Let $f, g \in G$ such that $C \cap \{f = 0\} \subset \{g = 0\}$. Then there exists $f' \in G$ such that

a) $f' = f$ on C, $f' = g$ on $\{f = 0\}$.
b) $\langle f, g \rangle = \langle f', f'fg \rangle$.

In many applications of **HL** only part *a)* is used, but for the proof of the most important results in Chapter V the information of part *b)* will be essential.

Note that both axioms hold trivially for a prespace of orderings, since then the Zariski topology is trivial.

Proposition 1.13   *Suppose that (X, G) is associated to a commutative ring A with unit. Then **PE** and **HL** hold for (X, G).*

Proof.   Apply the Hörmander-Łojasiewicz inequality (Proposition II.1.16) and take $f' = (1 + h^2)f + g^l$.   □

Next we draw some consequences from **PE** and **HL**.

Proposition 1.14   *Suppose that **PE** and **HL** hold for the prespace of signs (X, G). Let $x, y \in X$ be independent, that is, neither $x \to y$ nor $y \to x$. Then there exists $h \in G$ with $h(x) = +1$, $h(y) = -1$.*

Proof.   By assumption, we find $f \in G$ with $f(x) \geq 0$, $f(y) = -1$. If $f(x) = +1$ we are done. So assume $f(x) = 0$. Similarly, we find $g \in G$ with $g(x) = +1$, $g(y) = 0$. We choose b with $\{b = 0\} = \{f = 0, g = 0\}$, so that $x, y \notin \{b = 0\}$. Now, replace f by fb^2, g by gb^2, and apply **HL** to f, g and $C = \{g = 0\}$. We find $f' \in G$ with $f'(x) = +1$ and $f'(y) = -1$. So $h = f'$ does the job.   □

Corollary 1.15   *Suppose that **PE** and **HL** hold for the prespace of signs (X, G). Let $x \in X$. Then the specializations of x form a chain. Moreover, this chain contains a unique closed point x_0.*

Proof.   Assume that $x \to y$, $x \to z$ and y, z are independent. Choose $g \in G$ with $g(y) = +1, g(z) = -1$. Then $+1 = g(x) = -1$, contradiction. Now consider $\bigcap_{y \in \mathrm{Adh}(x)} \mathrm{Adh}(y)$. By compactness, this intersection is not empty. Now, if x_1, x_2 belong to that intersection, the intersection itself is contained in $\mathrm{Adh}(x_1)$

and $\mathrm{Adh}(x_2)$. Consequently, $x_1 \to x_2$ and $x_2 \to x_1$, that is, $x_1 = x_2$. Whence, our intersection consists of a single point, which is the x_0 we sought. $\square$

Now, we consider the set $X_{\max}$ of all closed points of X. The corresponding real space $(X_{\max}, G|X_{\max})$ is compact with respect to the Harrison topology, and has sometimes advantages as we will see in Section 4. However, as we saw in the case of real spectra, it may also be very important to have non closed points at disposal. Furthermore, $(X_{\max}, G|X_{\max})$, in general, is not a prespace of signs, since $X_{\max}$ may not be compact with respect to the constructible topology.

Proposition 1.16   *Suppose that* **PE** *and* **HL** *hold for the prespace of signs* (X, G). *Let* $x_1, \ldots, x_n$ *be closed points in* X *such that no more than three of them have the same Zariski closure. Let* $h : \{x_1, \ldots, x_n\} \to \{+1, -1\}$ *be any function. Then there exists* $a \in G$ *such that* $h(x_i) = a(x_i)$, $i = 1, \ldots, n$.

Proof.   Let $V_1, \ldots, V_m$ be the distinct Zariski closures of $x_1, \ldots, x_n$. We proceed by induction on m, the case $m = 1$ being a straightforward computation. Let V_m be minimal among the V_i, that is, $V_i \not\subset V_m$ for $i \neq m$ (but it may happen that $V_m \subset V_i$ for $i \neq m$). Let $\mathrm{Adh}_Z(x_i) \neq V_m$ for $i = 1, \ldots, k$ and $\mathrm{Adh}_Z(x_i) = V_m$ for $i > k$. By induction we find $f, g \in G$ such that $h(x_i) = f(x_i)$ for $i \leq k$ and $h(x_i) = g(x_i)$ for $i > k$. Now, for each $i = 1, \ldots, k$, $X \setminus \{x_i\}$ is an open neighborhood of $\{f = 0\} \cup V_m$; since this set is quasicompact, we find a closed constructible set C_i such that $x_i \in C_i$ and $C_i \cap (\{f = 0\} \cup V_m) = \emptyset$. Let $C = C_1 \cup \cdots \cup C_k$. By compactness and **PE**, we also find $\ell \in G$ with $\{\ell = 0\} \cap C = \emptyset$ and $V_m \subset \{\ell = 0\}$. Now by **HL** we get $f' \in G$ such that $f' = \ell f$ on C and $f' = g$ on $\{\ell f = 0\}$, and this element $a = f'$ does the job. $\square$

Proposition 1.17   *Let* (X, G) *be a prespace of signs and let* $(Y, G|Y)$ *be a subspace. If* **PE** *and* **HL** *hold for* (X, G), *then they do so for* $(Y, G|Y)$.

Proof.   Obviously, **PE** goes over to $(Y, G|Y)$. For **HL** first consider the case that Y is of the form

$$Y = \bigcap_{h \in G_1} \{h \geq 0\} , \ G_1 \subset G .$$

We choose $C' \subset X$ closed and constructible with $C = Y \cap C'$. Then, in general, we will not have

$$C' \cap \{f = 0\} \subset \{g = 0\}$$

but by compactness this holds if we replace C' by $C' \cap T$ where $T = \{h_1 \geq 0, \ldots, h_r \geq 0\}$ for suitable $h_1, \ldots, h_r \in G_1$. Then we can apply **HL** in X and so we get it for Y by restriction. Thus, we may assume that Y is of the form

$$Y = \bigcap_{h \in G_2} \{h > 0\}, \ G_2 \subset G .$$

We choose C' as before and again by compactness (with respect to the constructible topology) we get

$$U \cap C' \cap \{f = 0\} \subset \{g = 0\},$$

where $U = \{h_1 > 0, \ldots, h_r > 0\}$ for suitable $h_1, \ldots, h_r \in G_2$. So we may replace Y by the basic open set U and C by $C' \cap Y$. But still C is not closed in X. Therefore we replace C' by the closed constructible set $C \cup \{h_1 = 0\} \cup \cdots \cup \{h_r = 0\}$. Also $C' \cap Y = C$. Now we can apply **HL** to C', $fh_1 \cdots h_r$ and $gh_1 \cdots h_r$ in X and so we get it for Y by restriction. $\qquad\square$

We next introduce a further axiom which is the key to make the abstract theory work.

MM: Let $\rho = \langle a_1, \ldots, a_n \rangle$, $\tau = \langle b_1, \ldots, b_m \rangle$ be forms over X, and let $h \in G$ be such that

$$\rho + \tau = \langle h, c_2, \ldots, c_{n+m} \rangle,$$

with $c_2, \ldots, c_{n+m} \in G$. Let $x \in X$ be a point such that

$$h(x) \neq 0, \ a_i(x) \neq 0, \ b_j(x) \neq 0, \ c_k(x) \neq 0$$

for all i, j, k. Then, there exist forms

$$\rho' = \langle f, a_2', \ldots, a_n' \rangle \ \text{ and } \ \tau' = \langle g, b_2', \ldots, b_m' \rangle,$$

an element $c \in G$, and a Zariski neighbourhood U of x such that, for all $y \in U$,

    i) $a_i'(y) \neq 0$, $b_j'(y) \neq 0$, $f(y) \neq 0$, $g(y) \neq 0$, $c(y) \neq 0$ for all i, j.
    ii) $\rho(y) = \rho'(y)$, $\tau(y) = \tau'(y)$,
    iii) $\langle f, g \rangle(y) = \langle h, c \rangle(y)$.

Remarks and Notations 1.18 In the above situation, we say that the form $\rho + \tau$ *represents* the element h over X. In general, we say that a form λ over X of dimension k represents an element $h \in G$ over a subset $C \subset X$ if there are $c_2, \ldots, c_k \in G$ such that $\lambda(x) = \langle h, c_2, \ldots, c_k \rangle(x)$ for all $x \in C$. The set of all the h's represented by λ over C is denoted by $D_C(\lambda)$; we write $D(\lambda)$ instead of $D_X(\lambda)$ if there is no risk of confusion. With this terminology, conditions *ii)* and *iii)* above read as $f \in D_U(\rho)$, $g \in D_U(\tau)$ and $h \in D_U(\langle f, g \rangle)$.

On the other hand, condition *i)* expresses that the *restrictions* of ρ and τ to U are *regular*, since they are defined by units. This condition is essential in general spaces of signs, but is trivially fulfilled if G is a group. Hence, if that is the case, **MM** means that for any two forms ρ, τ over X and any $h \in D(\rho + \tau)$, there are $f \in D(\rho)$ and $g \in D(\tau)$ such that $h \in D(\langle f, g \rangle)$.

Note also that condition *b)* in **HL** means that $\langle f, g \rangle$ represents f'.

If one looks at the usual diagonal non degenerate quadratic forms $\rho = \langle a_1, \ldots, a_n \rangle$ over a field K of characteristic $\neq 2$, and reads "ρ represents h" as "*there exist* $t_1, \ldots, t_n \in K$ *such that* $h = t_1^2 a_1 + \cdots + t_n^2 a_n$", it is fairly clear where **MM** comes from. $\qquad\square$

We are thus ready to define:

Definitions 1.19 *A prespace of signs (X, G) is called a space of signs if **PE** and **HL** hold for (X, G) and **MM** holds for any Zariski closed subset of (X, G).*

A space of signs (X, G) is called a space of orderings if G is a group.

Example 1.20   Let (X, G) be a space of signs, and let $V \subset X$ be a subvariety. According to 1.9, consider V^* and $G(V^*) = (G|V^*)^* = (G|V^*)\backslash\{0\}$. Then $(V^*, G(V^*))$ is a space of orderings.

Indeed, **MM** holds for V and we have to prove it for V^*. To that end, we extend data from V^* to V in the following way. Suppose, for instance, that we are given two forms ρ and ρ' over V with $\rho(x) = \rho'(x)$ for all $x \in V^*$, and we want this for all $x \in V$. Then, the set

$$C = \{x \in V \mid \rho(x) \neq \langle a_1, \ldots, a_n\rangle(x)\}$$

is constructible, and

$$\emptyset = C \cap V^* = C \cap V \cap \bigcap_{\substack{g \in G \\ V \not\subset \{g=0\}}} \{g \neq 0\}.$$

By compactness

$$C \cap V \cap \{g_1 \neq 0, \ldots, g_r \neq 0\} = \emptyset,$$

where the g_i's do not vanish on V. Since V is irreducible, we deduce that their product $g = g_1 \cdots g_r$ does not vanish either. Hence, $C \subset V \cap \{g = 0\}$, with $g \in G|V \setminus \{0\}$. This latter condition means that $g^2(x) = +1$ for all $x \in V^*$, and, consequently, $g^2\rho = g^2\rho'$. Thus, it is clear how to deduce **MM** for V^*.

$\square$

One of the main results of this chapter is the following:

Proposition 1.21   *The real space associated to a commutative ring with unit (resp. to a field) is a space of signs (resp. of orderings).*

This will be shown in Section 5.

2. Forms

For the sequel, we assume that (X, G) is a space of orderings. We are going to investigate the most important properties of forms over X.

Proposition and Definition 2.1   *Let* $g_1, h_1, \ldots, g_n, h_n \in G$ *be such that* $\langle g_1, \ldots, g_n\rangle = \langle h_1, \ldots, h_n\rangle$. *Then* $\prod_{i=1}^n g_i = \prod_{i=1}^n h_i$.
This product is called the discriminant *of* $\rho = \langle g_1, \ldots, g_n\rangle$.

Proof.   Fix $x \in X$ and let q be the number of g_i's for which $g_i(x) = -1$. Then q is also the number of h_i's for which $h_i(x) = -1$ and

$$\prod_{i=1}^n g_i(x) = (-1)^q = \prod_{i=1}^n h_i.$$

$\square$

Corollary 2.2   *Let $\rho = \langle a, b \rangle$, $a, b \in G$, and let $g \in D(\rho)$. Then $\rho = \langle g, abg \rangle$.*

Proposition 2.3   *Let ρ be a form over G and $a \in G$. Then*

    a) $D(\langle a \rangle) = \{a\}$
    b) $D(a\rho) = aD(\rho)$
    c) $D(m \times \rho) = D(\rho)$ for all $m \in \mathbb{N}$.

Proof. *a)* and *b)* are obvious. We show *c)* first for $m = 2$. Here we proceed by induction on $n = \dim(\rho)$.

Let $n = 1$: Assume first $\rho = \langle 1 \rangle$. Let $g \in D(\langle 1, 1 \rangle)$, so $\langle 1, 1 \rangle = \langle g, g \rangle$, thus $\langle g, g \rangle(x) = 2$ for all $x \in X$. Hence $g = 1 \in D(\langle 1 \rangle)$. Now let $\rho = \langle a \rangle$. Then $D(\rho + \rho) = aD(\langle 1, 1 \rangle) = aD(\langle 1 \rangle) = D(\langle a \rangle) = D(\rho)$.

Let $n > 1$: Write $\rho = \langle b \rangle + \tau$ and let $g \in D(\rho + \rho) = D(\langle b, b \rangle + 2 \times \tau)$. By **MM** we find $c \in D(\langle b, b \rangle)$ and $d \in D(2 \times \tau)$ such that $g \in D(\langle c, d \rangle)$. But by induction $c \in D(\langle b \rangle)$ and $d \in D(\tau)$, thus $g \in D(\langle b \rangle + \tau) = D(\rho)$.

Now let $m > 2$. By induction $D(2^k \times \rho) = D(\rho)$, but clearly $D(m \times \rho) \subset D(n \times \rho)$ for $n \geq m$. Thus $D(\rho) = D(m \times \rho)$ for $m \geq 1$. $\square$

Proposition 2.4   *For a form ρ over X the following are equivalent:*

    a) ρ is isotropic.
    b) For every $g \in G$ there is a form τ such that $\rho = \langle g, -g \rangle + \tau$.
    c) $D(\rho) = G$

Proof.   *a)* $\Rightarrow$ *b)* If ρ is isotropic, $\rho \sim \tau$ with $\dim(\rho) < \dim(\tau)$. Clearly, $\dim(\rho) = \dim(\tau')$ mod 2, from which this implication follows at once.

b) $\Rightarrow$ *c)* Obvious.

c) $\Rightarrow$ *a)* Assume $D(\rho) = G$. Then $\dim(\rho) > 1$ by Proposition 2.3 *a)*. Now we write $\rho = \langle 1 \rangle + \tau$. By **MM** we find $t \in D(\tau)$ with $-1 \in D(\langle 1, t \rangle)$, that is, $\langle 1, t \rangle = \langle -1, -t \rangle$ by Corollary 2.2. Then $\langle 1, 1 \rangle = \langle -t, -t \rangle$. Therefore, $-t = 1$ and $t = -1$. Hence $\rho = \langle 1 \rangle + \langle -1 \rangle + \tau_1$ is isotropic. $\square$

Corollary 2.5   *If the form ρ over X is anisotropic, then so is $m \times \rho$ for all $m \in \mathbb{N}$.*

Proof. We have $D(\rho) \neq G$, so $D(m \times \rho) = D(\rho) \neq G$. $\square$

Recall that a form φ over X of the type $\varphi = \langle 1, g_1 \rangle \otimes \cdots \otimes \langle 1, g_n \rangle$ is denoted by $\langle\langle g_1, \ldots, g_n \rangle\rangle$ and called an n-fold Pfister form. The unique form φ' for which $\varphi = \langle 1 \rangle + \varphi'$ is called the *pure part* of φ.

Proposition 2.6   *Let $C = \{g_1 > 0, \ldots, g_n > 0\}$ be basic in X and let $\varphi = \langle\langle g_1, \ldots, g_n \rangle\rangle$. Then*

    a) $C = \{x \in X \mid \varphi(x) = 2^n\} = \{x \in X \mid \varphi(x) \neq 0\}$.
    b) C can be generated by less elements, $C = \{h_1 > 0, \ldots, h_{n-1} > 0\}$, if and only if there exists a Pfister form ψ with $\varphi = 2 \times \psi$.

c) $D(\varphi) = \{g \in G \,|\, g(x) > 0 \text{ for all } x \in C\} = \{g \in G \,|\, g\varphi = \varphi\}$. *In particular, $D(\varphi)$ is a subgroup of G.*

d) *If φ is isotropic, then $\varphi(x) = 0$ for all $x \in X$, and $C = \emptyset$.*

e) *Let ρ be a form over X. Then $D_C(\rho) = D(\rho \otimes \varphi)$.*

Proof. a) *and* b) *are obvious.*

c) Suppose that $g(x) > 0$ for all $x \in C$. Then $\varphi + g\varphi = \langle 1, g\rangle \otimes \varphi = 2 \times \varphi$. Hence $g \in D(2 \times \varphi) = D(\varphi)$ by Proposition 2.3. Conversely, let $g \in D(\varphi)$. Then $\varphi = \langle a_1, \ldots, a_{2^n}\rangle$ for $a_1 = g$ and suitable $a_2, \ldots, a_{2^n} \in G$. For $x \in C$ we have

$$2^n = \sum_{i=1}^{2^n} a_i(x),$$

hence $a_1(x) = +1$. This proves the first equality and the second is evident.

d) By c) we have $\varphi = -\varphi$.

e) Suppose $f \in D_C(\rho)$, and let $m = \dim(\rho)$. Then, there are $f_2, \ldots, f_m \in G$ such that $\rho(x) = \langle f, f_2, \ldots, f_m\rangle(x)$ for all $x \in S$. Hence, by a),

$$(\rho \otimes \varphi)(x) = 2^n \times \langle f, f_2, \ldots, f_m\rangle(x)$$

for all $x \in X$, and $f \in D(\rho \otimes \varphi)$. Conversely, suppose $f \in D(\rho \otimes \varphi)$. Let $\rho = \langle a_1, \ldots, a_m\rangle$, so that $\rho \otimes \varphi = a_1\varphi + \cdots + a_m\varphi$. By **MM** and Proposition 2.3 b), we find $d_1, \ldots, d_m \in D(\varphi)$ such that $f \in D(\langle a_1 d_1, \ldots, a_m d_m\rangle)$. Thus, there are $c_2, \ldots, c_m \in G$ such that $\langle f, c_2, \ldots, c_m\rangle(x) = \langle a_1 d_1, \ldots, a_m d_m\rangle(x)$ for all $x \in X$. But, by c), if $x \in C$ we have $d_1(x) = \cdots = d_m(x) = +1$, and so $\langle f, c_2, \ldots, c_m\rangle(x) = \langle a_1, \ldots, a_m\rangle(x) = \rho(x)$. Thus, $f \in D_S(\rho)$. $\square$

Proposition 2.7   *Let φ be an n-fold Pfister form with pure part φ' and let $g \in D(\varphi')$. Then there exists an $(n-1)$-fold Pfister form ψ with $\varphi = \langle 1, g\rangle \otimes \psi$.*

Proof.   Write $\varphi = \langle 1, a\rangle \otimes \rho$ for $a \in G$ and an $(n-1)$-fold Pfister form ρ. Thus $\varphi' = \rho' + a\rho$. By **MM** and Proposition 2.3 b), we find $c \in D(\rho')$ and $d \in D(\rho)$ such that $g \in D(\langle c, ad\rangle)$. Since $d \in D(\rho)$ we have $d\rho = \rho$ by Proposition 2.6 c), so $\varphi = \rho + a\rho = \rho + ad\rho = \langle 1, ad\rangle \otimes \rho$. Replacing a by ad we may assume $d = 1$. By induction on n we find an $(n-2)$-fold Pfister form τ with $\rho = \langle 1, c\rangle \otimes \tau$. So $\varphi = \langle 1, a\rangle \otimes \langle 1, c\rangle \otimes \tau$ and $g \in D(\langle a, c\rangle)$. Then $\langle a, c\rangle = \langle g, gac\rangle$ and $\langle 1, a\rangle \otimes \langle 1, c\rangle = \langle 1, a, c, ac\rangle = \langle 1, g, gac, ac\rangle = \langle 1, g\rangle \otimes \langle 1, gac\rangle$. $\square$

Corollary 2.8   *Let $g_1, h_1, \ldots, g_n, h_n \in G$ and $k_1 \in D(\langle g_1, h_1\rangle)$ be such that*

$$\varphi = \langle\!\langle g_1, \ldots, g_n\rangle\!\rangle = \langle\!\langle h_1, \ldots, h_n\rangle\!\rangle.$$

Then, there exist $k_2, \ldots, k_n \in G$ such that $\varphi = \langle\!\langle k_1, \ldots, k_n\rangle\!\rangle$.

Proof.   We have $g_1, h_1 \in D(\varphi')$. Therefore, using Proposition 2.3, $k_1 \in D(\varphi' + \varphi') = D(\varphi')$ and we may apply Proposition 2.7. $\square$

Corollary 2.9   *Let φ, ψ be forms over X such that $\varphi = 2 \times \psi$ and φ is a Pfister form. Then ψ is a Pfister form too.*

Proof.   One gets $1 \in D(2 \times \psi)$ hence $1 \in D(\psi)$ by Proposition 2.3, thus $1 \in D(\varphi')$, that is $\varphi = 2 \times \rho$ for a Pfister form ρ. But $\rho = \psi$. $\square$

Proposition 2.10   *For a form $\rho = \langle g_1, \ldots, g_n \rangle$ over X the following are equivalent*

    a) $g\rho = \rho$ for all $g \in D(\rho)$.
    b) $\rho = m \times \varphi$ for some $m \in \mathbb{N}$ and a Pfister form φ.

Proof.   *b) $\Rightarrow$ a)* Let $g \in D(\rho)$. Then $g \in D(\varphi)$ by Proposition 2.3. By Proposition 2.6 *c)* we get $g\rho = \rho$.

    a) $\Rightarrow$ b) Let $C = \{x \mid \rho(x) \neq 0\}$. Then $C \subset \{g > 0\}$ for all $g \in D(\rho)$. If $C = \emptyset$, *b)* is surely true. Otherwise we have $\rho(x) = \dim(\rho)$ for all $x \in C$. Thus $C = \{g_1 > 0, \ldots, g_n > 0\}$. From ρ and $\psi = \langle\langle g_1, \ldots, g_n \rangle\rangle$ we get a form φ with $\varphi(x) = 2^k$ for $x \in C$ and $\varphi(x) = 0$ outside, where $2^k = \gcd(2^n, n)$ (namely, if $2^k = r2^n + sn$, take $\varphi = r \times \psi + s \times \rho$). Choose φ anisotropic, so that $2^{n-k} \times \varphi$ is anisotropic. Since ψ is also anisotropic, $\psi = 2^{n-k} \times \varphi$. So φ is a Pfister form by Corollary 2.9. Let $m \cdot 2^k = n$. Since $\rho(x) = n$ for $x \in C$, also ρ is anisotropic. Thus $\rho = m \times \varphi$. $\square$

3. SAP-Spaces and Fans

In view of Proposition 1.16, one may ask what happens if four of the points x_i have the same Zariski closure. For this we can restrict ourselves to spaces of orderings, what we shall do for all this section. So, let (X, G) be a space of orderings. We want to consider two types of spaces of orderings, which are extremal in some sense. In case (X, G) is finite,

$$2\#(X) \leq \#(G) \leq 2^{\#(X)}.$$

In fact, on the one hand we have $X \subset \hat{G}$ and $X \cap (-1)X = \emptyset$, and on the other $G \subset \mathrm{Cont}(X, \{+1, -1\})$. Here, the extreme spaces appear if the left or the right hand equality holds.

Proposition 3.1   *Let (X, G) be a space of orderings. Then $s(X) \leq 1$ (resp. $w(X) \leq 1$, $l(X) \leq 2$) if and only if each constructible set $C \subset X$ is principal. If that is the case, $t(X) \leq 1$.*

Proof.   We may suppose $\#(X) > 1$. Then, let $s(X) = 1$ and $C = \{g_1 > 0\} \cup \cdots \cup \{g_k > 0\}$ for some $g_1, \ldots, g_k \in G$. Thus, $X \setminus C = \{-g_1 > 0, \ldots, -g_k > 0\} = \{-h > 0\}$ for some $h \in G$, since $s(X) = 1$. Whence $C = \{h > 0\}$. The converse implication is obvious.

Now, suppose $w(X) = 1$ and let $C \subset X$ be constructible. Then there is a form $\rho = \langle f_1, \ldots, f_n \rangle$ such that $\rho(x) = 2$ for $x \in C$, $\rho(x) = 0$ otherwise. Hence n is even, and we put $f = f_1 \cdots f_n$ if $n \equiv 2 \mod 4$, $f = -f_1 \cdots f_n$ if $n \equiv 0 \mod 4$. One easily checks that $C = \{f > 0\}$.

Finally, if $l(X) = 2$, we have $w(X) = 1$. $\qquad\qquad\qquad\qquad\square$

We shall see later that $t(X) \geq s(X)$, $w(X) = s(X)$ and $l(X) \geq 2^{s(X)}$ for all spaces of orderings (Corollary IV.7.5). In particular, we also get that $t(X) \leq 1$ if and only if every constructible set is principal.

Corollary 3.2   *One has $s(X) \leq 1$ if and only if $G = \mathrm{Cont}(X, \{+1, -1\})$. If X is finite, then $s(X) \leq 1$ if and only if $\#(G) = 2^{\#(X)}$.*

In particular, there is a unique space (X, G) with $\#(X) = n$ and $s(X) = 1$ for all $n \in \mathbb{N}$, $n > 1$. This will be denoted by $n \times E$. The background for this notation will appear in Section IV.2.

Examples 3.3   Let K be a number field and (X, G) the space of orderings associated to K. Then (X, G) is finite and $s(X) \leq 1$. (In fact, the elements $\alpha \in X$ correspond to the embeddings $K \to \mathbb{R}$. So X is finite, and by the elementary approximation theorem for finitely many spots one gets $s(X) \leq 1$.) Now let $K \supset \mathbb{R}$ be a formally real function field of dimension 1 and let (X, G) be associated to K. Here (X, G) is infinite, but again $s(X) = 1$.

This two facts will be shown in Section VI.5.

Remark 3.4   In view of Proposition 3.1 and Corollary 3.2, a space (X, G) for which $s(X) \leq 1$ is called *SAP-space*. These initials stand for Strong Approximation Property. $\qquad\qquad\qquad\qquad\square$

Now we are going to look at the other end of the scene. To do that we shall use the embedding $X \subset \widehat{G}$. We first present a characterization of X in $\widehat{G}$.

Proposition 3.5   *Let (X, G) be a space of orderings and let $1 \neq \sigma \in \widehat{G}$. Then $\sigma \in X$ if and only if $D(\langle 1, g \rangle) \subset \mathrm{Ker}(\sigma)$ for every $g \in \mathrm{Ker}(\sigma)$.*

Proof.   The "*only if*" part is clear. So, let $\sigma \in \widehat{G}$ be such that $D(\langle 1, g \rangle) \subset \mathrm{Ker}(\sigma)$ and $\sigma \neq 1$. We start by proving that $D(\varphi) \subset \mathrm{Ker}(\sigma)$ for any Pfister form $\varphi = \langle\langle g_1, \ldots, g_k \rangle\rangle$, where $g_1, \ldots, g_k \in \mathrm{Ker}(\sigma)$. This is done by induction on k, the case $k = 1$ being true by assumption. Let $k > 1$ and $\psi = \langle\langle g_2, \ldots, g_k \rangle\rangle$. Assume $a \in D(\varphi)$. We have $\varphi = \psi + g_1 \psi$. By **MM** we find $c, d \in D(\psi)$ such that $a \in D(\langle c, g_1 d \rangle)$. Then $ca \in D(\langle 1, g_1 cd \rangle)$. Now $g_1 \in \mathrm{Ker}(\sigma)$ and by induction also $c, d \in \mathrm{Ker}(\sigma)$. Thus $ca \in \mathrm{Ker}(\sigma)$ by hypothesis, and finally $a = c(ca) \in \mathrm{Ker}(\sigma)$.

Now, since $\sigma \neq 1$, for all $g_1, \ldots, g_k \in \mathrm{Ker}(\sigma)$ we have $D(\varphi) \neq G$, and, by Proposition 2.6 c), $\{g_1 > 0, \ldots, g_k > 0\} \neq \emptyset$. Hence, by compactness of X, we have

$$X \supset \bigcap_{g_1,\ldots,g_k \in \mathrm{Ker}(\sigma)} \{g_1 > 0, \ldots, g_k > 0\} \neq \emptyset.$$

This means that there is $\tau \in X$ such that $G \supset \mathrm{Ker}(\tau) \supset \mathrm{Ker}(\sigma)$. Since $G/\mathrm{Ker}(\sigma) \simeq \{+1, -1\}$, it follows either $G = \mathrm{Ker}(\tau)$ or $\mathrm{Ker}(\tau) = \mathrm{Ker}(\sigma)$. As $\tau \neq 1$, we get $\mathrm{Ker}(\tau) = \mathrm{Ker}(\sigma)$, that is, $\sigma = \tau \in X$. $\qquad\square$

After this result we introduce a special type of spaces of orderings:

Proposition and Definition 3.6   *For a space of orderings (X, G) the following properties are equivalent:*

a) $X = \{\sigma \in \widehat{G} \mid \sigma(-1) = -1\}$.
b) For all $\sigma_1, \sigma_2, \sigma_3 \in X$, the product $\sigma_1\sigma_2\sigma_3 \in \widehat{G}$ lies in X.
c) For all $g \in G$, $g \neq -1$: If $\{g > 0\} \subset \{f > 0\}$, then $f = 1$ or $f = g$.
d) For all $g \in G$, $g \neq -1$: $D(\langle 1, g \rangle) = \{1, g\}$.

If these properties hold, (X, G) is called a fan.

Proof.   Obviously *a)* $\Rightarrow$ *b)*. So assume *b)* and let $f, g \in G$ be such that $\emptyset \neq \{g > 0\} \subset \{f > 0\} \neq X$. If $\{g > 0\} \neq \{f > 0\}$, there are $\sigma_1 \in X \setminus \{f > 0\}$, $\sigma_2 \in \{f > 0\} \setminus \{g > 0\}$ and $\sigma_3 \in \{g > 0\}$, and by hypothesis, $\sigma = \sigma_1\sigma_2\sigma_3 \in X$. But then, $f(\sigma) < 0$ and $g(\sigma) > 0$, contradiction. Consequently, $\{g > 0\} = \{f > 0\}$, and so $f = g$. This proves *c)*. Now the implication *c)* $\Rightarrow$ *d)* is immediate from Proposition 2.6 *c)*. Finally, suppose *d)*, and let $\sigma \in \widehat{G}$, $\sigma(-1) = -1$. Then $\sigma \neq 1$, and if $g \in \mathrm{Ker}(\sigma)$, then $g \neq -1$. Thus, by *d)*, $D(\langle 1, g \rangle) \subset \mathrm{Ker}(\sigma)$, hence from Proposition 3.5 we get $\sigma \in X$. $\qquad\square$

Let us look at finite fans. For every $n \in \mathbb{N}$ we have a unique fan (X, G) with $\#(X) = 2^n$ (and $\#(G) = 2^{n+1}$). This fan will be denoted by F_n or $E[\mathbb{Z}_2^n]$ or $(2 \times E)[\mathbb{Z}_2^{n-1}]$. The latter notations will be introduced in Chapter IV.2.

Example 3.7   Let $\mathbb{R}((x_1)) \ldots ((x_n)) = K$ be the field of iterated power series over $\mathbb{R}$ and let (X, G) be the real space associated to K. Then (X, G) is the fan F_n. $\qquad\square$

The following properties of finite fans are easily verified:

Proposition 3.8   *Let (X, G) be a finite fan, and let $\#(X) = 2^n$. Then*

a) If $C \subset X$ is principal, then $\#(C) = 0$ or 2^{n-1} or 2^n.
b) If $C \subset X$ is basic, say $C = \{g_1 > 0, \ldots, g_k > 0\} \neq \emptyset$ with k minimal, then $\#(C) = 2^{n-k}$, $k \leq n$.
c) $s(X) = n$, $t(X) \geq n$, $w(X) = n$, $l(X) \geq 2^n$.
d) For a form ρ over X one has

$$\sum_{x \in X} \rho(x) \equiv 0 \bmod 2^n.$$

Proof.   Consider X as an affine hyperplane and the elements $g \in G$ as linear forms of the $\mathbb{F}_2$-vector space $\hat{G}$. Thus, a basic set $C = \{g_1 > 0, \ldots, g_r > 0\}$ is an affine subvariety, and $C = \{g_{i_1} > 0, \ldots, g_{i_k} > 0\}$, where $k \leq n$ is the maximal number of the mod (-1) independent elements among $g_1, \ldots, g_r$. We get $\#(C) = 2^{n-k}$. In particular, $s(X) \leq n$ (the first inequality by I.3.8). Now, the description of a singleton requires n elements g_i, and we get $s(X) = n$. The lower bound for $t(X)$ follows by complementation of a singleton. So we have proved *a)*, *b)*, and half of *c)* and for *d)* it is sufficient to consider 1–dimensional forms, for which the claim follows from *a)*. Now, let ρ define a singleton $\{x\}$. We have $2^n | \rho(x)$ by *d)*, and so $2^n \leq \rho(x) \leq \dim(\rho)$. Hence, $w(X) \geq n$ and $l(X) \geq 2^n$. Since $n = s(X) \geq w(X)$ we are done.   □

Property *c)* in the definition of fan leads to a further invariant for a space of orderings (X, G), concerning chains of principal sets.

Proposition 3.9   *We have:*

a) Let $X = \{a_1 > 0\} \cup \cdots \cup \{a_k > 0\}$ be a disjoint union of principal sets. Then every partial union is also principal.

b) Let $\{a > 0\} \subset \{b > 0\} \subset X$ be principal sets. Then also $\{b > 0\} \setminus \{a > 0\}$ is principal.

Proof.   *a)* Let $\{b_1, \ldots, b_m\} \subset \{a_1, \ldots, a_k\}$. Then

$$\{b_1 > 0\} \cup \cdots \cup \{b_m > 0\} = \{(-1)^{m+1} b_1 \cdots b_m > 0\}.$$

b) $\{b > 0\} \setminus \{a > 0\} = \{-ab > 0\}$.   □

Definition 3.10   *The* chain length *of (X, G), denoted by $cl(X)$, is the maximal number $\ell \leq \infty$ such that X is a union of ℓ mutually disjoint principal sets, or, equivalently, the maximal number $\ell \leq \infty$ such that there is an increasing chain $\emptyset \subsetneq \{a_1 > 0\} \subsetneq \cdots \subsetneq \{a_\ell > 0\} = X$.*

This invariant could be defined for arbitrary real spaces but in the geometrical examples it makes minor sense. However, for spaces of orderings it is of high technical importance (Section IV.4). Immediately from the definitions we get:

Proposition 3.11   *A space of orderings (X, G) is a fan if and only if $cl(X) \leq 2$.*

Apparently, fans are very special spaces of orderings. However, for an arbitrary space of orderings (X, G) any subspace is again a space of orderings, as we shall see in Section 6, and in general many of these subspaces are fans. It will turn out that most of the properties of (X, G) depend on the fans in X. So the notion of fan is the most important one in the theory of spaces of orderings.

The concept of fan is generalized as follows

Definition 3.12   *Let (X, G) be a space of signs and let $(F, G|F)$ be a subspace. If $(F, (G|F) \setminus \{0\})$ is a fan, then F is called a* fan *in X.*

Remarks 3.13   *a)* A fan F in a space of signs (X, G) is always a subspace of a space of orderings $(V^*, G(V^*))$ for some subvariety V of X (Example 1.20). In fact, if Y is a subspace of X such that $(G|Y) \setminus \{0\}$ is a group, then any $f \in G$ vanishing at some point of Y must vanish on the whole of Y. Hence all points $x \in Y$ have the same Zariski closure V in X, and $Y \subset V$.   □

b) Let C be a basic open set of a space of signs (X, G), say $C = \{g_1 > 0, \ldots, g_k > 0\}$. It follows readily from Proposition 3.8 that, for every finite fan F in X,

$$\#(F) \equiv 0 \bmod \#(C \cap F) \quad \text{and} \quad 2^k \#(C \cap F) \equiv 0 \bmod \#(F).$$

In fact, one main point of the theory is to show that, conversely, these congruences imply $C = \{g_1 > 0, \ldots, g_k > 0\}$ for suitable g_i's. This is the content of the so-called *generation formulae* (Theorems IV.7.3 and V.1.4).

SAP-spaces on the one hand and fans on the other are extremal types at both ends in the collection of all spaces of orderings. We already remarked that if the space of orderings (X, G) is finite,

$$2\#(X) \leq \#(G) \leq 2^{\#(X)}.$$

Then X is a fan if and only if the left hand equality holds, while X is a SAP–space if and only if the right hand equality holds. In general, X is neither a SAP–space nor a fan, but has many fans as subspaces. Later on (Section IV.5) we shall classify finite spaces of orderings and more general ones. Let us look here at the cases where $\#(X) \leq 4$:

Remark 3.14   Let (X, G) be a space of orderings and put $n = \#(X)$. If $n = 1$ then (X, G) is the atomic space E. If $n = 2, 3$, (X, G) is $n \times E$. One has $E = F_0$ and $2 \times E = F_1$; F_0 and F_1 are called *trivial* fans. For $n = 4$ one has two different spaces, namely, $4 \times E$ and F_2. The first space which is not a fan but for which $s(X) > 1$ appears for $n = 5$. Here $\#(G) = 16$. This space is given by the constructions in Section IV.2.

We end this section with the following fancy "non–example", which will be crucial later:

Lemma 3.15 (Miraculous Lemma)   *Let $n \geq 5$ be odd, and let $G = \mathbb{Z}_2^n$. Let $Z = \{\sigma_1, \ldots, \sigma_n, \sigma_1 \cdots \sigma_n, \tau_1, \ldots, \tau_m\} \subset \widehat{G}$, where $\sigma_1, \ldots, \sigma_n$ are linearly independent and $\tau_1, \ldots, \tau_m$ are elements (possibly none) among $\sigma_1\sigma_3\sigma_4$, $\sigma_1\sigma_3\sigma_5$, $\sigma_2\sigma_3\sigma_4$, $\sigma_2\sigma_3\sigma_5$. Then (Z, G) is not a space of orderings.*

Proof.   Let $g_1, \ldots, g_n \in G$ be such that $\sigma_i(g_j) = +1$ for $i \neq j$ and $\sigma_i(g_i) = -1$. Consider the forms

$$\rho = \langle 1, g_1 g_2 g_3 g_4, g_1 g_2 g_3 g_5 \rangle \quad \text{and} \quad \tau = \langle g_1 g_2 g_4 g_5, g_2 g_3, g_3 g_1 \rangle.$$

One easily checks that $\rho = \tau$. Assume that Z is a space of orderings. Then by **MM** there exists

$$h \in D(\langle g_1 g_2 g_3 g_4, g_1 g_2 g_3 g_5 \rangle) = g_1 g_2 g_3 g_4 D(\langle 1, g_4 g_5 \rangle)$$

such that $\langle 1, h \rangle$ represents $g_1 g_2 g_4 g_5$. Now

$$\{g_4 g_5 > 0\} \supset \{\sigma_1, \sigma_2, \sigma_3, \sigma_6, \ldots, \sigma_n, \sigma_1 \cdots \sigma_n\},$$

a collection of $n - 1$ independent orderings. Hence, $\{g_4 g_5 > 0\}$ is a linear subspace of codimension 1 of $\widehat{G}$, and, by Proposition 2.6 c), $\langle 1, g_4 g_5 \rangle$ only represents 1 and $g_4 g_5$, hence $h = g_1 g_2 g_3 g_4$ or $g_1 g_2 g_3 g_5$. Now $\sigma_5(g_1 g_2 g_3 g_4) = +1$, $\sigma_5(g_1 g_2 g_4 g_5) = -1$. Thus we have $g_1 g_2 g_4 g_5 \notin D(\langle 1, g_1 g_2 g_3 g_4 \rangle)$. A similar argument gives $\sigma_4(g_1 g_2 g_3 g_5) = +1$, $\sigma_4(g_1 g_2 g_4 g_5) = -1$, and thus $g_1 g_2 g_4 g_5 \notin D(\langle 1, g_1 g_2 g_3 g_5 \rangle)$. Contradiction. $\qquad\qquad \square$

4. Local Spaces of Signs

Let (X, G) be a prespace of signs. Then any Z–closed set $W \subset X$ contains a minimal non-empty Z–closed set V. This is easily seen by Zorn's Lemma, using the compactness of X in the Zariski topology. Clearly, V is a subvariety. This gives rise to the following definition:

Definition 4.1   *Let (X, G) be a prespace of signs. Then (X, G) is called* local, *if X contains a unique minimal non-empty Z–closed set V.*

In this situation V is contained in any Z–closed subset $W \neq \emptyset$ of X. Moreover, $g \in G^*$ if and only if $V \not\subset \{g = 0\}$.

Example 4.2   Let (X, G) be a prespace of orderings. Then (X, G) is local, with $V = X$.

(4.3) Localization.   Let (X, G) be a prespace of signs and let $V \subset X$ be a subvariety. We set

$$X_V = \bigcap_{\substack{g \in G \\ V \not\subset \{g=0\}}} \{g \neq 0\}$$

and

$$G_V = G | X_V$$

Then (X_V, G_V) is a subspace of (X, G), and consequently a prespace of signs. This prespace of signs is, in fact, local. The unique minimal non-empty Z–closed set of X_V is $V^* = V \cap X_V$. This local prespace of signs is called the localization of (X, G) at V.

Examples 4.4   *a)* Suppose X itself is irreducible, and let $V = X$. Then (X_V, G_V) is just the prespace of orderings $(V^*, G(V^*))$ we considered in 1.9.

b) Suppose that (X, G) is associated to a commutative ring with unit A and let $\mathfrak{p} \subset A$ be a real prime ideal. Moreover, let V be the zero set of $\mathfrak{p}$ in X. Then (X_V, G_V) is associated to the localization $A_{\mathfrak{p}}$.

Proposition 4.5   *Let (X, G) be a local prespace of signs. Let $X_{\max}$ be the set of closed points of X, and consider the group of units $G^* \subset G$. If* **HL** *holds for (X, G), the pair $(X_{\max}, G^*)$ is a prespace of orderings. If, moreover,* **MM** *holds for (X, G) then $(X_{\max}, G^*)$ is a space of orderings.*

Proof.   First we have to show that $(X_{\max}, G^*)$ is a real space. So let $g \in G^*$ such that $g(x) = +1$ for all $x \in X_{\max}$. Since g is a unit in G, by continuity $g(x) = +1$ for all $x \in X$, and $g = 1$.

Now the constructible and the Harrison topologies coincide for $(X_{\max}, G^*)$. Next we show at a time that G^* separates points in $X_{\max}$ and that $X_{\max}$ is a Hausdorff space. For this let $x \neq y \in X_{\max}$ and choose an element z in the minimal non-empty Z-closed set V of X. By the minimality of V, $V = \mathrm{Adh}_Z(z)$, and we can choose z closed. Hence, using **HL**, we find $g \in G$ with $g(x) = +1$, $g(y) = -1$ and $g(z) \neq 0$. In particular, $V \not\subset \{g = 0\}$, and, as already remarked, $g \in G^*$. We are done.

The compactness of $X_{\max}$ with respect to the topology defined by G^* follows from the fact that $X_{\max}$ is a compact subset of X with respect to the Harrison topology.

Finally, it is easily seen that **MM** goes over from (X, G) to $(X_{\max}, G^*)$. $\square$

Remark 4.6   We considered on $X_{\max}$ the topology $\mathcal{T}^*$ coming from the space of signs $(X_{\max}, G^*)$. Let $\mathcal{T}$ be the topology on $X_{\max}$ which is induced by the Harrison topology of X. Then $\mathcal{T} \supset \mathcal{T}^*$ and $X_{\max}$ is compact with respect to both. Hence $\mathcal{T} = \mathcal{T}^*$. However, $X_{\max}$ admits two different structures as a real space, coming from G^* and $G|X_{\max}$ respectively, and so there are more constructible sets in $X_{\max}$ with respect to the latter. $\square$

As a corollary of Proposition 4.5 we get:

Proposition 4.7   *Let (X, G) be a space of signs and let $V \subset X$ be a subvariety. Then (X_V, G_V) is a space of signs and $((X_V)_{\max}, G_V{}^*)$ is a space of orderings.*

Proof.   By Proposition 4.5 it suffices to prove the assertion for the localization (X_V, G_V). But, by Proposition 1.17, we know that **PE** and **HL** hold for (X_V, G_V), and consequently it remains to show that **MM** holds for every Z-closed subset Z of (X_V, G_V).

To prove that, we write $Z = X_V \cap Z'$ for some Z-closed subset Z' of X. Then $Z' \supset V$, and $Z = Z'_V$. Consequently, we have to show that **MM** goes over from Z' to Z. To do it, we can suppose $Z' = X$ and $Z = X_V$. Then, like in Example 1.20, we extend data from X_V to X as follows. Let ρ be a form over X

and $x \in X_V$ such that for suitable $a_1, \ldots, a_n \in G$ it holds $\rho(y) = \langle a_1, \ldots, a_n \rangle(y)$ for all $y \in X_V$ and $a_1(x) \neq 0, \ldots, a_n(x) \neq 0$. We then consider the constructible set $C = \{y \in X \mid \rho(y) \neq \langle a_1, \ldots, a_n \rangle(y)\}$, so that $C \cap X_V = \emptyset$. By definition of X_V and compactness of the constructible topology, we get $g_1, \ldots, g_s \in G$, not vanishing on V, such that $C \cap \{g_1 \neq 0, \ldots, g_s \neq 0\} = \emptyset$. Since V is irreducible, $g = g_1 \cdots g_s$ does not vanish on V either, and clearly $C \subset \{g = 0\}$. We conclude that $g \in G_V{}^*$, and $g^2 \rho = \langle g^2 a_1, \ldots, g^2 a_n \rangle$. Using this method the conclusion is immediate. $\qquad\square$

We conclude this section by an important converse of the preceding proposition.

Proposition 4.8   *Let (X, G) be a prespace of signs for which* **PE** *and* **HL** *hold. If for all subvarieties $V \subset X$* **MM** *holds for the pair $((X_V)_{\max}, G_V{}^*)$, then (X, G) is a space of signs.*

Proof.   We have to show that **MM** holds for every Z-closed set Z of (X, G). However, it will be clear from the argument below that it is enough to treat the case $Z = X$. So, let $\rho = \langle a_1, \ldots, a_n \rangle$, $\tau = \langle b_1, \ldots, b_m \rangle$, $h, c_2, \ldots, c_{n+m} \in G$, $x \in X$ in the hypotheses of **MM**, and consider $V = \mathrm{Adh}_Z(x)$. Then $h|X_V, a_i|X_V, b_j|X_V, c_k|X_V$ belong to $G_V{}^*$. Since **MM** holds for $((X_V)_{\max}, G_V{}^*)$, we find $\rho' = \langle f, a_2', \ldots, a_n' \rangle$, $\tau' = \langle g, b_2', \ldots, b_m' \rangle$ and $\lambda' = \langle h, c \rangle$, such that,

$$a_i'(y) \neq 0, \ b_j'(y) \neq 0, \ f(y) \neq 0, \ g(y) \neq 0, \ c(y) \neq 0$$

for all i, j and $y \in X_V$, and

$$\rho(y) = \rho'(y), \ \tau(y) = \tau'(y), \ \langle f, g \rangle(y) = \langle h, c \rangle(y)$$

for all $y \in (X_V)_{\max}$. Now we need this to hold for all y in some Zariski neighbourhood of x. This U is found as follows. First, let $\ell \in G$ be the product of h, f, g, c and all the a_i's, a_i''s, b_j's, b_j''s, c_k's, and c_k''s, and consider the Zariski open set $U_1 = \{\ell \neq 0\}$. Clearly, $x \in X_V \subset U_1$. Now, let

$$C = \{y \in X \mid \rho(y) \neq \rho'(y)\}$$

and $C' = C \cap X_V$. Then C and C' are constructible in X and X_V respectively. By assumption

$$C' \cap (X_V)_{\max} = C \cap (X_V)_{\max} = \emptyset.$$

Since ρ and ρ' are generated by units of G_V, C' is open and closed in X_V. It follows $C' = \emptyset$.

Hence, $C \subset X \setminus X_V$ is covered by the Zariski closed sets

$$\{u = 0\}, \ u \in G, \ V \not\subset \{u = 0\}.$$

Since C is compact in the constructible topology, $C \subset W$, where W is Zariski closed and $V \not\subset W$. Then $U = U_1 \setminus W$ is a Zariski neighbourhood of x and $\rho(y) = \rho'(y)$ for all $y \in U$. Clearly, the conclusion follows by repeating the argument for the other forms. $\qquad\square$

5. The Space of Signs of a Ring

We prove here that the real space associated to a ring (resp. a field) is a space of signs (resp. of orderings), as was stated in Proposition 1.21. Clearly, it is enough to prove the assertion for rings, so let A be a commutative ring with unit, $X = \mathrm{Spec}_r(A)$ and (X, G) the associated real space (II.1.5). We already know that (X, G) is a prespace of signs (Example 1.2) and that **PE** and **HL** hold for (X, G) (Proposition 1.13). Then, by Proposition 4.8, we have to show that for every subvariety $V \subset X$, **MM** holds for the pair $((X_V)_{\max}, G_V^*)$. But by 1.9 and Example 4.4 $b)$, the subvariety V is the zero set of a real prime ideal $\mathfrak{p} \subset A$, and (X_V, G_V) is the real space associated to the localization $A_{\mathfrak{p}}$. In conclusion we are reduced to prove the following:

Proposition 5.1   *Let A be a local ring with formally real residue field, and let (X, G) be the real space associated to A. Then* **MM** *holds for* $(X_{\max}, G^*)$.

Now note that the group G^* consists of all functions $\mathrm{sign}[f]$, $f \in A$, such that $f(x) \neq 0$ for all prime cones $x \in X = \mathrm{Spec}_r(A)$. Since A is a local ring whose maximal ideal $\mathfrak{m}$ has formally real residue field, this condition is equivalent to $f \notin \mathfrak{m}$. In other words, G^* consists of the functions $\mathrm{sign}[f]$ for all units f of A. To simplify the notation, we shall write $\bar{f}$ instead of $\mathrm{sign}[f]$.

The proof of Proposition 5.1 requires a special characterization of the elements represented by forms over $(X_{\max}, G^*)$. For this we consider the set

$$\Sigma = \{t \in A \mid t(x) \geq 0 \text{ for all } x \in X\}$$

(see Corollary II.1.15). Then, for any given $a_1, \ldots, a_n \in A$ we put:

$$d(a_1, \ldots, a_n) = \{t_1 a_1 + \cdots + t_n a_n \mid t_1, \ldots, t_n \in \Sigma\}.$$

With this terminology, we have the following elementary lemma:

Lemma 5.2   *Let $u, a_1, \ldots, a_n, b_1, \ldots, b_m \in A$ be units such that*

$$u \in d(a_1, \ldots, a_n, b_1, \ldots, b_m).$$

Then there are units $v, w \in A$ such that $u = v + w$, and

$$v \in d(a_1, \ldots, a_n) \qquad w \in d(b_1, \ldots, b_m).$$

Proof.   By hypothesis, $u = a + b$ with $a \in d(a_1, \ldots, a_n)$ and $b \in d(b_1, \ldots, b_m)$. Then we write $a_1/u = s - t$ with $s = 1 + a_1/u + (a_1/u)^2$, $t = 1 + (a_1/u)^2$. Clearly $s, t \in \Sigma$, and since A is local with formally real residue field, both s and t are units. Then we have $u = a' + b'$ with $a' = (at + a_1)/s$, $b' = bt/s$, and:

   $i)$ If a is not a unit, $at + a_1$ is a unit.

   $ii)$ If b is a unit, bt/s is a unit.

This shows how to replace at least one summand by a unit, and then the two of them at a time by two units. $\qquad\square$

The previous lemma is clearly patterned upon Axiom **MM**, and in fact we have:

Proposition 5.3   *Let $a_1, \ldots, a_n, b \in A$ be units. Then $b \in d(a_1, \ldots, a_n)$ if and only if $\bar{b} \in D(\langle \bar{a}_1, \ldots, \bar{a}_n \rangle)$.*

Now, to prove this, we have to recall some basic facts on Σ-modules.

Definition 5.4   *A subset $M \subset A$ is called a Σ-module if $M + M \subset M$, $\Sigma \cdot M \subset M$ and $1 \in M$. A given Σ-module M is called* proper *if $-1 \notin M$. A proper Σ-module M is called* prime *if $A = M \cup -M$ and $M \cap -M$ is a prime ideal.*

Of course, the Σ-modules we shall deal with are the sets $d(a_1, \ldots, a_n)$ defined above. In general, we have:

Proposition 5.5   *Let M be a prime Σ-module, $\mathfrak{p} = M \cap -M$, and $t_1, \ldots, t_n \in \Sigma$, $s_1, \ldots, s_n \in M$. If $s_1 t_1 + \cdots + s_n t_n \in -M$, then for each $i = 1, \ldots, n$ either $s_i \in \mathfrak{p}$ or $t_i \in \mathfrak{p}$. In particular, $\mathfrak{p}$ is a real prime ideal.*

Proof.   By assumption $-s_1 t_1 - \cdots - s_n t_n \in M$, and clearly $s_2 t_2 + \cdots + s_n t_n \in M$. Hence $-s_1 t_1 \in M$, and we get $s_1 t_1 \in M \cap -M = \mathfrak{p}$. Since $\mathfrak{p}$ is prime, $s_i \in \mathfrak{p}$ or $t_1 \in \mathfrak{p}$. Finally, setting $s_1 = \cdots = s_n = 1 \in M$ we see that $\mathfrak{p}$ is real. $\qquad\square$

Proposition 5.6   *Every proper Σ-module M is contained in a prime Σ-module N.*

Proof.   By Zorn's lemma, there is a maximal proper Σ-module $N \supset M$. We shall show in five steps that N is prime.

i) The aditive group $A/N \cap -N$ has no element of order 2. Let $a \in A$ be such that $2a \in N$. If $a \notin N$, then $N + a\Sigma$ is a larger Σ-module, hence there are $s \in N$ and $t \in \Sigma$ with $-1 = s + at$. Thus, $-2 = 2s + 2at \in N$ and also $-1 = -2 + 1 \in N$. Contradiction.

ii) $N \cap -N$ is an ideal of A. Let $a \in N \cap -N$ and $b \in A$. We have $4b = (b+1)^2 - (b-1)^2$, hence $4ba = (b+1)^2 a + (b-1)^2(-1)a \in N \cap -N$. By *i)* we conclude $ba \in N \cap -N$.

iii) If $b \in A$ and $b^2 \in N \cap -N$, then $b \in N \cap -N$. Suppose that $b \notin N$ and consider the Σ-module $N + b\Sigma$. Then $-1 = s + bt$ for some $s \in N, t \in \Sigma$. Hence $4s = -(4 + 4bt) = -(2 + bt)^2 - (-b^2)t^2 \in -N$. Thus $4s \in N \cap -N$, and by *i)* we get $s \in N \cap -N$. This means that $-1 = bt \mod N \cap -N$, and consequently $1 = b^2 t^2 = 0 \mod N \cap -N$, a contradiction. Whence $b \in N$. Replacing b by $-b$ we conclude $-b \in N$, whence $b \in N \cap -N$ as wanted.

iv) $N \cap -N$ is a prime ideal. Let $a, b \in A$ be such that $ab \in N \cap -N$. As above, we consider the Σ-module $N + b\Sigma$. Then either $b \in N$ or $-1 = s + bt$ with $s \in N, t \in \Sigma$. In the latter case, $-a^2 = a^2 s + a^2 bt = a^2 s + (at)(ab) \in N$

by *ii)*. Hence $a^2 \in N \cap -N$, and $a \in N \cap -N$ by *iii)*. Replacing b by $-b$ we see that either $-b \in N$ or $a \in N \cap -N$. We are done.

v) $N \cup -N = A$. Assume that there is $a \in A \setminus N \cup -N$. Then $-1 \in (N + a\Sigma) \cap (N - a\Sigma)$, say $-1 = s_1 + at_1 = s_2 - at_2$. By *ii)*, $a, t_1, t_2 \notin N \cap -N$. Now, multiplying these equations by t_2 and t_1 respectively, and adding we get $t_2 s_1 + t_1 s_2 = -(t_1 + t_2) \in -\Sigma$. Thus, by Proposition 5.5, $s_1, s_2 \in N \cap -N$. But then $-a^2 t_1 t_2 = (at_1)(-at_2) = (1 + s_1)(1 + s_2) \in N \cap -N$, and by *iv)*, either of a, t_1, t_2 is in $N \cap -N$. Contradiction. $\qquad\square$

Finally:

Proof of Proposition 5.3. Suppose first that $b \in d(a_1, \ldots, a_n)$. Then $b = t_1 a_1 + \cdots + t_n a_n$, with $t_1, \ldots, t_n \in \Sigma$. By Lemma 5.2 we may assume that $t_1 a_1$ and $b' = t_2 a_2 + \cdots + t_n a_n$ are units. Then, by induction on n, we find units $b_3, \ldots, b_n \in A$ such that $\langle \bar{a}_2, \ldots, \bar{a}_n \rangle = \langle \bar{b}', \bar{b}_3, \ldots, \bar{b}_n \rangle$. Hence,

$$\langle \bar{a}_1, \ldots, \bar{a}_n \rangle = \langle \bar{t}_1 \bar{a}_1, \bar{b}', \bar{b}_3, \ldots, \bar{b}_n \rangle.$$

Now, we replace $t_1 a_1$ and b' by $b = t_1 a_1 + b'$ and $b_2 = (t_1 a_1 + b') t_1 a_1 b'$ to get

$$\langle \bar{a}_1, \ldots, \bar{a}_n \rangle = \langle \bar{b}, \bar{b}_2, \bar{b}_3, \ldots, \bar{b}_n \rangle.$$

Thus, the proof of the only if part is complete.

Conversely, let $\bar{b} \in D(\langle \bar{a}_1, \ldots, \bar{a}_n \rangle)$. Then, for suitable units $b_2, \ldots, b_n \in A$, we have $\langle \bar{a}_1, \ldots, \bar{a}_n \rangle = \langle \bar{b}, \bar{b}_2, \bar{b}_3, \ldots, \bar{b}_n \rangle$. We claim that the Σ-module $M = d(-a_1 b^{-1}, \ldots, -a_n b^{-1}, 1)$ is not proper.

Indeed, otherwise by Proposition 5.6 there is a prime Σ-module $N \supset M$, and we consider the real prime ideal $\mathfrak{p} = N \cap -N$ (Proposition 5.5). Now we consider the following diagonal quadratic forms over the residue field k of $\mathfrak{p}$:

$$q = \langle -a_1 b^{-1}, \ldots, -a_n b^{-1}, 1 \rangle \bmod \mathfrak{p}, \quad q' = \langle 1, -b_2 b^{-1}, \ldots, -b_n b^{-1}, -1 \rangle \bmod \mathfrak{p}.$$

After multiplying by b we see immediately from the hypothesis that q and q' have the same signature in all orderings of the field k. Then, by Pfister's local global principle ([Pr2 §7] or [Schl2 Ch.2, §7]), the forms $l \times q$ and $l \times q'$ are congruent for some $l \in \mathbb{N}$. Now, it is obvious that $l \times q'$ has non trivial zeroes in k^{ln}, and consequently $l \times q$ has them too. Such a non trivial zero gives sums of squares $t_1, \ldots, t_{n+1}$ not all in $\mathfrak{p}$ such that

$$t_1(-a_1 b^{-1}) + \cdots + t_n(-a_n b^{-1}) + t_{n+1} \in \mathfrak{p}.$$

Since all the $-a_i b^{-1}$'s are units, this contradicts Proposition 5.5, and our claim is proved.

The fact that M is not proper means that $-1 \in d(-a_1 b^{-1}, \ldots, a_n b^{-1}, 1)$, hence that $b \in d(a_1, \ldots, a_n, -b)$. Thus

$$b = t_1 a_1 + \cdots + t_n a_n - tb,$$

for suitable $t, t_1, \ldots, t_n \in \Sigma$. Since the residue field of A is formally real and $t \in \Sigma$, $1 + t$ is a unit. Hence

$$b = (1 + t)^{-1}t_1 a_1 + \cdots + (1 + t)^{-1}t_n a_n - tb.$$

$\square$

As explained before, Proposition 5.3 shows that Proposition 5.1 is equivalent to Lemma 5.2, and this completes the proof of Proposition 1.21.

6. Subspaces

This section is devoted to the proof of the following important fact:

Theorem 6.1   *Let (X, G) be a space of signs (resp. of orderings) and let $(Y, G|Y)$ be a subspace. Then $(Y, G|Y)$ is again a space of signs (resp. of orderings).*

Proof.   It is enough to prove the assertion for spaces of signs. By Propositions 1.8 and 1.17, **PE** and **HL** hold for $(Y, G|Y)$. Then, by Proposition 4.8, it remains to show that for every subvariety V of Y, **MM** holds for the pair $((Y_V)_{\max}, (G|Y)_V{}^*)$. Let

$$Y = \bigcap_{g \in G_1} \{g \geq 0\} \cap \bigcap_{g \in G_2} \{g > 0\}, \quad G_1, G_2 \subset G.$$

We shall argue in three steps.

1st *step: Reduction to the case when (X, G) is a space of orderings.* Denote by W the Zariski closure of V in X. We first remark that $X_W \cap Y = Y_V$, and so $(X_W)_{\max} \cap Y \subset (Y_V)_{\max}$. We claim that, in fact,

$$(X_W)_{\max} \cap Y = (Y_V)_{\max}.$$

To see it, suppose that $y \in Y_V$ and $y \to x$ in X_W. For all $g \in G_1$, we have $g(y) \geq 0$, and by continuity $g(x) \geq 0$. If $g \in G_2$, $W \not\subset \{g = 0\}$ and consequently $g(x) \neq 0$; but $g(y) > 0$ implies $g(x) \geq 0$ by continuity, and we conclude $g(x) > 0$. Hence, $x \in X_W \cap Y = Y_V$. This shows that $(Y_V)_{\max} \subset (X_W)_{\max}$ and our claim is proved. On the other hand, let $g \in G$ be such that $g(y) \neq 0$ for all $y \in Y_V$. Then, g does not vanish on the non-empty set $V^* = V \cap Y_V \subset W$, and consequently, $X_W \subset \{g \neq 0\}$. All this means that $((Y_V)_{\max}, (G|Y)_V{}^*)$ is a subspace of $((X_W)_{\max}, G_W{}^*)$, which is a space of orderings by Proposition 4.5. This completes the step.

2nd *step: Reduction to the case when Y is a basic open set.* This follows if we prove that given two forms ρ, ρ' over X such that $\rho(x) = \rho'(x)$ for all $x \in Y$ there are $g_1, \ldots, g_s \in G_1$ such that $\rho(x) = \rho'(x)$ for all $x \in \{g_1 >$

$0, \ldots, g_s(x) > 0\}$. But given such forms, Y does not meet the constructible set $C = \{x \in X \mid \rho(x) \neq \rho'(x)\}$, that is,

$$\emptyset = C \cap Y = C \cap \bigcap_{g \in G_1} \{g > 0\}.$$

Then, by compactness of the constructible topology, there are $g_1, \ldots, g_s \in G_1$ such that $\{g_1 > 0, \ldots, g_s > 0\} \cap C = \emptyset$. These g_i are the functions we sought.

3rd *step: Conclusion.* By the preceding reductions, we can suppose that (X, G) is a space of orderings and $Y = \{g_1 > 0, \ldots, g_s > 0\}$. In order to prove **MM** for $(Y, G|Y)$ consider $\rho = \langle a_1, a_2, \ldots, a_n \rangle$, $\tau = \langle b_1, b_2, \ldots, b_m \rangle$, $h, c_2, \ldots, c_{n+m} \in G$ such that $(\rho + \tau)(x) = \langle h, c_2, \ldots, c_{n+m} \rangle(x)$ for all $x \in Y$. Then, consider the Pfister form $\varphi = \langle\!\langle g_1, \ldots, g_s \rangle\!\rangle$ and the products $\rho \otimes \varphi$, $\rho' \otimes \varphi$. Clearly,

$$\langle h, c_2, \ldots, c_{n+m} \rangle \otimes \varphi = \langle h, d_2, \ldots, d_l \rangle$$

for suitable $d_2, \ldots, d_l \in G$, and by Proposition 2.6 *a)*,

$$(\rho \otimes \varphi + \tau \otimes \varphi)(x) = \langle h, d_2, \ldots, d_l \rangle(x)$$

for all $x \in X$. Thus, by **MM** for (X, G), we find $f, g \in G$ such that $f \in D(\rho \otimes \varphi), g \in D(\tau \otimes \varphi)$ and $h \in D(\langle f, g \rangle)$. Finally, by Proposition 2.6 *e)*, we have $f \in D_Y(\rho)$, $g \in D_Y(\tau)$, which completes the proof. $\square$

Notes

The idea of studying the space of all orderings of an algebraic structure from an axiomatic viewpoint is due to Marshall, who was motivated by the case of spaces of orderings of fields ([Mr 1-5]). Later, it turned out that his axioms also applied to skew fields ([Ts], [Cr]), to local rings ([Kn1]), and even to ternary fields ([Kl]). The abstract setting for real spectra of rings that is presented here is new. Our axioms imitate more or less well known facts of rings. The meaning of the initials, surely guessed by the reader, are the following: **PE**=positive equation, **HŁ**=Hörmander-Łojasiewicz, and **MM**=Murray Marshall. In fact, Marshall discovered that all the pleasant properties of spaces of orderings of fields depend just on these axioms (which simplify considerably in the field case). The axiom **MM** is related Pfister's local-global principle for fields ([Pf]), plus an almost trivial local-global principle for rings ([Br4]; see Section 5, in particular Proposition 5.6, and also [Mr-Wa]). Recently, Marshall has found a different approach, enough to deal with spectra of rings ([Mr9]), by using one single axiom which implies **PE**, **HŁ** and **MM**; it has been shown later that this single axiom is in fact equivalent to the other three.

Most of the material of Section 1 is standard, but Proposition 1.16 seems to be new even for real spectra. The content of section 2 can be found in [Mr1], except for Proposition 2.10. The important Proposition 2.7 is due to

W. Scharlau ([Schl1,2]). For spaces of orderings of fields the notion of SAP space was introduced in [El-Lm]. These spaces were characterized by Prestel in the very important paper [Pr1]. The notion of fan was introduced in [Be-Kö]; however its importance was not yet clear at that time. The transcription of these notions to spaces of orderings is straightforward. Chain length was first defined in [Mr4], and is crucial for the proof of local-global principles in spaces of orderings (see Sections IV.5-6). Also the miraculous lemma is Marshall's contribution; it is the key for the classification of finite spaces of orderings in the axiomatic context. The material of Section 4 is new in the general case, but for real spectra of local rings it can be found already in [Kn1]. Concerning spaces of orderings, the result of Section 6 is in [Mr1].

Chapter IV. Spaces of Orderings

Summary. This chapter contains a fully detailed presentation of the theory of spaces of orderings. In Section 1 we reformulate the notion of space of orderings to stress the connection with the duality of topological groups and the theory of reduced quadratic forms. Section 2 is devoted to sums and extensions of spaces of orderings and their basic properties. In Section 3 we introduce spaces of finite type and their trees, which support the use of induction in many proofs. The fundamental fact that the chain length of a space of orderings is bigger than or equal to that of any subspace is proved in Section 4. Also in this section we define solid fans, impervious fans, and places, which are essential to prove in Section 5 that finite chain length is equivalent to finite type. This is the key technical result of the theory. We prove in Section 6 the local-global principle that reduces problems on forms from the whole space to its finite subspaces. In Section 7 we draw the consequences: the representation theorem, the generation formula and the stability formula. Using these, we bound each invariant s, $\bar{s}$, t, $\bar{t}$, w, l in terms of any of the others. In particular, all of them are finite or infinite at a time. The final result of the section and the chapter is a local-global separation principle.

1. The Axioms Revisited

In this chapter we shall use equivalent definitions of prespace and space of orderings. The reader will soon realize that several hidden references to this alternative presentation have been scattered in the preceding chapters.

(1.1) Marshall's Axioms. Let G be a (multiplicative) abelian group *of exponent* 2, that is, such that $g^2 = 1$ for all $g \in G$, and let $\widehat{G} = \mathrm{Hom}(G, \mathbb{C})$ the topological dual group of G for the discrete topology on G (see [Mo]). Then, $\sigma(g) \in \{+1, -1\}$ for all $\sigma \in \widehat{G}$ and $g \in G$, so that $\widehat{G} = \mathrm{Hom}(G, \{+1, -1\})$. Furthermore, let us be given a *distinguished element* $-1 \in G$, $-1 \neq 1$, and a subset X of $\widehat{G}$. The pair (X, G) is a prespace of orderings if the following conditions hold:

$\mathbf{O_1}$: X is closed in $\widehat{G}$.

O_2: $\sigma(-1) = -1$ for all $\sigma \in X$.

O_3: The unique element $g \in G$ such that $\sigma(g) = +1$ for all $\sigma \in X$ is $g = 1$.

Having in mind that $\widehat{G}$ is compact the equivalence of this definition and ours should be immediately clear. The only difference is that there we write $g(x)$ for $g \in G$ and $x \in X$ and here we write $\sigma(g)$ for $\sigma \in X$ and $g \in G$. The present notation, specially greek letters for the elements of $\widehat{G}$ and in particular for the points of $X \subset \widehat{G}$ is used to stress the fact that they are seen as homomorphisms $G \to \{+1, -1\}$. Thus, for instance, for $g_1, \ldots, g_k \in G$ we have

$$\{x \in X \mid g_1(x) > 0, \ldots, g_k(x) > 0\} = \{\sigma \in X \mid \sigma(g_1) = +1, \ldots, \sigma(g_k) = +1\}$$

In this setting, an *isomorphism* from a prespace of orderings (X_1, G_1) onto another (X_2, G_2) is a group isomorphism $\mu : G_1 \to G_2$ such that the induced dual isomorphism

$$\widehat{\mu} : \widehat{G}_2 \to \widehat{G}_1 \, ; \, \sigma \mapsto \sigma \circ \mu$$

maps X_2 onto X_1.

Let (X, G) be a prespace of orderings. Of course, we have forms over X and all related notions; in particular, we have the signature $\sigma(\rho) = \sigma(g_1) + \cdots + \sigma(g_n)$ of a form $\rho = \langle g_1, \ldots, g_n \rangle$ at a point $\sigma \in X$. Then, (X, G) is a space of orderings if also the following condition holds:

O_4: For any two forms ρ, τ on X and $a \in D(\rho + \tau)$ there exist $g \in D(\rho)$, $h \in D(\tau)$ such that $a \in D(\langle g, h \rangle)$.

As was already explained (Remarks and Notations 1.18), this is nothing but Axiom **MM**. In addition, here we have a very important new equivalent axiom:

O_4': Let ρ, τ be forms on X such that $\rho + \tau$ is isotropic. Then there exists $g \in D(\rho) \cap -D(\tau)$.

Proof of the equivalence: First we see that $O_4 \Rightarrow O_4'$. Let $\rho + \tau$ be isotropic. If $\dim(\tau) = 1$ we set $-a = g$ and are done. Otherwise, we write $\tau = \langle a \rangle + \tau_1$. Then $\rho + \tau = \langle a, -a \rangle + \psi$. We get $\rho + \tau_1 = \langle -a \rangle + \psi$ hence $-a \in D(\rho + \tau_1)$. By O_4 we have $g \in D(\rho)$, $h \in D(\tau_1)$ with $-a \in D(\langle g, h \rangle)$, that is $\langle g, h \rangle = \langle -a, b \rangle$ for some $b \in G$ or $\langle a, h \rangle = \langle -g, b \rangle$. But $D(\langle a, h \rangle) \subset D(\tau)$, hence $-g \in D(\tau)$.

For the converse implication $O_4' \Rightarrow O_4$, let $a \in D(\rho + \tau)$. Then $\rho_1 + \tau$ is isotropic for $\rho_1 = \langle -a \rangle + \rho$. By O_4' we find $h \in D(\tau)$ with $-h \in D(\rho_1)$. Then $\rho + \langle -a, h \rangle$ is isotropic. Again by O_4' we find $g \in D(\rho)$ with $-g \in D(\langle -a, h \rangle)$. So $\langle -a, h \rangle = \langle -g, b \rangle$ for some $b \in G$ or $\langle a, b \rangle = \langle g, h \rangle$, hence $a \in D(\langle g, h \rangle)$ with $g \in D(\rho)$ and $h \in D(\tau)$. $\qquad\qquad \square$

It is clear that a prespace of orderings which is isomorphic to a space of orderings is a space of orderings.

Finally, for arbitrary subsets $Z \subset \widehat{G}$ and $H \subset G$ we write:

$$Z^{\perp} = \{h \in G \mid \sigma(h) = +1 \text{ for all } \sigma \in Z\},$$
$$H^{\perp} = \{\sigma \in X \mid \sigma(h) = +1 \text{ for all } h \in H\} = \bigcap_{h \in H} \{h > 0\}.$$

(1.2) Subspaces of a Space of Orderings. Let (X, G) be a prespace of orderings and let $Y \subset X$ be a subspace, namely,

$$Y = \bigcap_{g \in G_1} \{g > 0\}, \quad \text{with } G_1 \subset G.$$

Then $G_1 \subset Y^{\perp}$ and $Y = \bigcap_{h \in Y^{\perp}} \{h > 0\}$, that is, $Y = Y^{\perp\perp}$. Consequently, $G|Y$ can be identified with $G/Y^{\perp}$.

Conversely, by the very definition, any $Y \subset X$ such that $Y = Y^{\perp\perp}$ is a subspace. For instance, for any $H \subset G$, $Y = H^{\perp}$ is a subspace.

Let $(Y, G/Y^{\perp})$ be a subspace of (X, G). For $g \in G$, let $\bar{g} = g|Y \equiv gY^{\perp} \in G/Y^{\perp}$ and for a form $\rho = \langle g_1, \ldots, g_n \rangle$ over X, let $\bar{\rho} = \langle \bar{g}_1, \ldots, \bar{g}_n \rangle$ be the corresponding form over Y. So $\bar{\rho}$ is the *restriction of ρ to Y*. $\qquad\square$

For the sequel, we fix a space of orderings (X, G).

Remarks and Notations 1.3 *a)* For $Z \subset X$ let $\mathrm{Lin}(Z)$ be the group generated by Z in $\widehat{G}$. Then

$$Z^{\perp\perp} = X \cap \mathrm{Adh}(\mathrm{Lin}(Z)) \supset \mathrm{Adh}(X \cap \mathrm{Lin}(Z)).$$

In general, this inclusion is strict. Of course, $Z^{\perp\perp}$ is a subspace of X. This is called the *subspace generated by* Z. A subset Z generates X if and only if $\mathrm{Lin}(Z)$ is dense in $\widehat{G}$.

Indeed, this follows from the facts that $\mathrm{Adh}(\mathrm{Lin}(Z))$ consists of the σ's in $\widehat{G}$ such that $\sigma(g) = +1$ for all $g \in Z^{\perp}$ [Mo Ch.7], and that $X^{\perp} = \{1\}$. $\qquad\square$

b) Conversely, let H be a subgroup of G and $Y = H^{\perp}$. Then

$$Y^{\perp} = \mathrm{Sat}(H) =$$
$$\{a \in G \mid \text{there are } d_1, \ldots, d_k \in H \text{ such that } a \in D(\langle\!\langle d_1, \ldots, d_k \rangle\!\rangle)\} \ .$$

Indeed, $a \in Y^{\perp}$ if and only if $\{a > 0\} \supset \bigcap_{d \in H} \{d > 0\}$, and, by compactness, if and only if $\{a > 0\} \supset \{d_1 > 0, \ldots, d_k > 0\}$ for some $d_1, \ldots, d_k \in H$. Hence, the assertion follows from Proposition III.2.6 *c)*.

The subgroup $\mathrm{Sat}(H)$ is called the *saturation of H*. Thus $(Y, G/\mathrm{Sat}(H))$ is a subspace. $\qquad\square$

Example 1.4 Let K be a formally real field and let (X, G) be the space of orderings associated to K, which in Marshall's presentation has the group

$G = K^*/\Sigma^*$, where $\Sigma = \sum K^2$. Let $T \subset K$ be a precone, that is, $T + T \subset T$, $T \cdot T \subset T$, $K^2 \subset T$ and $-1 \notin T$; in particular, $T \supset \Sigma$. Let

$$X/T = \{\sigma \in X \mid \sigma(t) = 1 \text{ for all } t \in T^*\}, \quad G/T = K^*/T^*.$$

Then $(X/T, G/T)$ is a subspace of (X, G). In fact, the first observation in the theory of formally real fields says that $(X/T)^\perp = G/T$ again. Conversely, each subspace Y of (X, G) has this form, with $T = \{t \in K \mid t = 0 \text{ or } t\Sigma^* \subset Y^\perp\}$.
$\square$

We now turn to fans in (X, G).

Proposition 1.5   *We have:*

    a) The group L generated in $\widehat{G}$ by finitely many fans $F_1, \ldots, F_n \subset X$ is closed, and the subspace generated by those fans in X is $Y = X \cap L$.
    b) A subset F of X is a fan in X if and only if F is closed in X and $F^3 \subset F$.

Proof.   *a)* Here L consists of the products of at most $2n$ elements of the F_i's. Whence L is closed, and by Remarks and Notations 1.3 *a)* we get $Y = X \cap L$.

b) The necessity is clear. Conversely, suppose that $F \subset X$ is closed, and that $F^3 \subset F$. We must see that F is a subspace of X. But by *a)*, $\mathrm{Lin}(F)$ is closed and its intersection with X is F. Hence, by Remarks and Notations 1.3 *a)*, $F^{\perp\perp} = F$.
$\square$

Of course, any $F \subset X$ with $\#(F) = 1$ or 2 is a fan in X, which is called *trivial*. Next, the subspace generated by three elements $\sigma_1, \sigma_2, \sigma_3 \in X$ is

$$Y = \{\sigma_1, \sigma_2, \sigma_3, \sigma_1\sigma_2\sigma_3\} \cap X,$$

which is a (non trivial) fan if and only if $\sigma_1\sigma_2\sigma_3 \in X$.

Example 1.6   Let K be a field with a valuation ring $V \subset K$ and let k_v the residue field of V. Let $\overline{F}$ be a fan in the space of orderings associated to k_v and let

$$F = \{\sigma \in \mathrm{Spec}_r(K) \mid \sigma \text{ is compatible with } V \text{ and } \overline{\sigma} \in \overline{F}\}$$

where $\overline{\sigma}$ is the ordering induced by σ in k_v. Then F is a fan in the space of orderings associated to K

Indeed, this is an easy exercise on real valuations, which belongs to Section II.3 and is left to the reader (but see Theorem VI.1.3).
$\square$

Next we look at the question of how to recognize a subspace Y of a space of orderings (X, G). The following result gives a partial answer:

Proposition 1.7   *Let (X, G) be a space of orderings and $Y \subset X$ such that $\#(F \cap Y) \neq 3$ for all fans $F \subset X$ with $\#(F) = 4$. Then $Y = \mathrm{Lin}(Y) \cap X$.*

Proof. Assume $\sigma \in \mathrm{Lin}(Y) \cap X$ but $\sigma \notin Y$. Then $\sigma = \sigma_1 \cdots \sigma_n$, n odd, with $\sigma_i \in Y$, $i = 1, \ldots, n$. Choose σ and the σ_i's such that n is the smallest possible. By hypothesis $n \geq 5$ and no partial product of the σ_i lies in X. In particular, $\sigma_1, \ldots, \sigma_n$ are independent in $\widehat{G}$. Let $Z = \{\sigma_1, \ldots, \sigma_n, \sigma_1 \cdots \sigma_n\}$. Then $Z = \mathrm{Lin}(Z) \cap X$. But $\mathrm{Lin}(Z)$ is closed, since it is finite. Thus Z is a subspace, hence $(Z, G/Z^{\perp})$ is a space of orderings. This contradicts the Miraculous Lemma (Lemma III.3.15). $\qquad\square$

Corollary 1.8 *Let $Y \subset X$ be a finite subset such that $\#(F \cap Y) \neq 3$ for all fans $F \subset X$ with $\#(F) = 4$. Then Y is a subspace.*

A corresponding result holds if Y is constructible but this requires deeper insight into the theory (Theorem 7.2 *b)*). Anyway, we already see fans at work.

2. Basic Constructions

In all this section (X, G) will be a prespace of orderings and, with few exceptions, even a space of orderings. First we shall introduce two basic constructions, which allow to build up new spaces of orderings from given ones. Later we shall see the importance of the spaces of orderings that can be constructed in this way, starting from the atomic space E.

Definition 2.1 (Addition) *Let (X_1, G_1), (X_2, G_2) be prespaces of orderings. Let*

$$(X, G) = (X_1 \cup X_2, G_1 \times G_2)$$

with the distinguished element $(-1, -1) \in G$ and the action

$$\sigma_1(g_1, g_2) = \sigma_1(g_1), \quad \sigma_2(g_1, g_2) = \sigma_2(g_2)$$

for $\sigma_i \in X_i$, $i = 1, 2$, and $(g_1, g_2) \in G_1 \times G_2$. Then (X, G), which is again a prespace of orderings, is called the sum *of (X_1, G_1) and (X_2, G_2), and denoted by $(X_1, G_1) + (X_2, G_2)$.*

Proposition 2.2 *Let (X_1, G_1) and (X_2, G_2) be two spaces of orderings and let $(X, G) = (X_1, G_1) + (X_2, G_2)$. Then:*

a) (X, G) is a space of orderings.
b) For subsets $C_i \subset X_i$, $i = 1, 2$, or $C \subset X$, respectively, the following properties are preserved under the union

$$(C_1, C_2) \mapsto C_1 \cup C_2 \subset X_1 \cup X_2$$

and the intersection

$$C \subset X_1 \cup X_2 \mapsto C_i = C \cap X_i, i = 1, 2 :$$

open, closed, constructible, principal, basic, subspace.

c) Let Y be a subspace of X. Then

$$(Y, G/Y^\perp) = (Y_1, G_1/Y_1^\perp) + (Y_2, G_2/Y_2^\perp),$$

where $Y_i = Y \cap X_i$ for $i = 1, 2$.

d) $s(X) = \max\{s(X_1), s(X_2)\}$ and $t(X) = \max\{t(X_1), t(X_2)\}$ unless $X_1 = X_2 = E$. Here one has $s(E) = t(E) = 0$ but $s(E + E) = t(E + E) = 1$.

e) $cl(X) = cl(X_1) + cl(X_2)$.

f) Let $F \subset X$ be a fan with $\#(F) \geq 4$. Then $F \subset X_1$ or $F \subset X_2$. Conversely, any fan in X_i, $i = 1, 2$, is a fan in X.

g) The addition of spaces of orderings is associative.

The proof of these facts is rather obvious.

Remarks 2.3 For the atomic space E let

$$n \times E = \underbrace{E + \cdots + E}_{n} .$$

Then $s(n \times E) = 1$. Thus, $n \times E$ is in fact the unique space (X, G) with $\#(X) = n$ and $s(X) = 1$. We find two different spaces with 5 elements, namely $5 \times E$ and $F_2 + E$ (where F_2 is the fan with $4 = 2^2$ elements).   □

The sum of two spaces of orderings can be characterized as follows.

Proposition 2.4   Let (X, G) be a space of orderings and let $Y, Z \subset X$ be subspaces such that $X = Y \cup Z$, $Y \cap Z = \emptyset$ and G is generated by $Y^\perp \cup Z^\perp$. Then

$$(X, G) = (Y, Z^\perp) + (Z, Y^\perp).$$

Proof.   Clearly $Y^\perp \cap Z^\perp = X^\perp = \{1\}$. Thus $G = Y^\perp \times Z^\perp$. The projection of $-1 \in G$ onto the factors gives us the distinguished elements in $Y^\perp$ and $Z^\perp$. Now one easily sees that the claim holds.   □

Definition 2.5   The space (X, G) is called indecomposable, *if it is not isomorphic to a sum of non trivial spaces. Otherwise* (X, G) *is called* decomposable.

For instance, by Proposition 2.2 *f)*, fans with more than 2 elements are indecomposable.

Proposition 2.6   Let (X, G) be a space of orderings.

a) Let $W \subset X$ be an indecomposable subspace and let $X = Y \cup Z$ be a decomposition of X. Then $W \subset Z$ or $W \subset Y$.

b) A decomposition of X into a sum of finitely many indecomposable spaces is unique.

Proof. First we prove *a)*. Suppose we have $W \cap Y \neq \emptyset$ and $W \cap Z \neq \emptyset$. Then by Proposition 2.4 W would be decomposable. Now, *b)* follows immediately from *a)*. $\qquad\square$

Definition 2.7 (Extension) *Let (X', G') be a prespace of orderings and H a group of exponent 2, endowed with the discrete topology. Let*

$$(X, G) = (\widehat{H} \times X', H \times G')$$

with distinguished element $(1, -1) \in G$ and the action

$$(\alpha, \sigma)(h, g) = \alpha(h)\sigma(g)$$

$(\alpha, \sigma) \in \widehat{H} \times X'$ and $(h, g) \in H \times G'$. Then (X, G), which is again a prespace of orderings, is called the extension *of (X', G') by H, and denoted by $(X', G')[H]$.*

Example 2.8 Let K' be a formally real field and (X', G') the space of orderings associated to K' (Theorem III.6.1). Then, the space of orderings (X, G) associated to $K = K'((X))$ is isomorphic to the extension $(X', G')[\mathbb{Z}_2]$. This example generalizes to henselian valued fields (Remark VI.1.5) $\qquad\square$

Now we shall show that if (X', G') is a space of orderings, so is $(X', G')[H]$. For this we introduce a construction which allows us to recognize whether a space is an extension. On the way, we exhibit further properties which recall the situation in valued fields.

(2.9) Residue Space. Let (X, G) be a prespace of orderings. The closed subgroup of $\widehat{G}$,
$$\mathrm{Tr}(X, G) = \{\tau \in \widehat{G} \mid \tau X = X\},$$

is called the *translation group of (X, G)*. Let T be a closed subgroup of $\mathrm{Tr}(X, G)$, $G' = T^{\perp} \subset G$ and let X' be the image of X in $\widehat{G}'$ via restriction. Then (X', G') is a prespace of orderings which is called the *residue space of (X, G) with respect to T.*

Now choose a complement H of G' in G: $G = H \times G'$. Then $\widehat{G} = \widehat{H} \times \widehat{G}'$ and, By Remark 1.3 *a)*, we may identify T with $\widehat{H}$. Write $\sigma \in X$ as $\sigma = \alpha\sigma'$ with $\alpha \in T = \widehat{H}$ and $\sigma' \in \widehat{G}'$. Since $\alpha X = X$ we get $\sigma' \in X'$. Thus we may identify X' with $X \cap \widehat{G}'$ and $X = \widehat{H} \times X'$. In conclusion,

$$(X, G) = (X', G')[H].$$

Thus we get:

Proposition 2.10 *A prespace of orderings is a proper extension if and only if its translation group is non trivial.*

We keep this situation in mind and consider a form $\rho = \langle h_1 g_1, \ldots, h_n g_n \rangle$ over X, $h_i \in H$ and $g_i \in G'$. We may reorder the coefficients $h_i g_i$ and write

$$\rho = \sum_{h \in H} h \rho_h$$

where ρ_h is a well-defined form over X' for finitely many h's, and vanishes for the others.

Definition 2.11   *The family $(\rho_h)_{h \in H}$ is called a* system of residue forms of ρ.

In general, the residue forms of ρ depend not only on the class of ρ, but on the special n-tuple which presents ρ. However, one has:

Proposition 2.12   *With the preceding notations, let ρ and τ be forms over $X = \widehat{H} \times X'$. We have:*

a) $\rho \sim \tau$ *if and only if ρ_h vanishes when τ_h is not defined, τ_h vanishes when ρ_h is not defined and $\rho_h \sim \tau_h$ when both ρ_h and τ_h are defined.*

b) ρ *is anisotropic if and only if all ρ_h's which are defined are anisotropic.*

c) If ρ is anisotropic, then $\rho = \tau$ if and only if ρ_h and τ_h are defined for the same h's, and for them $\rho_h = \tau_h$.

d) If ρ is anisotropic, $h \in H$ and $g \in G'$, then $hg \in D_X(\rho)$ if and only if ρ_h is defined and $g \in D_{X'}(\rho_h)$.

Proof.   *a)* Assume $\rho \sim \tau$ over $X = \widehat{H} \times X'$, the converse direction being obvious. We show that $\sigma(\rho_h - \tau_h) = 0$ for all $\sigma \in X'$, $h \in H$. Let $h_1, \ldots, h_m$ be the elements of H such that $\rho_{h_i} - \tau_{h_i}$ does not vanish. We proceed by induction on m, the case $m = 1$ being obvious. For $m > 1$ choose $\chi \in \widehat{H}$ which changes sign on the h_i's, say $\chi(h_i) = 1$ for $i = 1, \ldots, k$, $\chi(h_i) = -1$ for $i = k+1, \ldots, m$, where $1 \leq k < m$. Let

$$\pi = \sum_{i=1}^{k} h_i (\rho - \tau)_{h_i}$$

and

$$\omega = \sum_{i=k+1}^{m} h_i (\rho - \tau)_{h_i},$$

so that $\rho - \tau = \pi + \omega$. Let $\sigma \in X$. Then

$$0 = \sigma(\rho - \tau) = \sigma(\pi) + \sigma(\omega) ; \qquad 0 = \chi\sigma(\rho - \tau) = \sigma(\pi) - \sigma(\omega).$$

Thus $\sigma(\pi) = \sigma(\omega) = 0$ for all $\sigma \in X$ and the induction works.

b) If ρ is anisotropic, then clearly all the ρ_h's are anisotropic. Conversely, suppose ρ_h is anisotropic for all $h \in H$. If ρ were isotropic, there would be a form τ over X such that $\dim(\tau) < \dim(\rho)$ and $\sigma(\rho) = \sigma(\tau)$ for all $\sigma \in X$. Thus, $\dim(\tau_h) < \dim(\rho_h)$ for at least one $h \in H$, and, by *a)*, $\sigma'(\tau_h) = \sigma'(\rho_h)$ for all $\sigma' \in X'$. Contradiction.

c) and d) follow directly from *a)* and *b)*. $\square$

Proposition 2.13   *We have:*

a) *Let (X, G) be a prespace of orderings and let T be a closed subgroup of $\mathrm{Tr}(X, G)$. Then $\mathbf{O}_4$ holds for the residue space (X', G') with respect to T if and only if $\mathbf{O}_4$ holds for (X, G). Thus, extensions and residue spaces of spaces of orderings are again spaces of orderings.*

b) *Let (X_1, G_1) and (X_2, G_2) spaces of orderings, which are not extensions, that is, $\mathrm{Tr}(X_i, G_i) = 1$ for $i = 1, 2$ and let H_i, $i = 1, 2$, be groups of exponent 2. If $(X_1, G_1)[H_1]$ and $(X_2, G_2)[H_2]$ are isomorphic, then so are (X_1, G_1) and (X_2, G_2) and hence H_1 and H_2.*

Proof.   a) Instead of $\mathbf{O}_4$ we show $\mathbf{O}'_4$. Let ρ, τ be forms over X such that $\rho + \tau$ is isotropic and assume that $\mathbf{O}'_4$ holds for (X', G'). We look for $a \in D_X(\rho)$ with $-a \in D_X(\tau)$. We choose a complement H of $T^\perp$ in G. Now by Proposition 2.12 b), for at least one $h \in H$ the residue form $(\rho + \tau)_h$ over G' is isotropic from which we get the claim immediately. Conversely, assume $\mathbf{O}'_4$ holds for (X, G) and let ρ, τ be forms over X' such that $\rho + \tau$ is isotropic. Again we look for $a \in D_{X'}(\rho)$ with $-a \in D_{X'}(\tau)$. This is easily found if ρ or τ is isotropic. Otherwise consider ρ and τ as forms over X. These have a single residue form $\rho = \rho_1$ and $\tau = \tau_1$ respectively. Choose $a \in D_X(\rho)$ with $-a \in D_X(\tau)$. By Proposition 2.12 c), also $a \in D_{X'}(\rho)$ and $-a \in D_{X'}(\tau)$. Thus we are done.

b) Obvious, since (X_i, G_i) is the residue space of $(X_i, G_i)[H_i]$ with respect to the full translation group $\mathrm{Tr}((X_i, G_i)[H_i]) = \widehat{H}_i$ for $i = 1, 2$. $\square$

Proposition 2.14   *Let (X, G) be a space of orderings and let H, H' be groups of exponent 2.*

a) *$((X, G)[H])[H'] \cong (X, G)[H \times H']$.*

b) *$(\{1\} \times X)^\perp = H \times \{1\}$ and $\{1\} \times X = (H \times \{1\})^\perp$. Thus (X, G) embeds naturally as a subspace in $(X, G)[H]$.*

c) *For subsets C of either X or $\widehat{H} \times X$ respectively, the following properties are preserved under the lift*

$$C \subset X \mapsto \widehat{H} \times C \subset \widehat{H} \times X$$

and the intersection
$$C \subset \widehat{H} \times X \mapsto C \cap X :$$

open, closed, constructible, principal, basic, subspace, fan.

d) *Let Y be a subspace of $(X, G)[H]$. Then Y is (non canonically) isomorphic to $(X', G')[H']$ where X' is a subspace of X and H' is a subgroup of H.*

e) *$s(\widehat{H} \times X) = s(X) + \dim_{\mathbb{F}_2}(H)$.*

f) *$cl(\widehat{H} \times X) = cl(X)$ unless $X = E$ and $H \neq \{1\}$. In that case, $cl(E) = 1$ but $cl(\widehat{H} \times E) = 2$.*

g) *Let $H \neq \{1\}$. Then $(X, G)[H]$ is decomposable if and only if $(X, G) = E$ and $H = \mathbb{Z}_2$.*

h) *$E[\mathbb{Z}_2^n] \cong F_n$. More generally, fans are exactly the spaces of the form $E[H]$*

i) *The projection into X of a fan in $\widehat{H} \times X$ is again a fan.*

Proof. One easily shows *a)*, *b)*, *c)* and *i)*.

d) Consider the projection $\pi : H \times G \to H$ and choose a complement H_1 of $\pi(Y^\perp)$ in H. Set

$$G_1 = \{g \in G \,|\, (1, g) \in Y^\perp\}.$$

Then

$$\mu : H_1 \times (G/G_1) \to (H \times G)/Y^\perp$$
$$(h_1, \bar{g}) \mapsto \overline{(h_1, g)}$$

is an isomorphism. Now, a carefull computation using the various identifications involved shows that the dual isomorphism $\hat{\mu}$ maps Y onto $\widehat{H_1} \times X'$ where $X' = G_1^\perp$. Note that $X'^\perp = G_1$ by Remarks and Notations 1.3 *b)*.

e) We first see the inequality $\geq$. Let φ be an s-fold Pfister form over X which cannot be written as $\varphi = 2 \times \psi$ for an $(s - 1)$-fold Pfister form ψ over G and let $h_1 \ldots, h_k \in H$ be linearly independent. Let $H' \subset H$ be the group generated by $\{h_1, \ldots, h_k\}$. Consider the form

$$\rho = \sum_{h \in H'} h\varphi = \langle\!\langle h_1, \ldots, h_k \rangle\!\rangle \otimes \varphi$$

as a form over $\widehat{H} \times X$. Then all residue forms of ρ coincide, namely, $\rho_h = \varphi$ for all h. By Propositions 2.12 *d)* and III.2.7, the pure part ρ' of ρ does not represent 1. Thus ρ is not of the form $\rho = 2 \times \tau$ for any Pfister form τ. Now the inequality follows from Proposition III.2.6 *b)*.

Let us prove to the converse inequality. It is trivial if H is not finite. Thus, by *a)*, we may assume that $H = \mathbb{Z}_2$. Consider a basic set $C \subset \widehat{H} \times X$ which is described by m elements , say $f_1, \ldots, f_m \in H \times G$, and which cannot be described by fewer. Then $f_1, \ldots, f_m$ are linearly independent. Let

$$\varphi = \langle\!\langle f_1, \ldots, f_m \rangle\!\rangle = \langle 1, b_2, \ldots, b_{2^m} \rangle.$$

Then $D = \{1, b_2, \ldots, b_{2^m}\}$ is the group generated by $f_1, \ldots, f_m$ in $H \times G$. If D is a subgroup of G, we have $C = \widehat{H} \times \{\sigma \in X \,|\, \sigma(f_1) = +1, \ldots, \sigma(f_m) = +1\}$, and $m \leq s(X)$. Otherwise $G \cap D$ is a subgroup of index 2 in D. Let $k_1, \ldots, k_{m-1}$ be generators of $G \cap D$ and let $\psi = \langle\!\langle k_1, \ldots, k_{m-1} \rangle\!\rangle$. Thus ψ is a Pfister form over X and $\varphi = \psi + h\psi$ for some $h \in D \setminus G$. If $m > s(X) + 1$ we get $\psi = 2 \times \psi_1$ for an $(m - 2)$-fold Pfister form ψ_1 over X. Then $\varphi = 2 \times \langle 1, h \rangle \otimes \psi_1$. Thus m was not minimal.

f) Let $(h, g) \in H \times G$ with $h \neq 1$. Direct inspection shows that the chain $\emptyset \subset \{(h, g) > 0\} \subset \widehat{H} \times X$ can not be refined by other principal sets. From this and *c)* we get the assertion.

g) Suppose $(X, G)[H] = (X_1, G_1) + (X_2, G_2)$. We have just seen that there is a principal set $C \subset X$ such that the chain $\emptyset \subset C \subset X$ cannot be refined by other principal sets. On the other hand, by Proposition 2.2 *b)*, such a principal set C is a union $C = C_1 \cup C_2$, where $C_i = C \cap X_i$ is principal in X_i, $i = 1, 2$. Therefore, unless $X_1 = X_2 = E$, by Proposition 2.2 *b)* again, the chain $\emptyset \subset C \subset X$ can be refined.

h) Take H to be the translation group of that fan. Then the residue space is E. $\qquad\qquad\square$

After these constructions have been described, the reader is invited to produce all spaces of orderings of small cardinalities (see Remark III.3.14).

3. Spaces of Finite Type

Starting with the atomic space E the basic constructions lead to an important class of spaces of orderings:

Definition 3.1   *A space of orderings (X, G) is called a* space of finite type, *if it is built up, starting from atomic spaces, by finitely many additions and extensions.*

In the following sections we shall show that (X, G) is of finite type if and only if $cl(X, G) < \infty$. This is in fact the key technical result of the whole theory.

Note that by Propositions 2.6, 2.13 *b)* and 2.14 *a)*, *f)*, the construction of a space of finite type is essentially unique. The only ambiguity lies in the isomorphism

$$E + E \cong E[\mathbb{Z}_2]$$

and the derived ones, as

$$(E + E)[H] \simeq E[\mathbb{Z}_2 \times H],$$

or

$$E + E + (X, G) \simeq E[\mathbb{Z}_2] + (X, G).$$

We are going to attach a graph to each space of finite type. In order to get a bijection we fix the following conventions.

(3.2) Construction of Weighted Trees. Let (X, G) be a space of finite type.

a) For each space (X_i, G_i) which appears in the construction of (X, G) set a point i. In particular, the atomic space E is denoted just by a single point.

b) Denote the sum $(X_i, G_i) + (X_j, G_j) = (X_k, G_k)$ by

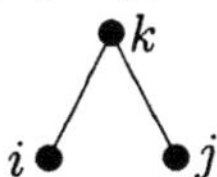

c) Consecutive sums collapse associatively:

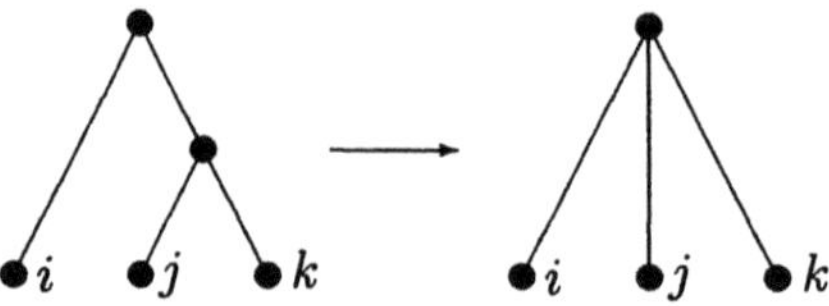

d) Denote the extension $(X_i, G_i)[H] = (X_k, G_k)$ by

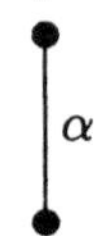

where $1 \leq \alpha = \dim_{\mathbb{F}_2}(H)$ is called the *weight*.

e) Consecutive extensions collapse to a single one:

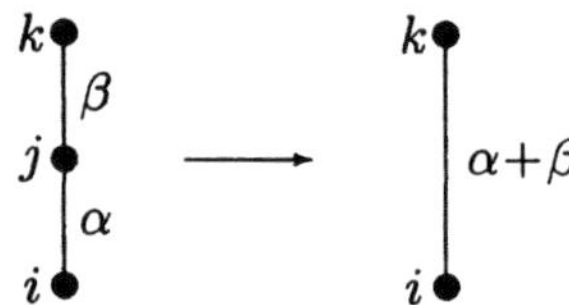

f) Pure extensions of atomic spaces are not allowed, thus they are replaced as follows:

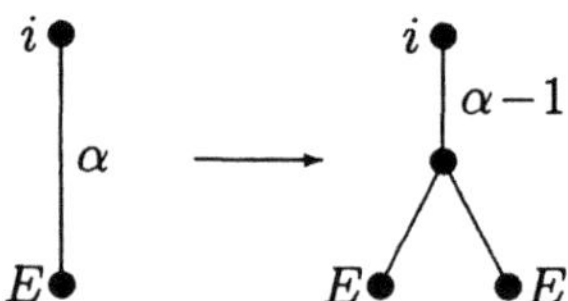

Consequently, no diagram admits a vertical line at the bottom. The final diagram obtained this way is called the *weighted tree* of (X, G). Typical examples of weighted trees look like

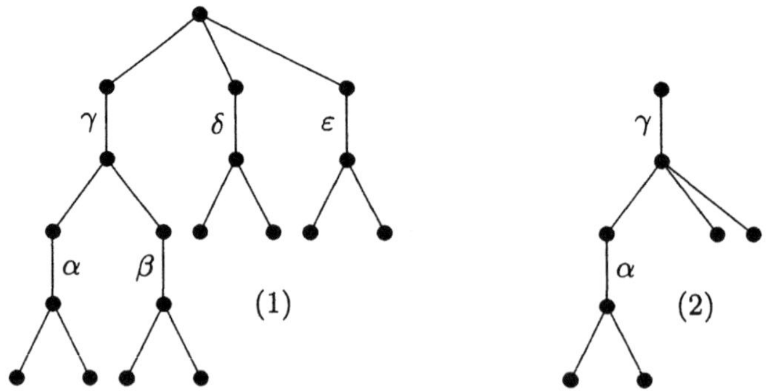

where the lowest points denote atomic spaces and the greek letters are weights. In (1) the whole space is a sum while in (2) the whole space is an extension. The reader is invited to write down the spaces corresponding to the trees (1) and (2).

Summarizing we get

Theorem 3.3   *Construction 3.2 defines a bijection between isomorphism classes of spaces of finite type and weighted trees.*

This bijection allows us to show properties of spaces of finite type by *induction along the tree.*

We next introduce the notion of a subtree.

(3.4) Subtrees. Given a weighted tree, its *subtrees* are obtained by applying the following rules.

a) Cutting off a summand, like

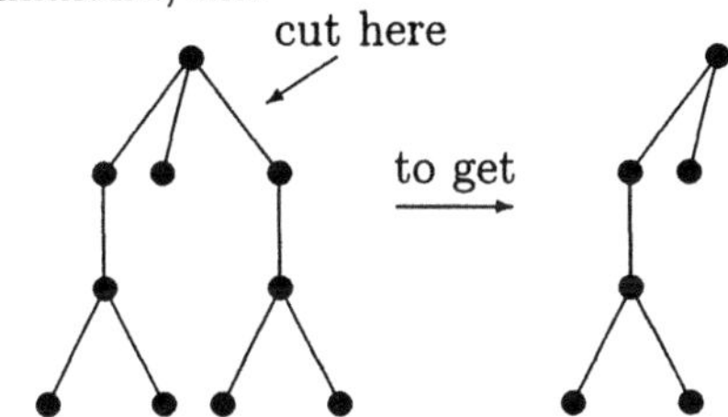

b) Lowering weights at vertical edges, like

with $0 \le \alpha' \le \alpha$.

c) Collapsing vertical edges of weight zero and meaningless non vertical edges.

d) Normalizing according to the rules 3.2 *c)*, *e)* and *f)*.

For instance, the tree (2) in the above example is the subtree of (1) obtained by cutting off the second and third summand and lowering β to zero.

Proposition 3.5   *Let (X, G) be a space of finite type. Then the tree of a subspace of X is a subtree of the tree of X, and vice versa.*

Proof.   By induction along the tree, using Propositions 2.2 *c)* and 2.14 *d)*.   □

Proposition 3.6   *A non trivial space of finite type (X, G) is a fan if and only if its tree looks like*

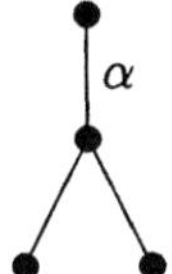

Proof. By Proposition 2.14 *h)*, $(X, G) = E[H]$. Now use rule 3.2 *f)*.   □

Now one can nicely read some invariants of a space of finite type from its tree.

Proposition 3.7   *Let (X, G) be a space of finite type.*

   a) The chain length $cl(X)$ is the number of bottom points of the tree.
   b) The stability index $s(X)$ is computed as follows. If there is a subtree like

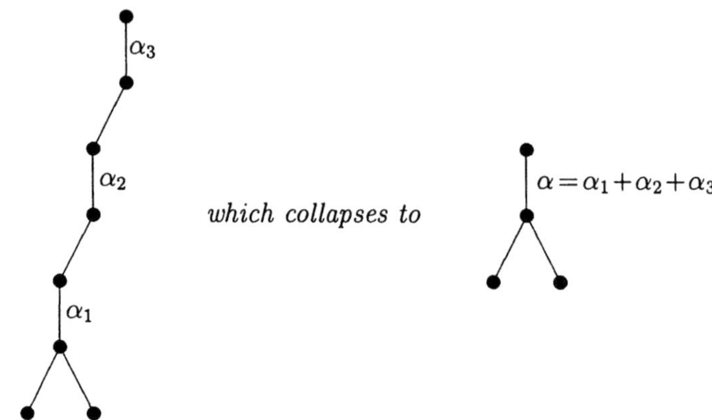

which collapses to

there is one with α maximal, and then $s(X) = \alpha + 1$. If no such a subtree exists, then $s(X) = 0$.
   c) If $X \neq E$, then $\#(X) \leq cl(X)2^{s(X)-1}$.

Proof.   By induction along the tree using Proposition 2.2 *d)*, *e)* and 2.14 *e)*, *f)*, taking into account rule 3.2 *f)*.   □

As a consequence, we get the following results for a space of finite type (X, G).

(3.8) Let Y be a subspace of X. Then $cl(Y) \leq cl(X)$.

(3.9) $s(X) = \sup\{s(F) \,|\, F$ is a fan in $X\}$.

Contrarily, there is no obvious formula for $t(X)$ depending on the tree.

In the next section we shall show that 3.8 holds in general. Also 3.9 is true in general. This is the *stability formula*, which will be proved in Section 7.

4. Spaces of Finite Chain Length

Proposition 4.1   *Let (X, G) be a space of orderings and let Y be a subspace. Then $cl(Y) \leq cl(X)$.*

Proof. We prove this non trivial result in several steps.

Step 1: Let Y_λ be a descending chain of subspaces of X such that $cl(Y_\lambda) \leq n < \infty$ for all λ, and let $Y_0 = \bigcap_\lambda Y_\lambda$. Then $cl(Y_0) \leq n$. In fact, Y_0 is a subspace. Let $a_1, \ldots, a_k \in G$ be such that

$$\{a_i > 0\} \cap Y_0 \cap \{a_j > 0\} = \emptyset \ for \ i \neq j$$

and

$$Y_0 \subset \bigcup_{i=1}^{k} \{a_i > 0\}.$$

Then the same holds if we replace Y_0 by a suitable Y_λ, since $\{a > 0\}$ is open and closed for any $a \in G$ and X is compact.

Step 2: Now assume that $cl(X) = n < \infty$. Otherwise there is nothing to do. Let Y_λ be the family of all subspaces with $X \supset Y_\lambda \supset Y$ and $cl(Y_\lambda) \leq n$. By *Step 1*, we may apply Zorn's lemma and thus find a minimal subspace Y_1 in that family. Replace X by Y_1.

Step 3: If $Y = X$ we are done. Otherwise, since $cl(X) < \infty$, we find a chain

$$Y \subset \{a_1 > 0\} \subset \cdots \subset \{a_m > 0\} \subset X$$

which cannot be refined. By the above constructions we have $cl(\{a_m > 0\}) > n$. Moreover, by Proposition III.2.6 c), a_m is *rigid*, that is, $D(\langle 1, a_m \rangle) = \{1, a_m\}$. Altogether we are reduced to the next step.

Step 4: Let $b \in G$ be rigid and let $a_0, \ldots, a_k \in G$ be such that

$$\{b > 0, a_0 > 0\} \subset \cdots \subset \{b > 0, a_k > 0\}.$$

Then, there exist $a'_0, \ldots, a'_k \in G$ such that

$$\{a'_0 > 0\} \subset \cdots \subset \{a'_k > 0\}$$

and

$$a'_i \equiv a_i \quad \mod \ D(\langle 1, b \rangle),$$

that is,

$$\{b > 0, a_i > 0\} = \{b > 0, a'_i > 0\},$$

for $i = 1, \ldots, k$. We show *Step 4* by induction on k.

Step 5, $k = 1$: In the subspace $Y = \{b > 0\}$ we have $a_1 \in D_Y(\langle 1, a_0 \rangle)$, thus $a_1 \in D(\langle\langle a_0, b \rangle\rangle)$ by Proposition III.2.6 e). But

$$\langle\langle a_0, b \rangle\rangle = \langle 1, b \rangle + a_0 \langle 1, b \rangle$$

and b is rigid, thus $a_1 \in D(\langle u, a_0 v \rangle)$ where $u, v \in \{1, b\}$. Then $a_1 u \in D(\langle 1, a_0 u v \rangle)$. Now choose $a'_0 = a_0 u v$ and $a'_1 = a_1 u$.

Step 6, $k \geq 3$ (the case $k = 2$ will be shown in Step 7): By induction we have $\tilde{a}_i, a^*_i$ and $a'_i \in G$ such that

$$\tilde{a}_i, a^*_i, a'_i \equiv a_i \quad \mod \ D(\langle 1, b \rangle)$$

and

$$\{\tilde{a}_0 > 0\} \subset \cdots \subset \{\tilde{a}_{k-1} > 0\},$$
$$\{a_1^* > 0\} \subset \cdots \subset \{a_k^* > 0\},$$
$$\{a_0' > 0\} \subset \{a_2' > 0\} \subset \cdots \subset \{a_k' > 0\}.$$

If $\tilde{a}_i = a_i^*$ for some $i = 1, \ldots, k-1$, or $\tilde{a}_i = a_i'$ for some $i = 2, \ldots, k$, we can splice the first and the second (resp. first and third) chains together to get the chain required in *Step 4*. Thus we assume that all $\tilde{a}_i \neq a_i^*$ and $\tilde{a}_i \neq a_i'$, that is, $\tilde{a}_i = a_i^* b$ and $\tilde{a}_i = a_i' b$. But then $a_i^* = a_i'$ for $i = 2, \ldots, k-1$. We claim that $\{a_0' > 0\} \subset \{a_1^* > 0\}$, which allows us to splice the second and the third chain together.

To prove the claim, let $\sigma \in \{a_0' > 0\}$. If $\sigma(b) = +1$ then

$$\sigma \in \{a_0' > 0, b > 0\} = \{a_0 > 0, b > 0\} \subset \{a_1^* > 0\}.$$

So assume $\sigma(b) = -1$. Now

$$\sigma(a_0') = +1 \;\Rightarrow\; \sigma(a_{k-1}') = +1$$
$$\Rightarrow\; \sigma(\tilde{a}_{k-1}) = \sigma(a_{k-1}' b) = -1$$
$$\Rightarrow\; \sigma(a_1^* b) = \sigma(\tilde{a}_1) = -1$$
$$\Rightarrow\; \sigma(a_1^*) = +1.$$

Step 7, $k = 2$: By *Step 5* we may assume that

$$a_0 \neq a_1 \neq a_2$$

and
$$\text{I}: \qquad \{a_0 > 0\} \subset \{a_1 b > 0\}, \quad \{a_1 > 0\} \subset \{a_2 > 0\}.$$

Moreover, we may also assume

$$\text{II}: \qquad \begin{cases} \{a_0 > 0\} \not\subset \{a_1 > 0\}, & \{a_1 b > 0\} \not\subset \{a_2 > 0\}, \\ \{a_0 b > 0\} \not\subset \{a_1 > 0\}, & \{a_1 b > 0\} \not\subset \{a_2 b > 0\}. \end{cases}$$

Indeed, any inclusion here would give by splicing the chain wanted in *Step 4*. From I and II we shall deduce a contradiction: we look for $\sigma_1, \sigma_2, \sigma_3, \sigma_4 \in \widehat{G}$ with the following table of signs:

	b	a_0	a_1	a_2
σ_1	$-$	$+$	$-$	$-$
σ_2	$-$	$-$	$-$	$+$
σ_3	$-$	$+$	$-$	$+$
σ_4	$-$	$-$	$-$	$-$

Inspection of the conditions I and II shows that in each set $\{\sigma_1, \sigma_3\}$, $\{\sigma_1, \sigma_4\}$, $\{\sigma_2, \sigma_4\}$, $\{\sigma_2, \sigma_3\}$ at least one of the two elements can be realized in X.

For instance, since $\{a_0 > 0\} \not\subset \{a_1 > 0\}$ we find $\sigma \in X$ with $\sigma(a_0) = +1$,

$\sigma(a_1) = -1$. If $\sigma(b) = +1$ we get $\sigma(a_1 b) = -1$. Contradiction to $\{a_0 > 0\} \subset \{a_1 b > 0\}$. Thus $\sigma(b) = -1$ and σ is like σ_1 or σ_3. The argument for the other pairs runs correspondingly.

We obtain that both σ_1 and σ_2 or both σ_3 and σ_4 exist in X.

Consider now the Pfister forms $\varphi = \langle\!\langle a_1, a_0 a_2 \rangle\!\rangle$ and $\psi = \langle\!\langle b, a_0 \rangle\!\rangle$. For $\sigma \in X$ we get by I:

$$\sigma(b) = \sigma(a_0) = +1 \quad \Leftrightarrow \quad \sigma(a_1) = \sigma(a_0) = +1 \quad \Leftrightarrow \quad \sigma(a_1) = \sigma(a_0 a_2) = +1,$$

thus $\varphi = \psi$. By cancellation of $\langle 1 \rangle$ we see that $a_0 a_2, a_0 a_1 a_2 \in D(\psi')$, $\psi' = \langle b \rangle + a_0 \langle 1, b \rangle$. But b is rigid. Thus

$$D(\psi') = D(\langle b, a_0 \rangle) \cup D(\langle b, a_0 b \rangle),$$

hence

$$\text{III}: \qquad a_0 a_1 a_2, \ a_0 a_2 \in D(\langle b, a_0 \rangle) \cup D(\langle b, a_0 b \rangle).$$

If σ_1 and σ_2 exist, then $\sigma_2(b) = \sigma_2(a_0) = -1$, $\sigma_1(b) = \sigma_1(a_0 b) = -1$, but $\sigma_1(a_0 a_1 a_2) = \sigma_2(a_0 a_1 a_2) = +1$. Contradiction to III. Similarly, if both σ_3 and σ_4 exist, then $\sigma_4(b) = \sigma_4(a_0) = -1$, $\sigma_3(b) = \sigma_3(a_0 b) = -1$, but $\sigma_3(a_0 a_2) = \sigma_4(a_0 a_2) = +1$, which again contradicts III and completes the proof of the proposition. $\qquad\square$

The main technical result of the whole theory says that a space of orderings has finite chain length if and only if it is a space of finite type. In order to show this we first try to mimic, in spaces of finite chain length, real places as they occur in spaces of orderings of fields. This requires some technical notions.

Notation 4.2   Let (W, D) be a fan and Y a subspace of W. Then Y is again a fan and Y, W can be considered as affine subspaces of $\widehat{G}$. For $k =$ codimension of Y in W, we set $(W : Y) = 2^k$ and call this number the *index of Y in W*.
$\qquad\square$

Definition 4.3   *Let (X, G) be a space of orderings and F a fan in X.*

   a) *The fan F is called* solid *in X if $(W : F \cap W) \leq 2$ for all fans $W \subset X$ with $F \cap W \neq \emptyset$.*
   b) *The fan F is called* impervious *in X if F is maximal with respect to the property of being solid.*

Clearly, a solid fan has codimension 1 in any other bigger fan. Hence, every solid fan is contained in an impervious one. Note also that the property "solid" is preserved under intersections with subspaces of X.

Definition 4.4   *Let (X, G) be a space of orderings and $\sigma \in X$. Then σ is called* archimedean *in X if σ is not contained in any 4-element fan of X.*

We shall show in Lemma VI.1.7 that for spaces of orderings of fields, archimedean orderings are archimedean elements in this sense, but, in general, not vice versa (consider for instance the space of orderings of the field $\mathbb{R}((t))$).

In the sequel let (X, G) be a space of finite type. Let us see how the above definitions are related to the tree of X.

Remark 4.5   The element $\sigma \in X$ is archimedean if and only if $(X, G) = (X', G') + E$ and σ is the element of E.

This follows by induction along the tree, using Proposition 2.2 $f)$ and 2.14 $c)$. $\square$

Examples 4.6   Consider the following trees for (X, G)

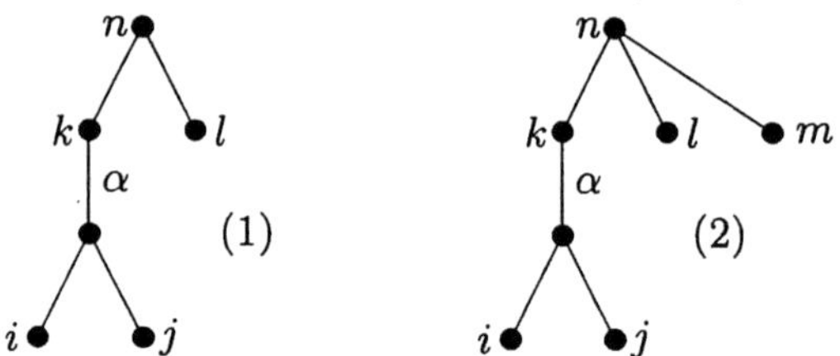

where $\alpha \geq 2$. Then the elements of the atomic spaces l, m in (1) and (2) are archimedean. The fan

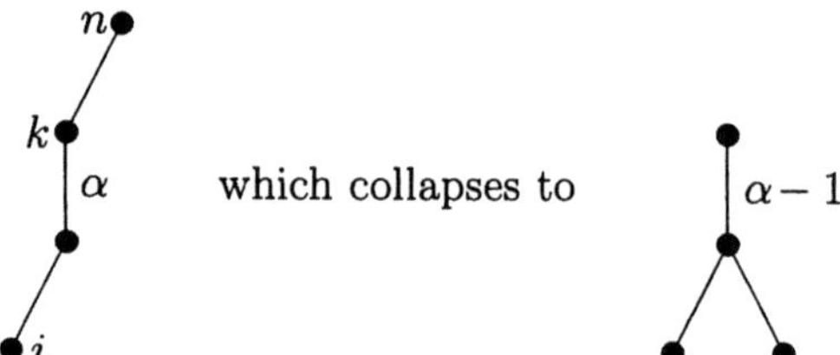

in (1) and (2) is solid. Moreover, the fan

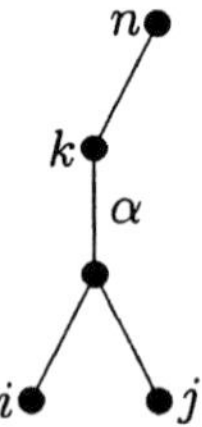

is impervious in (1) and in (2). In both cases, there is exactly one impervious fan more, namely, the singleton l in (1) (which is not a maximal fan) and the 2-element fan

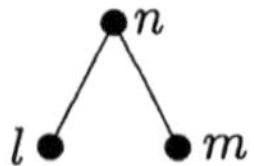

in (2). Note, that any pair $\{\sigma, \tau\} \subset X$, where σ is archimedean and $\tau \in k$ forms a maximal fan, which is not impervious. Thus, if α is not finite, one has an infinity of maximal fans. $\qquad\square$

The essential fact is that the impervious fans on the one hand cover X, while on the other hand their number is bounded by a function of $cl(X)$. In order to show this we shall characterize impervious fans in the context of the tree of (X, G). We start with the following easy result:

Proposition 4.7

    a) *Let $(X, G) = (X_1, G_1) + (X_2, G_2)$ and let $F \subset X_1$ be a fan. Then F is solid in X_1 if and only if F is solid in X.*

    b) *Let $(X, G) = (X_1, G_1)[H]$, let $F_1 \subset X_1$ be a fan and $F = \widehat{H} \times F_1$. Then F_1 is solid in X_1 if and only if F is solid in X.*

Proof. This follows directly from the definition, part *a)* using Proposition 2.2 *f)*, and part *b)* using Proposition 2.14 *c)* and *i)*. $\qquad\square$

Now, we can characterize impervious fans in a space of finite type by induction along the tree:

Proposition 4.8   *Let (X, G) be a space of finite type, and let $F \subset X$ be a fan.*

    a) *Let (X, G) be the atomic space. Then $F = X$ is impervious in X.*

    b) *Let $(X, G) = (X_1, G_1) + \cdots + (X_k, G_k)$, where (X_i, G_i) is indecomposable and $k \geq 2$. Then, F is impervious in X if and only if one of the following conditions holds:*

        i) *$\#(F) \geq 2$, $F \subset X_i$ for some $i = 1, \ldots, k$ (which holds automatically if $\#(F) \geq 4$), and F is impervious in X_i.*

        ii) *$F = \{\sigma, \tau\}$, $\sigma \neq \tau$, where σ and τ are archimedean in X. Note that then σ, τ lie in different spaces X_i and X_j which must be atomic.*

        iii) *$F = \{\sigma\}$, where σ is archimedean and there is no other archimedean element in X. Note that then σ lies in the unique atomic space among the X_i's.*

    c) *Let $(X, G) = (X_1, G_1)[H]$. Then F is impervious in X if and only if $F = \widehat{H} \times F_1$, where F_1 is an impervious fan in X_1.*

Proof. *a)* is obvious.

    b) Part *i)* follows readily from Propositions 2.2 *f)* and 4.7.

    For *ii)*, let $F = \{\sigma, \tau\}$ with $\sigma \in X_i$ and $\tau \in X_j$. If σ is not archimedean in X, it belongs to a 4-element fan W, which must be completely contained in X_i. Hence, $F \cap W = \{\sigma\}$ and $(W : F \cap W) > 2$, so that F is not solid. Analogously, if τ is not archimedean, F is not solid. Conversely, if both σ and τ are archimedean, no 4-element fan meets F, and F is impervious. That X_i and X_j are atomic follows from Remark 4.5.

Finally, let $F = \{\sigma\}$. If F is impervious, arguing as above we see that σ must be archimedean. If there were other archimedean element τ, the fan $\{\sigma, \tau\}$ would be solid (by $ii)$), and F would not be impervious. Conversely, it is obvious that any archimedean singleton is impervious.

$c)$ First, let $F \subset X$ be a solid fan and let F_1 be the projection of F into X_1, which by Proposition 2.14 $i)$ is a fan. We claim that F_1 is solid in X_1. Indeed, otherwise we would have a fan $W_1 \subset X_1$ with $W_1 \cap F_1 \neq \emptyset$ and $(W_1 : W_1 \cap F_1) > 2$. Let $W = \widehat{H} \times W_1$. Then $W \cap F \neq \emptyset$ and $(W : W \cap F) > 2$. Contradiction. Now, by part $b)$ of the previous lemma, $\widehat{H} \times F_1$ is solid in X.

Next, suppose that F is impervious. Since $F \subset \widehat{H} \times F_1$, we get $F = \widehat{H} \times F_1$. Again by the previous lemma, it follows that F_1 is impervious in X_1.

Conversely, let F_1 be an impervious fan in X_1, and set $F = \widehat{H} \times F_1$. By the previous lemma, F is solid. Let F' be a solid fan in X that contains F. By the first remark, the projection F'_1 of F' into X_1 is solid. Since F'_1 contains F_1, which is impervious in X_1, we conclude $F'_1 = F_1$, so that

$$F \subset F' \subset \widehat{H} \times F_1 = F.$$

$\square$

Note that the definition of impervious fan is intrinsic, while the latter characterization leads to the following bound:

Proposition 4.9   *Let (X, G) be a non atomic space of finite type. Then, the number of impervious fans in X is bounded by $\frac{1}{2}cl(X)(cl(X) - 1)$.*

Proof.   By induction along the tree once again, using the characterization of the previous proposition and Propositions 2.2 $e)$ and 2.14 $f)$. $\square$

Definition 4.10   *Let (X, G) be a space of finite type. A fan $P \subset X$ is called a place of X if it is either*

$p_1:$ *a non-empty intersection of two distinct impervious fans, or*
$p_2:$ *an impervious fan which does not intersect any other impervious fan.*

By induction along the tree, using Proposition 4.8, we immediately get:.

Proposition 4.11   *Let (X, G) be a space of finite type and let $P \subset X$ be a fan.*

a) Let $(X, G) = (X_1, G_1) + \cdots + (X_k, G_k) + (X_{k+1}, G_{k+1}) + \cdots + (X_{k+l}, G_{k+l})$, where (X_i, G_i) is an extension for $1 \leq i \leq k$ and (X_i, G_i) is atomic for $k + 1 \leq i \leq k + l$. Then P is a place in X if and only if it is of the following type:

    i) $P \subset X_i$, $i \leq k$, and P is a place of X_i.
    ii) $l \neq 2$, $P = X_i$, $i > k$, that is, P is an archimedean singleton.
    iii) $l = 2$, $P = X_{k+1} \cup X_{k+2}$, that is, P consists of the two archimedean singletons.

b) *Let $(X, G) = (X_1, G_1)[H]$. Then P is a place in X if and only if $P = \widehat{H} \times P_1$ where P_1 is a place in X_1.*

Using this result, one obtains:

Proposition 4.12   *Let (X, G) be a space of finite type.*

   a) *Each place $P \subset X$ is solid.*
   b) *The places form a partition of X.*
   c) *An impervious fan is either a union of two places of type p_1 or a single place of type p_2.*
   d) *$\frac{1}{2} cl(X) \leq \#\{places\} \leq cl(X)$.*
   e) *Let $X = F_1 \cup \cdots \cup F_m$ be a covering of X by impervious fans such that m is minimal, and let $F \subset F_1$ be a fan $(\neq F_1)$. If $F, F_2, \ldots, F_m$ still cover X, then $F_1 = F \cup F'$ where F, F' are places.*

Proof.   *a)-d)* follow easily by induction along the tree.

   e) For the proof, we can assume that no F_i is a place of type p_2. Then, by *c)*, for every i we can write $F_i = P_i \cup P_i'$, where P_i and P_i' are places of type p_1. By *b)*, places are disjoint, and since m is minimal and $F \neq F_1$, we may assume that $P_1 \subset F$ and $P_1' = P_2'$. Finally, by *a)*, places are solid fans, and from $P_1 \subset F \subset F_1$ and $F \neq F_1$ we conclude $P_1 = F$. $\square$

5. Finite Type = Finite Chain Length

Here we shall prove the result that is the heart of the whole theory:

Theorem 5.1   *For a space of orderings (X, G) the following properties are equivalent:*

   a) *(X, G) is a space of finite type.*
   b) *X is a union of finitely many impervious fans.*
   c) *X is generated by finitely many fans.*
   d) *$cl(X) < \infty$.*

Proof.   We shall argue along the scheme $a) \Rightarrow b) \Rightarrow c) \Rightarrow a) \Rightarrow d) \Rightarrow a)$.

   $a) \Rightarrow b)$ By Propositions 4.12 *b)*, 4.9, 4.12 *a)* and the fact that every solid fan is contained in an impervious one.

   $b) \Rightarrow c)$ Obvious.

   $c) \Rightarrow a)$ We postpone this to the end of the proof.

   $a) \Rightarrow d)$ By Proposition 3.7 *a)*.

   $d) \Rightarrow a)$ We assume the equivalence of *a)*, *b)* and *c)*. Let Y_λ, $\lambda \in \Lambda$, be an ascending chain of subspaces of X such that all Y_λ are spaces of finite type.

Claim. *There is a subspace Y of X such that Y is a space of finite type, and $Y_\lambda \subset Y$ for all $\lambda \in \Lambda$.*

From the claim, by Zorn's lemma, we get a maximal subspace Y_0 of X which is a space of finite type. If $\sigma \in X \setminus Y_0$, finitely many fans of Y_0 and the fan $\{\sigma\}$ generate a larger subspace of finite type by *c)*. This proves *a)*.

Proof of the claim. By Propositions 4.1, 4.9 and 4.12, we may assume that $cl(Y_\lambda) = c$ and that a minimal covering of Y_λ by impervious fans consists of exactly m fans for all $\lambda \in \Lambda$. Let $\lambda > \mu$ and let $F_1, \ldots, F_m$ be impervious fans or places of Y_λ which cover Y_λ. Then the fans $F_i \cap Y_\mu$ are solid in Y_μ and they cover Y_μ. Thus by Proposition 4.12 *e)*, the fans $F_i \cap Y_\mu$ are impervious fans or places of Y_μ. So, for each $\lambda \in \Lambda$ we have the finite family C_λ of coverings of Y_λ by m fans, which are impervious fans or places, together with restriction maps $C_\lambda \to C_\mu$ for $\mu < \lambda$.

Since $\lim_\lambda C_\lambda \neq \emptyset$ [Bk TE III, §7.4, Th.1, Ex.I, p.III.58-60], we find for each $i = 1, \ldots, m$, an ascending chain of fans, $(F_{\lambda,i})_{\lambda \in \Lambda}$, such that

$$Y_\lambda \subset \bigcup_{i=1}^{m} F_{\lambda,i}$$

for all $\lambda \in \Lambda$. It is easily seen using Proposition 1.5 *b)*, that the closure F_i of $\bigcup_{\lambda \in \Lambda} F_{\lambda,i}$ is a fan in X.

By condition *c)* the claim holds for the subspace Y of X generated by $F_1, \ldots, F_m$. This completes the proof of the claim.

Now, it remains the implication *c)* $\Rightarrow$ *a)*, whose proof fills the rest of this section and requires several steps. $\square$

For the following lemmas (X, G) is an arbitrary space of orderings.

Lemma 5.2   *Let S be a subset of X which generates X and let $\alpha \in \widehat{G}$ such that $\alpha S \subset X$. Then $\alpha X = X$.*

Proof. For $\sigma \in S$ we have $\sigma(-1) = \alpha\sigma(-1) = -1$, hence $\alpha(-1) = 1$. Next, we see that for $h \in G$ with $\alpha(h) = -1$ it holds

$$D(\langle 1, h \rangle) = \{1, h\}.$$

Indeed, let $f \in D(\langle 1, h \rangle)$, $f \neq 1$; this means that $\tau(f) = +1$ for all $\tau \in X$ such that $\tau(h) = +1$. Now, since S generates X, there is $\sigma \in S$ such that $\sigma(f) = -1$. Hence $\sigma(h) = -1$. Consequently, $\alpha\sigma(h) = \alpha(h)\sigma(h) = +1$, and, since $\alpha\sigma \in X$, $\alpha\sigma(f) = +1$. We conclude that $\alpha(f) = -1$. Now, let $\tau \in S$. If $\tau(h) = +1$, we have $\tau(f) = +1$ and $\tau(hf) = +1$. If $\tau(h) = -1$, then $\alpha\tau(h) = +1$ and $\alpha\tau(f) = +1$; thus, $\tau(f) = -1$ and again $\tau(hf) = +1$. Since S generates X, it follows that $hf = 1$, and $f = h$.

After this preparation, we show that $\alpha X = X$. Clearly, it is enough to see that $\alpha X \subset X$. To that end, according to Proposition III.3.5, we see that given

$\tau \in X$ and $g \in \mathrm{Ker}(\alpha\tau)$, it is $D(\langle 1, g\rangle) \subset \mathrm{Ker}(\alpha\tau)$. If $\alpha(g) = -1$, we just apply the previous remark to $h = g$. Now, assume $\alpha(g) = +1$ and let $f \in D(\langle 1, g\rangle)$. There are two possibilities:

$\alpha(f) = -1$: Then $\alpha(-f) = \alpha(-1)\alpha(f) = -1$ and the previous remark with $h = -f$ gives $D(\langle 1, -f\rangle) = \{1, -f\}$. Since $-g \in D(\langle 1, -f\rangle)$, it follows $f = g \in \mathrm{Ker}(\alpha\tau)$.

$\alpha(f) = +1$: Since $\alpha(g) = \alpha\tau(g) = +1$, we get $\tau(g) = +1$, and consequently $\tau(f) = +1$. Thus $\alpha\tau(f) = +1$.

This completes the proof. $\qquad\qquad\qquad\qquad\qquad\qquad\qquad\qquad\qquad\quad$ $\square$

Lemma 5.3   *For $\alpha \in \widehat{G}$ let*

$$X_\alpha = \{\sigma \in X \mid \alpha\sigma \in X\} = X \cap \alpha X.$$

Then X_α is a (possibly trivial) subspace of X.

Proof.   By definition, X_α generates $X_\alpha^{\perp\perp}$, and, clearly, $\alpha X_\alpha \subset X_\alpha \subset X_\alpha^{\perp\perp}$. Thus, by the previous lemma, $\alpha X_\alpha^{\perp\perp} \subset X_\alpha^{\perp\perp} \subset X$ and $X_\alpha^{\perp\perp} \subset X \cap \alpha X = X_\alpha$. $\qquad$ $\square$

Lemma 5.4   *Let $\sigma_1, \sigma_2, \sigma_3 \in X$ and $\alpha, \beta \in \widehat{G}$. Suppose that $\sigma_1 \in X_\alpha \cap X_\beta$, $\sigma_2 \in X_\alpha \setminus X_\beta$ and $\sigma_3 \in X_\beta \setminus X_\alpha$. Then*

a) $\sigma_1\alpha\beta \notin X$.

b) If $X_\alpha \cap X_\beta = \{\sigma_1\}$, then $\sigma_2 = \sigma_1\alpha$ or $\sigma_3 = \sigma_1\beta$.

Proof.   Consider the table of elements in $\widehat{G}$:

$$
\begin{array}{llll}
\sigma_1 & \sigma_2 & \sigma_3 & \sigma_1\sigma_2\sigma_3 \\
\sigma_1\alpha & \sigma_2\alpha & \sigma_3\alpha & \sigma_1\sigma_2\sigma_3\alpha \\
\sigma_1\beta & \sigma_2\beta & \sigma_3\beta & \sigma_1\sigma_2\sigma_3\beta \\
\sigma_1\alpha\beta & \sigma_2\alpha\beta & \sigma_3\alpha\beta & \sigma_1\sigma_2\sigma_3\alpha\beta
\end{array}
$$

a) Assume that $\sigma_1\alpha\beta \in X$. Then the elements of the first column are all in X. Also $\sigma_2, \sigma_2\alpha \in X$ but $\sigma_2\beta \notin X$ by assumption. Suppose $\sigma_2\alpha\beta \in X$. Let $\gamma = \sigma_1\sigma_2\alpha$. Then $\sigma_1\gamma = \sigma_2\alpha$, $\sigma_1\alpha\gamma = \sigma_2$, $\sigma_1\beta\gamma = \sigma_2\alpha\beta$. Thus, by Lemma 5.2, the space generated by $\sigma_1, \sigma_2, \sigma_1\alpha, \sigma_1\beta$ is invariant under γ. In particular $\sigma_1\alpha\beta\gamma = \sigma_2\beta \in X$. Contradiction. Thus $\sigma_2\alpha\beta \notin X$. Similarly we get $\sigma_3 \in X$, $\sigma_3\alpha \notin X$, $\sigma_3\beta \in X$ and $\sigma_3\alpha\beta \notin X$ in the third column. For the fourth column, if one of the elements $\sigma_1\sigma_2\sigma_3$, $\sigma_1\sigma_2\sigma_3\alpha$, $\sigma_1\sigma_2\sigma_3\beta$ is in X, we may replace in the table σ_2 by $\sigma_2\alpha$ or σ_3 by $\sigma_3\beta$, if necessary, and thus assume that $\sigma_1\sigma_2\sigma_3 \in X$. We have $\sigma_1, \sigma_2 \in X_\alpha$ and if also $\sigma_1\sigma_2\sigma_3 \in X_\alpha$ then $\sigma_3 = \sigma_1\sigma_2(\sigma_1\sigma_2\sigma_3) \in X_\alpha$ by Lemma 5.3. Contradiction. Thus $\sigma_1\sigma_2\sigma_3\alpha \notin X$ and similarly $\sigma_1\sigma_2\sigma_3\beta \notin X$. So the table above, the elements

$$
\begin{array}{lll}
\sigma_1 & \sigma_2 & \sigma_3 \\
\sigma_1\alpha & \sigma_2\alpha & \\
\sigma_1\beta & & \sigma_3\beta \\
\sigma_1\alpha\beta & &
\end{array}
\qquad \text{and possibly one or two of} \qquad
\begin{array}{l}
\sigma_1\sigma_2\sigma_3 \\
\\
\\
\sigma_1\sigma_2\sigma_3\alpha\beta
\end{array}
$$

are in X. Then these elements form the subspace Z generated by the elements $\sigma_1\alpha, \sigma_1\beta, \sigma_2\alpha, \sigma_2$ and σ_3, which are linearly independent. Indeed, any product of an odd number of them maps -1 to -1, hence cannot be 1. Then one checks case by case that none of the fifteen different products of two or four of them is 1. For instance:

$$(\sigma_1\alpha)\sigma_3 = 1 \;\Rightarrow\; \sigma_3\alpha = \sigma_1 \in X,$$

which is impossible, or

$$(\sigma_1\alpha)(\sigma_1\beta)(\sigma_2\alpha)(\sigma_2) = 1 \;\Rightarrow\; \beta = 1,$$

again absurd. The assumption $\sigma_1\alpha\beta \in X$ is used exactly in two cases, namely:

$$(\sigma_1\alpha)\sigma_2 = 1 \;\Rightarrow\; \sigma_2\beta = \sigma_1\alpha\beta \in X,$$

$$(\sigma_1\beta)\sigma_3 = 1 \;\Rightarrow\; \sigma_3\alpha = \sigma_1\alpha\beta \in X.$$

Once those five elements are independent, we have a contradiction by the miraculous lemma (Lemma III.3.15). Thus $\sigma_1\alpha\beta \notin X$.

b) Assume that $\sigma_2 \neq \sigma_1\alpha$ and $\sigma_3 \neq \sigma_1\beta$. Now, of our list, the elements in X are

$$
\begin{array}{lll}
\sigma_1 & \sigma_2 & \sigma_3 \\
\sigma_1\alpha & \sigma_2\alpha & \\
\sigma_1\beta & & \sigma_3\beta
\end{array}
\qquad \text{and possibly one or two of}
\qquad
\begin{array}{l}
\sigma_1\sigma_2\sigma_3 \\
\\
\sigma_1\sigma_2\sigma_3\alpha\beta
\end{array}
$$

First, $\sigma_2\beta, \sigma_3\alpha \notin X$ by assumption. Second, if $\sigma_2\alpha\beta$ were in X, then $\sigma_2\alpha \in X_\alpha \cap X_\beta$ and $\sigma_2\alpha \neq \sigma_1$, contradicting the assumption. Similarly $\sigma_3\alpha\beta \notin X$ and the argument for the fourth column corresponds to that in *a)*. Then $\sigma_1\alpha, \sigma_1\beta, \sigma_2\alpha, \sigma_2$ and σ_3 are linearly independent as one checks case by case as above. We remark here that the two cases settled in *a)* by the fact that $\sigma_1\alpha\beta \in X$ are settled here by the fact that $\sigma_2 \neq \sigma_1\alpha$ and $\sigma_3 \neq \sigma_1\beta$. Therefore, exactly as in *a)*, this leads to a contradiction by the miraculous lemma. $\square$

Lemma 5.5   *Let $\alpha, \beta \in \widehat{G} \setminus \{1\}$ such that $X_\alpha \cap X_\beta \neq \emptyset$ and $\#(X_\alpha) \geq 3$, $\#(X_\beta) \geq 3$. Then there exists $\gamma \in \widehat{G} \setminus \{1\}$ such that $X_\alpha, X_\beta \subset X_\gamma$.*

Proof. If $X_\alpha \subset X_\beta$ or $X_\beta \subset X_\alpha$ we are done. Otherwise, choose

$$\sigma_1 \in X_\alpha \cap X_\beta, \quad \sigma_2 \in X_\alpha \setminus X_\beta \quad \text{and} \quad \sigma_3 \in X_\beta \setminus X_\alpha.$$

By part *a)* of the previous lemma, $\sigma_1\alpha\beta \notin X$. Assume

$$X_\alpha \cap X_\beta = \{\sigma_1\}.$$

Then σ_2 and σ_3 can be chosen different from $\sigma_1\alpha$ and $\sigma_1\beta$, respectively, which contradicts part *b)* of the preceding lemma. Thus there exists $\sigma_4 \neq \sigma_1$ in $X_\alpha \cap X_\beta$. Take $\gamma = \sigma_1\sigma_4$. Then $\sigma_1, \sigma_1\alpha, \sigma_1\gamma = \sigma_4$ and $\sigma_1\alpha\gamma = \sigma_4\alpha$ are all in X. Hence by part *a)* of the preceding lemma either $X_\alpha \subset X_\gamma$ or $X_\gamma \subset X_\alpha$. Now $\sigma_1\beta\gamma = \sigma_4\beta \in X$, thus $\sigma_1\beta \in X_\gamma$. But, since $\sigma_1\beta\alpha \notin X$, $\sigma_1\beta \notin X_\alpha$. So $X_\alpha \subset X_\gamma$ and similarly one shows $X_\beta \subset X_\gamma$. $\square$

Proposition and Definition 5.6   *Two elements $\sigma_1, \sigma_2 \in X$ are called connected and we write $\sigma_1 \sim \sigma_2$, if either $\sigma_1 = \sigma_2$ or σ_1, σ_2 lie in a common fan F with $\#(F) \geq 4$, that is, there exists $\sigma \in X \setminus \{\sigma_1, \sigma_2\}$ such that $\sigma_1 \sigma_2 \sigma \in X$. Then:*

a) *This relation is an equivalence relation, whose classes are called connected components of X. If X consists of one single class, X is called connected.*

b) *A connected space is indecomposable.*

c) *If (X, G) is an extension, $(X, G) = (X_1, G_1)[H]$ with $\#(H) \geq 2$ and $\#(X_1) \geq 2$, then $\sigma_1 \sim \sigma_2$ for any two elements $\sigma_1, \sigma_2 \in X$.*

Proof.   c) Let $\sigma_1 \neq \sigma_2$ and $\sigma_i = \alpha_i \tau_i$, $\alpha_i \in \widehat{H}$, $\tau_i \in X_1$ for $i = 1, 2$. If $\tau_1 \neq \tau_2$ and $\alpha_1 \neq \alpha_2$ then σ_1, σ_2, $\sigma_3 = \alpha_1 \tau_2$ and $\sigma_4 = \alpha_2 \tau_1$ form a 4−element fan. If $\tau_1 = \tau_2$ choose $\tau \neq \tau_1$ and take $\{\sigma_1, \sigma_2, \sigma_3 = \alpha_1 \tau, \sigma_4 = \alpha_2 \tau\}$. Similarly for the case $\alpha_1 = \alpha_2$.

a) Let $\sigma_1, \sigma_2, \sigma_3 \in X$ be all different such that $\sigma_1 \sim \sigma_2$ and $\sigma_2 \sim \sigma_3$. We get fans

$$F = \{\sigma_1, \sigma_2, \sigma, \sigma_1 \sigma_2 \sigma\} \quad \text{and} \quad V = \{\sigma_2, \sigma_3, \tau, \sigma_2 \sigma_3 \tau\}$$

for suitable $\sigma, \tau \in X$. For $\alpha = \sigma_1 \sigma_2$ and $\beta = \sigma_2 \sigma_3$ we have $F \subset X_\alpha$, $V \subset X_\beta$. Thus $F \cup V \subset X_\gamma$ for some $\gamma \neq 1$ by Lemma 5.5. Now, γ belongs to the translation group (see 2.9) of the subspace X_γ, which consequently is not trivial. Thus, by Proposition 2.10, the subspace X_γ is a non trivial extension. Hence we may apply c) and get $\sigma_1 \sim \sigma_3$.

b) This follows from Proposition 2.2 f). $\square$

Lemma 5.7   *Let $\sigma_1, \ldots, \sigma_m \in X$ be linearly independent such that $\sigma = \sigma_1 \cdots \sigma_m \in X$. Then $\sigma \sim \sigma_i$ for $i = 1, \ldots, m$.*

Proof.   We show $\sigma \sim \sigma_1$ by induction on m. If $m = 3$ there is nothing to do. Assume $m \geq 5$. By the miraculous lemma once again, the subspace Y of X which is generated by $\sigma_1, \ldots, \sigma_m$ contains an element σ' which is a product of l elements of $\{\sigma_1, \ldots, \sigma_m\}$ where $3 \leq l \leq m - 2$.

Case I: σ_1 appears in σ', say $\sigma' = \sigma_1 \cdots \sigma_l$. By induction $\sigma_1 \sim \sigma'$ and also $\sigma' \sim \sigma$ since $\sigma = \sigma' \cdot \sigma_{l+1} \cdots \sigma_m$. Thus $\sigma_1 \sim \sigma$.

Case II: σ_1 does not appear in σ', say $\sigma' = \sigma_2 \cdots \sigma_{l+1}$. Then $\sigma = \sigma_1 \cdot \sigma' \cdot \sigma_{l+2} \cdots \sigma_m$. So by induction again $\sigma_1 \sim \sigma$. $\square$

Now we turn to the end of the proof of Theorem 5.1, the implication c) $\Rightarrow$ a). This follows from the next proposition by induction on the number N of non trivial fans among the generating fans. If $N = 0$ then X is finite, and one uses the next proposition and induction on $\#(X)$.

Proposition 5.8   *Suppose that X is generated by finitely many fans. Then:*

a) *X decomposes into finitely many connected components X_i, $i = 1, \ldots, s$.*

b) X_i is a subspace for $i = 1, \ldots, s$ and either $\#(X_i) = 1$ or $\#(X_i) \geq 4$ and X_i has a non trivial translation group.

c) $(X, G) = (X_1, G/X_1^{\perp}) + \cdots + (X_s, G/X_s^{\perp})$.

Proof. Let X be generated by n fans $F_1, \ldots, F_n$. By splitting each 2-element fan into two 1-element fans, we may assume that $\#(F_i) \neq 2$ for $i = 1, \ldots, n$. Now, by Proposition 1.5 *b)* and Remark and Notations 1.3 *a)*,

$$X \subset \mathrm{Lin}\left(\bigcup_{i=1}^{n} F_i\right) = \widehat{G}.$$

Let $\sigma \in X$. Write $\sigma = \sigma_1 \cdots \sigma_m$, where $\sigma_1, \ldots, \sigma_m \in \bigcup_{i=1}^{n} F_i$ and m is the smallest possible. Then, by the preceding lemma, $\sigma \sim \sigma_i$ for $i = 1, \ldots, m$. This proves *a)*. Since $\#(F_i) \neq 2$ each F_i lies in exactly one component X_j for $i = 1, \ldots, n$. Now, the group $\widehat{G}_j$ generated in $\widehat{G}$ by the F_i's contained in X_j is closed, and consequently the subspace generated by those F_i's in X is $X \cap \widehat{G}_j$. By the preceding lemma, $X_j = X \cap \widehat{G}_j$, and X_j is a subspace of X.

Assume $\#(X_j) \geq 2$; since X_j is connected, $\#(X_j) \geq 4$. Say $X_j = Y$ is generated by $F_1, \ldots, F_k$. We show that the translation group of X_j is not trivial. For this choose $\sigma_i \in F_i$, $i = 1, \ldots, k$, where we may assume $\sigma_i \neq \sigma_j$ for $i \neq j$. For each i take a 4-element fan $F_i' \subset X$ containing σ_i and σ_{i+1}. Clearly $F_i' \subset Y$. Since F_i, F_i' are fans there are α_i, α_i', different from 1, in $\widehat{G}$ such that

$$F_i \subset Y_{\alpha_i} \text{ for } i = 1, \ldots, k \text{ and } F_i' \subset Y_{\alpha_i'} \text{ for } i = 1, \ldots, k - 1.$$

If $\#(F_i) = 1$, we take $Y_{\alpha_i} = Y_{\alpha_i'}$ or $Y_{\alpha_i} = Y_{\alpha_{i+1}'}$, to guarantee that all Y_{α_i}'s and $Y_{\alpha_i'}$'s have at least three elements. Now by Lemma 5.5 there is $\beta_1 \neq 1$ with $Y_{\alpha_1} \cup Y_{\alpha_1'} \subset Y_{\beta_1}$. Correspondingly, there is $\beta_2 \neq 1$ with $Y_{\beta_1} \cup Y_{\alpha_2} \subset Y_{\beta_2}$, and so on. Ending up, we find $\gamma \neq 1$ with

$$F_1 \cup F_1' \cup F_2 \cup \cdots \cup F_{k-1}' \cup F_k \subset Y_{\gamma}$$

and by Lemma 5.2, γ belongs to the translation group of Y. This completes the proof of *b)*.

For *c)* note first that

$$\widehat{G} = \mathrm{Lin}(F_1 \cup \cdots \cup F_n) \subset \mathrm{Lin}(\widehat{G}_1 \cup \cdots \cup \widehat{G}_s).$$

We claim that in fact $\widehat{G} = \widehat{G}_1 \times \cdots \times \widehat{G}_s$ (direct product). Otherwise, we would have a relation

$$1 = \prod_{i,j} \sigma_{ij},$$

where $\sigma_{ij} \in X_i$ and $\prod_j \sigma_{ij} \neq 1$ for at least one i, say $i = 1$. Choose such a relation with the smallest possible number of factors. Then the elements σ_{ij}, $(i, j) \neq (1, 1)$, are linearly independent. Thus, by the preceding lemma, all these σ_{ij} lie in the same component. Contradiction. Now the decomposition $\widehat{G} = \widehat{G}_1 \times \cdots \times \widehat{G}_s$ yields a corresponding dual one ([Mo Ch.7])

$$G = G/G_1 \times \cdots \times G/G_s = G/X_1^{\perp} \times \cdots \times G/X_s^{\perp},$$

which shows *c)*. $\square$

6. Local-Global Principles

Throughout this section let (X, G) be a space of orderings. Let $(Y, G/Y^\perp)$ be a subspace. For a form ρ over G we have the restriction $\bar\rho$ of ρ to Y and

$$D_Y(\rho) \;=\; \left\{ g \in G \;\middle|\; \begin{array}{c} \text{there are } g_2, \ldots, g_n \in G \text{ such that} \\ \sigma(\rho) = \sigma(\langle g, g_2, \ldots, g_n \rangle) \text{ for all } \sigma \in Y \end{array} \right\} =$$
$$= \; \{ g \in G \mid \bar g \in D(\bar\rho) \}.$$

Clearly $D_Y(\rho) \supset D_X(\rho) = D(\rho)$. (See 1.2 and Remarks and Notations III.1.18.) The general local-global principle, on which all the others depend, is the following:

Theorem 6.1   *Let $\rho_1, \ldots, \rho_n$ be forms over X.*

 a) Assume that $\bigcap_{i=1}^n D(\rho_i) \neq \emptyset$ and there is $\sigma \in X$ such that $\bigcap_{i=1}^n D(\rho_i) \subset \mathrm{Ker}(\sigma)$. Then there is a finite subspace Y of X such that $\bigcap_{i=1}^n D_Y(\rho_i) \subset \mathrm{Ker}(\sigma)$. In particular, $\sigma \in Y$.
 b) Assume that $\bigcap_{i=1}^n D(\rho_i) = \emptyset$. Then there is a finite subspace Y of X such that also $\bigcap_{i=1}^n D_Y(\rho_i) = \emptyset$.

In both cases the finite subspace Y can be chosen subject to the additional property that $cl(Y) \leq r$ and $s(Y) \leq r^2$, where $r = \sum_{i=1}^n \dim(\rho_i)$. In particular, $\#(Y) \leq r \cdot 2^{r^2 - 1}$.

Note that if $\emptyset \neq \bigcap_{i=1}^n D_Y(\rho_i) \subset \mathrm{Ker}(\sigma)$, then automatically $\sigma \in Y$ (we have $gY^\perp \subset \mathrm{Ker}(\sigma)$ for some $g \in G$, thus $\sigma \in Y^{\perp\perp} = Y$).

Before proving the theorem we need some preliminary considerations. A form τ is a *subform* of a form ρ, if there is another form τ' with $\rho = \tau + \tau'$. Then, we define as usual the *Witt index* of ρ to be the maximal number k such that $\tau = k \times \langle 1, -1 \rangle$ is a subform of ρ. This is well defined by Proposition III.2.4.

Lemma 6.2   *Let ρ, τ be forms over X such that $\rho - \tau$ has Witt index $\geq k$ where $k \leq \dim(\rho), \dim(\tau)$. Then ρ and τ have a common subform of dimension k.*

Proof.   This follows easily by induction on k from Axiom $\mathbf{O}_4'$. $\square$

Lemma 6.3   *Let ρ, ρ_i, ρ_i' be forms over X such that $\rho = \rho_i + \rho_i'$ for $i = 1, \ldots, n$, and $\dim(\rho) > \sum_{i=1}^n \dim(\rho_i')$. Then $\rho_1, \ldots, \rho_n$ have a common subform β of dimension $\dim(\rho) - \sum_{i=1}^n \dim(\rho_i')$.*

Proof.   This is trivial for $n = 1$. Let $n = 2$. Then $\rho_1 + \rho_1' = \rho = \rho_2 + \rho_2'$, hence $\rho_1 - \rho_2 \sim \rho_2' - \rho_1'$. Comparing dimensions we see that the Witt index of $\rho_1 - \rho_2$ is at least $\dim(\rho) - \dim(\rho_1') - \dim(\rho_2')$ and the result follows from Lemma

6.2. Now assume, by induction, that $\rho_2, \ldots, \rho_n$ have a common subform τ of dimension $\dim(\rho) - \sum_{i=2}^{n} \dim(\rho_i')$. So $\rho = \tau + \tau'$ where $\dim(\tau') = \sum_{i=2}^{n} \dim(\rho_i')$. Applying the case $n = 2$ we get the desired common subform β of ρ_1 and τ with $\dim(\beta) = \dim(\rho) - \dim(\rho_1') - \dim(\tau') = \dim(\rho) - \sum_{i=1}^{n} \dim(\rho_i')$. $\square$

We separate the proof of the theorem in two steps.

1st *Step, reduction to spaces of finite chain length:* Assume that $\bigcap_{i=1}^{n} D(\rho_i) \neq \emptyset$. There exists a minimal subspace Y subject to the property that $\sigma(g) = 1$ for all $g \in \bigcap_{i=1}^{n} D_Y(\rho_i)$. In fact, using Zorn's lemma we mainly must show: Let Y_λ be a descending family of subspaces verifying the required property; then $Y_0 = \bigcap_\lambda Y_\lambda$ verifies it too. It is enough to see that $D_{Y_0}(\rho) = \bigcup_\lambda D_{Y_\lambda}(\rho)$ for any form ρ. For the non trivial inclusion "$\subset$" take $g \in D_{Y_0}(\rho)$. Then there are $g_2, \ldots, g_n \in G$ such that the constructible set

$$U = \{\sigma \in X \mid \sigma(\rho) = \sigma(\langle g, g_2, \ldots, g_n \rangle)\}$$

is an open neighbourhood of Y_0. By compactness $U \supset Y_\lambda$ for some λ, and $g \in D_{Y_\lambda}(\rho)$. Thus we get Y.

Now, as remarked above, $\sigma \in Y$, and we replace X by Y. We claim that $cl(X) \leq r = \sum_{i=1}^{n} \dim(\rho_i)$. By way of contradiction, suppose that we have $b_1, \ldots, b_k \in G$, $k > r$, such that $X = \{b_1 > 0\} \cup \cdots \cup \{b_k > 0\}$ is a non-trivial decomposition into disjoint principal sets. By the minimal choice of Y there are $c_j \in G$, $j = 1, \ldots, k$, with $\sigma(c_j) = -1$ and $c_j \in \bigcap_i D_{\{b_j < 0\}}(\rho_i)$, or, by Proposition III.2.6 e), $\rho_i \otimes \langle 1, -b_j \rangle = \langle c_j \rangle + \tau_{ij}$ for suitable forms τ_{ij} over X. One has $\sum_{j=1}^{k} \langle 1, -b_j \rangle = (k-1) \times \langle 1, 1 \rangle + \langle 1, -1 \rangle$, thus, after addition, $2(k-1) \times \rho_i + \rho_i - \rho_i = \langle c_1, \ldots, c_k \rangle + \sum_{j=1}^{k} \tau_{ij}$. In particular, $\dim(\sum_{j=1}^{k} \tau_{ij}) = 2k \dim(\rho_i) - k$. Therefore the Witt index of $2(k-1) \times \rho_i - \langle c_1, \ldots, c_k \rangle$ is at least $k - \dim(\rho_i)$. Now, by Lemma 6.2, $2(k-1) \times \rho_i$ and $\langle c_1, \ldots, c_k \rangle$ have a common subform γ_i of dimension $k - \dim(\rho_i)$. We get $\langle c_1, \ldots, c_k \rangle = \gamma_i + \gamma_i'$ where $\dim(\gamma_i') = \dim(\rho_i)$. This holds for $i = 1, \ldots, n$. Now, by assumption,

$$k - \sum_{i=1}^{n} \dim(\gamma_i') = k - \sum_{i=1}^{n} \dim(\rho_i) = k - r > 0,$$

and thus, by Lemma 6.3, $\gamma_1, \ldots, \gamma_n$ have a common subform of dimension $k - r$. In particular, there exists $c \in \bigcap_{i=1}^{n} D(\gamma_i)$. But γ_i is also a subform of $2(k-1) \times \rho_i$, hence, by Proposition III.2.3 c), $c \in D(\rho_i)$ for $i = 1, \ldots, n$. On the other hand, $c \in D(\langle c_1, \ldots, c_k \rangle)$ and since $\sigma(c_j) = -1$, $j = 1, \ldots, k$, we get $\sigma(c) = -1$. Contradiction.

Now consider the case that $D(\rho_1) \cap \cdots \cap D(\rho_n) = \emptyset$. Here we choose a subspace Y of X which is minimal subject to the property that $D_Y(\rho_1) \cap \cdots \cap D_Y(\rho_n) = \emptyset$. We use the same argumentation as above in order to show that $cl(Y) \leq \sum_{i=1}^{n} \dim(\rho_i)$. The only difference is that now there is no extra condition for the elements c_j. But here already the existence of some $c \in \bigcap_{i=1}^{n} D(\gamma_i)$ leads to a contradiction. $\square$

2nd Step: Assume that $cl(X) = k < \infty$. The theorem follows from the more general

Claim. *Let $\{\rho_{ij}\}$ be a system of forms over X, $i = 1, \ldots, m$, $j = 1, \ldots, n_i$ and let $\sigma \in X$. Assume that*

   a) ρ_{ij} is anisotropic for $i = 1, \ldots, m$; $j = 1, \ldots, n_i$.
   b) $\bigcap_{j=1}^{n_i} D(\rho_{ij}) = \emptyset$ for $i = 2, \ldots, l$.
   c) $\emptyset \neq \bigcap_{j=1}^{n_j} D(\rho_{ij}) \subset \mathrm{Ker}(\sigma)$ for $i = l+1, \ldots, m$.

Let $r = \sum_{i,j} \dim(\rho_{ij})$. Then there exists a finite subspace Y of X such that

   a) ρ_{ij} is anisotropic over Y for $i = 1, \ldots, m$; $j = 1, \ldots, n_i$.
   b) $\bigcap_{j=1}^{n_i} D_Y(\rho_{ij}) = \emptyset$, $j = 2, \ldots, l$.
   c) $\emptyset \neq \bigcap_{j=1}^{n_i} D_Y(\rho_{ij}) \subset \mathrm{Ker}(\sigma)$, $j = l+1, \ldots, m$.

Moreover, $s(Y) \leq rk$, and if the translation group $\mathrm{Tr}(X, G)$ is trivial, then even $s(Y) \leq r(k-1)$.

Proof of the claim. By induction along the tree of X.

   i) If (X, G) is the atomic space E, set $Y = X$. One has $\mathrm{Tr}(X, G) = \{1\}$. In fact, $0 = s(E) = r(cl(E) - 1)$.

   ii) Let $(X, G) \neq E$ such that $\mathrm{Tr}(X, G) = \{1\}$. Then we have a decomposition: $(X, G) = (X', G') + (X'', G'')$. Thus $k = cl(X) = k' + k''$, where $k' = cl(X')$, $k'' = cl(X'')$ (Proposition 2.2 e)). Let ρ be a form over X, say, $\rho = \langle a_1' a_1'', \ldots, a_n' a_n'' \rangle$. Set $\rho' = \langle a_1', \ldots, a_n' \rangle$, $\rho'' = \langle a_1'', \ldots, a_n'' \rangle$. Then ρ', ρ'' represent forms over X' and X'' respectively which do not depend on the special representation of ρ. One has $D(\rho) = D(\rho') \times D(\rho'')$. Thus $\bigcap_{j=1}^{n_j} D(\rho_{ij}) = \emptyset$ if and only if $\bigcap_{j=1}^{n_j} D_{X'}(\rho_{ij}') = \emptyset$ or $\bigcap_{j=1}^{n_j} D_{X''}(\rho_{ij}'') = \emptyset$. Now say $\sigma \in X'$. Then $\emptyset \neq \bigcap_{j=1}^{n_i} D(\rho_{ij}) \subset \mathrm{Ker}(\sigma)$ if and only if $\emptyset \neq \bigcap_{j=1}^{n_i} D_{X'}(\rho_{ij}') \subset \mathrm{Ker}(\sigma)$ and $\emptyset \neq \bigcap_{j=1}^{n_i} D_{X''}(\rho_{ij}'')$. Finally, ρ_{ij} is anisotropic over X if and only if ρ_{ij}' or ρ_{ij}'' are anisotropic. Thus we can split the conditions *a)*, *b)*, *c)* on the system of forms $\{\rho_{ij}\}$ into corresponding conditions *a')*, *b')*, *c')* and *a'')*, *b'')*, *c'')* on systems of forms $\{\rho_{ij}'\}$ and $\{\rho_{ij}''\}$ respectively. Note that we may drop among these all isotropic forms. Let $r' = \sum_{i,j} \dim(\rho_{ij}')$ and $r'' = \sum_{i,j} \dim(\rho_{ij}'')$. Then $r', r'' \leq r$. By induction we find finite subspaces Y' of X' and Y'' of X'' such that the conditions *a')*, *b')*, *c')* and *a'')*, *b'')*, *c'')* hold for the restrictions of the ρ_{ij}' to Y' and of the ρ_{ij}'' to Y'' respectively. Now $Y = Y' \cup Y''$ is a subspace of X which verifies the conditions of the claim. In fact, $(Y, G/Y^\perp) = (Y', G'/Y'^\perp) + (Y'', G''/Y''^\perp)$. Thus by induction and Proposition 2.2 *d)*, we also get $s(Y) \leq \max\{r'k', r''k''\} \leq r(k-1)$.

   iii) (X, G) has a non trivial translation group. Thus $(X, G) = (X', G')[H]$ where $H \neq \{1\}$ and $\mathrm{Tr}(X', G') = \{1\}$. By induction, we assume that the claim holds for (X', G'). Again, $k = cl(X')$ (unless X' is atomic). Let $H_1 \subset H$ be the subgroup generated by all $h \in H$ for which some residue form $(\rho_{ij})_h$ is defined. Clearly, $\dim_{\mathbb{F}_2}(H_1) \leq \sum_{i,j} \dim(\rho_{ij}) = r$. Now, by Proposition 2.12 *d)*, a condition $\bigcap_{j=1}^{n_i} D(\rho_{ij}) = \emptyset$ splits into the conditions $\bigcap_{j=1}^{n_i} D_{X'}((\rho_{ij})_h) = \emptyset$ with

$h \in H_1$ (such a condition is automatically fullfilled, if one of the residue forms $(\rho_{ij})_h$ vanishes).

Let $\sigma = \alpha\sigma'$ with $\alpha \in \widehat{H}$ and $\sigma' \in X'$. Then a condition $\emptyset \neq \bigcap_{j=1}^{n_i} D(\rho_{ij}) \subset \mathrm{Ker}(\sigma)$ splits into the conditions

$$\emptyset \neq \bigcap_{j=1}^{n_i} D_{X'}(\alpha(h)(\rho_{ij})_h) \subset \mathrm{Ker}(\sigma')$$

or

$$\bigcap_{j=1}^{n_i} D_{X'}((\rho_{ij})_h) = \emptyset,$$

with $h \in H_1$. Finally, the condition that ρ_{ij} is anisotropic, splits into the conditions that all $(\rho_{ij})_h$ are anisotropic over X'. So we get a system (ρ'_{ij}) of forms over X' with corresponding conditions $a')$, $b')$, $c')$. One obtains $r' = \sum_{i,j} \dim(\rho'_{ij}) \leq r$. By induction we find a finite subspace X_1 of X' such that the conditions $a')$, $b')$, $c')$ hold also for the restrictions of the forms ρ'_{ij} to X_1. Moreover, $s(X_1) \leq r'(cl(X') - 1) \leq r(k-1)$. Now let $H = H_1 \times H_2$ and $\widehat{H} = \widehat{H}_1 \times \widehat{H}_2$ correspondingly. We set $Y = \widehat{H}_1 \times X_1$. Then $Y^\perp = H_2 \times X_1^\perp$, thus $G/Y^\perp = H_1 \times (G'/X_1^\perp)$ and $(Y, G/Y^\perp) = (X_1, G'/X_1^\perp)[H_1]$. Hence, $s(Y) = s(X_1) + \dim_{\mathbb{F}_2}(H_1) \leq r(k-1) + r = rk$. By construction it is also clear that the conditions $a)$, $b)$, $c)$ hold for the restrictions of the ρ_{ij}'s to Y.
$\square$

Theorem 6.4   *Let ρ be an anisotropic form over X. Then there exists a finite subspace Y of X such that the restriction of ρ to Y is anisotropic. Moreover, Y can be chosen such that $cl(Y) \leq \dim(\rho)$ and $s(Y) \leq \dim(\rho)^2$.*

Proof.   This is clear if $\dim(\rho) = 1$. If $\dim(\rho) > 1$ write $\rho = \langle a \rangle + \rho'$. Then, by Axiom $\mathbf{O}'_4$, since ρ is anisotropic we have $D(\langle a \rangle) \cap D(\langle -\rho' \rangle) = \emptyset$. Thus the claim follows from Theorem 6.1 $b)$.
$\square$

Corollary 6.5   *Let ρ be a form over X such that for each finite subspace Y of X there exists a form τ with $\rho = 2 \times \tau$ over Y. Then $\rho = 2 \times \pi$ for a suitable form π over X.*

Proof.   By induction on n where $2n = \dim(\rho)$.

$n = 1$. Then $\rho = \langle a, b \rangle$ and $\sigma(a) = \sigma(b)$ for all $\sigma \in X$. Thus $a = b$ and $\rho = 2 \times \langle a \rangle$.

$n > 1$. Now let $\rho = \langle a_1, \ldots, a_{2n} \rangle = \langle a_1 \rangle + \rho'$. Over each finite subspace Y we have $\langle a_1 \rangle + \rho' = \tau + \tau$. Thus $a_1 \in D_Y(\tau + \tau) = D_Y(\tau)$ (Proposition III.2.3 $c)$). Hence, we get $\tau = \langle a_1 \rangle + \tau_1$ over Y. By cancellation, $\rho' = \langle a_1 \rangle + \rho_1$ with $\rho_1 = 2 \times \tau_1$, always over Y. Thus $\langle -a_1 \rangle + \rho'$ is isotropic over all finite subspaces Y. By Theorem 6.4, $\langle -a_1 \rangle + \rho'$ is isotropic over X, and by Axiom $\mathbf{O}'_4$, $a_1 \in D(\rho')$. Thus, $\rho' = \langle a_1 \rangle + \rho_1$ over X, and by induction, $\rho_1 = 2 \times \pi_1$ over X. Whence $\rho = 2 \times \pi$ for $\pi = \langle a_1 \rangle + \pi_1$.
$\square$

7. Representation Theorem and Invariants

Let (X, G) be a space of orderings. Every form ρ over X defines the signature map $\hat{\rho} : X \to \mathbb{Z}$; $\sigma \mapsto \sigma(\rho)$, which is continuous. Clearly, this map depends on the similarity class of the form, rather than on the form itself. Thus, $\text{Cont}(X, \mathbb{Z})$ contains the reduced Witt ring $W(X)$ of degenerate forms of X (I.3.6).

The first remark concerning this is the famous fact that $W(X)$ *is equal to* $\text{Cont}(X, \mathbb{Z})$ *up to 2-torsion.* Indeed, let $f : X \to \mathbb{Z}$ be continuous. Then, since X is compact, f takes finitely many values $m_1, \ldots, m_r$. Now, by Proposition and Definition I.3.7, for each $i = 1, \ldots, r$ there are a form τ_i and an integer k_i with $\sigma(\tau_i) = 2^{k_i}$ for $f(\sigma) = m_i$, $\sigma(\tau_i) = 0$ otherwise; obviously, we can assume all $k_i = k$. Then the form $\tau = \sum_{i=1}^{r} m_i \times \tau_i$ verifies $\hat{\tau} = 2^k f$.

However, such a result is not enough, as we need a full characterization of $W(X)$. This is done in the following fundamental result:

Theorem 7.1 (Representation Theorem)   *For a continuous function $f : X \to \mathbb{Z}$ the following conditions are equivalent:*

> a) *There is a form ρ over X such that $f = \hat{\rho}$.*
> b) *For each finite fan $F \subset X$ there is a form ρ over X such that $f|F = \hat{\rho}|F$.*
> c) $\sum_{\sigma \in F} f(\sigma) \equiv 0 \mod \#(F)$ *for each finite fan $F \subset X$.*

Proof.   Obviously *a)* $\Rightarrow$ *b)* $\Rightarrow$ *c)* (see Proposition III.3.8 *d)*).

c) $\Rightarrow$ *a)* As remarked before, we have $2^k f = \hat{\tau}$ for a suitable form τ over X and some k. So we may assume that *c)* holds for f and $2f = \hat{\tau}$ for a form τ over X. Suppose that τ is anisotropic. We prove the claim by showing that $\tau = 2 \times \rho$ and according to Corollary 6.5 we may assume that (X, G) is finite. Again we proceed by induction along the tree of (X, G).

i) (X, G) is the atomic space. Here the claim is obvious.

ii) $(X, G) = (X', G') + (X'', G'')$. For the form τ over G we have the components τ', τ'' where τ' is a form over G' and τ'' a form over G'' as in the proof of Theorem 6.1. In fact, τ' and τ'' are the restrictions of τ to X' and X'' respectively. One of these forms, say τ', is anisotropic. On the other hand, by Proposition III.2.4 we can write $\tau'' = k \times \langle 1, -1 \rangle + \tau_1$, where τ_1 is an anisotropic form over X''. By induction we have $\tau' = 2 \times \rho'$ and $\tau_1 = 2 \times \rho_1$. Let σ', σ'' belong to X', X'', respectively. By assumption *c)*, $\sigma'(\rho') \equiv \sigma''(\rho_1) \mod 2$, hence $\dim(\rho') \equiv \dim(\rho_1) \mod 2$ and $\dim(\tau') \equiv \dim(\tau_1) \mod 4$. Thus, $k \equiv 0 \mod 2$, and also $\tau'' = 2 \times \rho''$. These ρ', ρ'' are the components of the form ρ over X we sought.

iii) $(X, G) = (X', G')[\mathbb{Z}_2]$. Let $\mathbb{Z}_2 = \{1, a\}$ and $\widehat{\mathbb{Z}_2} = \{1, \alpha\}$. One has $\tau = \tau_1 + a\tau_a$ where τ_1 and τ_a are the residue forms of τ over X'. Let F' be a fan in X'. Then $F = F' \cup \alpha F'$ is a fan in X. We have

$$\sum_{\sigma \in F} \sigma(\tau) = \sum_{\sigma \in F'} \sigma(\tau_1) + \sum_{\sigma \in F'} \sigma(\tau_a) + \sum_{\sigma \in F'} \sigma(\tau_1) - \sum_{\sigma \in F'} \sigma(\tau_a) = 2 \sum_{\sigma \in F'} \sigma(\tau_1)$$

and by assumption

$$\sum_{\sigma \in F} \sigma(\tau) \equiv 0 \mod 2\#(F).$$

Thus $\sum_{\sigma \in F'} \sigma(\tau_1) \equiv 0 \mod \#(F)$, that is, $\equiv 0 \mod 2\#(F')$. Choose in particular $F' = \{\sigma\}$, $\sigma \in X'$. It follows that $\sigma(\tau_1)$ is even. Since this holds for all $\sigma \in X'$ we get $\widehat{\tau}_1 = 2g_1$ with $g_1 \in \mathrm{Cont}(X', \mathbb{Z})$. Correspondingly $\widehat{\tau}_a = 2g_a$. So by induction $\tau_1 = 2 \times \rho_1$ and $\tau_1 = 2 \times \rho_a$ for two forms ρ_1 and ρ_a over X'. Thus $\tau = 2 \times \rho$ for $\rho = \rho_1 + a\rho_a$ and the theorem is proved. $\square$

Theorem 7.2   *Let $C \subset X$ be a constructible set.*

   a) If $\#(C \cap F) \equiv 0 \bmod 2$ for all 4-element fans F in X, then C is principal.
   b) If $\#(C \cap F) \neq 3$ for all 4-element fans F in X, then C is basic.
   c) If for all 4-element fans F in X either $F \subset C$ or $F \subset D = X \setminus C$, then
   $$(X, G) = (C, G/C^\perp) + (D, G/D^\perp).$$

Proof.   *a)* Define $f \in \mathrm{Cont}(X, \mathbb{Z})$ by $f(\sigma) = 1$ for $\sigma \in C$ and $f(\sigma) = -1$ for $\sigma \notin C$. We show that $f = \widehat{\rho}$ for a form ρ over G, say $\rho = \langle a_1, \ldots, a_m \rangle$. Then m is odd and we set $g = \prod_{i=1}^m a_i$ if $m \equiv 1 \bmod 4$ and $g = - \prod_{i=1}^m a_i$ if $m \equiv 3 \bmod 4$ (compare with Proposition III.3.1). One easily sees that $C = \{g > 0\}$. Now, in order to show that $f = \widehat{\rho}$, by Theorem 7.1 we may assume that (X, G) is a finite fan, say $F_n = F_{n-1}[\mathbb{Z}_2]$ and we may proceed by induction on n. So we may write $f \mid F_{n-1} = \widehat{\rho}_1$. Let again $\mathbb{Z}_2 = \{1, a\}$ and $\widehat{\mathbb{Z}}_2 = \{1, \alpha\}$. For $\sigma, \tau \in F_{n-1}$, $\sigma \neq \tau$, consider the fan $F = \{\sigma, \tau, \alpha\sigma, \alpha\tau\}$. By assumption $f(\sigma)f(\alpha\sigma) = f(\tau)f(\alpha\tau)$. Now put $\rho = \rho_1$ if $f(\sigma) = f(\alpha\sigma)$ and $\rho = a\rho_1$ if $f(\sigma) = -f(\alpha\sigma)$ for one and thus for all $\sigma \in F_{n-1}$.

   b) Let f be the characteristic function of C. We have $2^k f = \widehat{\rho}$ for a suitable form ρ over X and some k. Say, $\rho = \langle a_1, \ldots, a_n \rangle$ is anisotropic. By Theorem 6.4 there is a finite subspace Y of X such that the restriction of ρ to Y is anisotropic. Now, by assumption and Corollary 1.8, $C \cap Y$ is a subspace. Thus $C \cap Y = Y \cap \{b_1 > 0, \ldots, b_r > 0\}$. Assume $r > k$. Then $\varphi = \langle\langle b_1, \ldots, b_r \rangle\rangle = 2^{r-k} \times \rho$ over Y. Then

$$1 \in D_Y(\varphi) = D_Y(2^{r-k} \times \rho) = D_Y(\rho),$$

the last equality by Proposition III.2.3 *c)*. It follows that $1 \in D_Y(\varphi')$, and by Proposition III.2.7 we get $\varphi = 2 \times \psi$ over Y, for another Pfister form ψ. Consequently, we can choose a smaller r. Thus we may assume $r \leq k$. Then $\dim(\varphi) = 2^r \leq n = \dim(\rho)$, and $\rho = 2^{k-r} \times \varphi$ over Y. Therefore, $n = 2^k$. But then $C = \{a_1 > 0, \ldots, a_n > 0\}$.

   c) Firstly, note that C is principal by *a)*, and so C and D are subspaces of X. By Proposition 2.4, we must see that $G = D^\perp \cdot C^\perp$. Let $g \in G$ and define $f_1, f_2 \in \mathrm{Cont}(X, \mathbb{Z})$ by $f_1(\sigma) = \sigma(g)$ for $\sigma \in C$, $f_1(\sigma) = 1$ for $\sigma \in D$; $f_2(\sigma) = 1$ for $\sigma \in C$, $f_2(\sigma) = \sigma(g)$ for $\sigma \in D$. By assumption $\#(F \cap f_i^{-1}(1)) \equiv 0 \bmod 2$ for all 4-element fan F in X. Thus, by *a)*, there exist $g_i \in G$, $i = 1, 2$, such that $f_i^{-1}(1) = \{g_i > 0\}$, that is $g = g_1 g_2$ with $g_1 \in D^\perp$ and $g_2 \in C^\perp$. We are done. $\square$

Here we have one of the generation formulae announced in Remarks III.3.13 *b)*:

Theorem 7.3 (Generation Formula for Spaces of Orderings) *Let $C \subset X$ be constructible and such that for all finite fans $F \subset X$,*

$$\#(F) \equiv 0 \mod \#(C \cap F)$$

and

$$2^m \#(C \cap F) \equiv 0 \mod \#(F).$$

Then there are $f_1, \ldots, f_m \in G$ such that $C = \{f_1 > 0, \ldots, f_m > 0\}$.

Proof. By Proposition 7.2 *b)*, C is basic. Thus $C = \{h_1 > 0, \ldots, h_k > 0\}$ and we must show that k can be chosen to be $\leq m$. For this consider the Pfister form $\varphi = \langle\!\langle h_1, \ldots, h_k \rangle\!\rangle$ with pure part φ' and assume $k > m$. By Theorem 7.1, there exists a form ρ over X such that $\sigma(\rho) = 2^m$ for all $\sigma \in C$ and $\sigma(\rho) = 0$ for all $\sigma \in X \setminus C$. Choose ρ anisotropic. Then $\varphi = 2^{k-m} \times \rho$. Arguing as in the preceding proof, we see that $1 \in D_Y(\varphi')$, then that $\varphi = 2 \times \psi$ for another Pfister form ψ, and finally that $C = \{g_1 > 0, \ldots, g_{k-1} > 0\}$ for suitable $g_i \in G$. So we finally find $f_1, \ldots, f_m \in G$ such that $C = \{f_1 > 0, \ldots, f_m > 0\}$. $\square$

Corollary 7.4 (Stability Formula for Spaces of Orderings)

$$s(X) = \sup\{k \mid \text{there is a fan } F \subset X \text{ with } \#(F) = 2^k\}.$$

Proof. First recall that $s(F) = k$ if F is a fan with 2^k elements (Proposition III.3.8 *c)*). Now, for the non trivial inequality "$\leq$" we may assume that the right hand side is finite, say m. Then any basic set $C \subset X$ verifies the hypotheses of the theorem for that m, and the conclusion follows. $\square$

A nice application of the previous results is that *the 2-torsion of $W(X)$ in* $\text{Cont}(X, \mathbb{Z})$ *is bounded by* $s = s(X)$. Indeed, by Theorem 7.1 and Corollary 7.4, for any continuous function $f : X \to \mathbb{Z}$ there is a form ρ with $\hat{\rho} = 2^s f$.

After the above computation of the stability of X, we shall discuss the relations among $s(X)$ and the other invariants $t(X)$, $w(X)$ and $l(X)$. First, we have:

Corollary 7.5 $s(X) \leq t(X)$, $s(X) = w(X)$, $2^{s(X)} \leq l(X)$.

Proof. This follows from the preceding corollary since the formulae hold for fans (Proposition III.3.8). $\square$

Next, we are going to bound $t(X)$ and $l(X)$ in terms of $s(X)$. We already know that $s(X) = 1$, if and only if $t(X) = 1$, if and only if $l(X) = 2$ (see Proposition III.3.1). We start with the following:

Lemma 7.6   *Let $(X, G) = (Y, H)[\mathbb{Z}_2]$ where $s(Y) = 1$ and let $C \subset X$ be constructible. Then there exists a form over X, $\beta = \langle b_1, b_2, ab_3, ab_4 \rangle$, where $b_i \in H$, $1 \leq i \leq 4$, and $1 \neq a \in \mathbb{Z}_2$, such that $\sigma(\beta) = 2$ for $\sigma \in C$ and $\sigma(\beta) = -2$ for $\sigma \in X \setminus C$.*

Proof.   Let $\mathbb{Z}_2 = \{1, a\}$ and $\widehat{\mathbb{Z}}_2 = \{1, \alpha\}$. We decompose Y into disjoint constructible sets Y_i, $1 \leq i \leq 4$ such that

$$C = (\{1\} \times Y_2) \cup (\{\alpha\} \times Y_3) \cup (\{1, \alpha\} \times Y_4).$$

Since $s(Y) = 1$ there exist elements $c_i \in H$ such that $Y_i = \{c_i > 0\}$. Now set $\beta = \langle -c_1 c_2 c_3, \ c_2 c_3 c_4, \ a c_2, \ -a c_3 \rangle$. $\qquad\qquad\square$

Proposition 7.7   *Let $1 \leq s = s(X) < \infty$ and let $C \subset X$ be constructible. Then there exists a form τ over X such that the following conditions hold:*

a) $\dim(\tau) \leq 4^{s-1}$
b) $\sigma(\tau) = 2^{s-1}$ for $\sigma \in C$ and $\sigma(\tau) = -2^{s-1}$ for $\sigma \in X \setminus C$.

Furthermore, $l(C) \leq 2^{w(C)-1}(1 + 2^{s-1})$ and, in particular, $l(X) \leq 2^{s-1} + 4^{s-1}$.

Proof.   Assume first that (X, G) is finite. We proceed by induction along the tree of (X, G).

   i) $s(X) = 1$. Here we set $\tau = \langle c \rangle$ where $C = \{c > 0\}$.

   ii) $(X, G) = (X', G') + (X'', G'')$ and the claim holds for (X', G') and (X'', G''). We have $C = C' \cup C''$ for which we find forms τ', τ'' over X' and X'', respectively, which are the components of the desired form τ.

   iii) $(X, G) = (Y, K)[\mathbb{Z}_2]$ and the claim holds for (Y, K) where $s(Y) = s - 1$. We provide Y with a second structure as a space of orderings. For this let $H = \mathrm{Cont}(Y, \{+1, -1\})$, thus, by Corollary III.3.2, $(Y, H) = u \times E$ where $u = \#(Y)$. Now choose a form over (Y, H), $\beta = \langle b_1, b_2, ab_3, ab_4 \rangle$, as in Lemma 7.6. By induction, for each $i = 1, \ldots, 4$ we find a form τ_i over (Y, K) such that $\sigma(\tau_i) = 2^{s-2}$ for $\sigma \in \{b_i > 0\} \subset Y$, $\sigma(\tau_i) = -2^{s-2}$ for $\sigma \in Y \setminus \{b_i > 0\}$ and $\dim(\tau_i) \leq 4^{s-2}$. Then set

$$\tau = \tau_1 + \tau_2 + a\tau_3 + a\tau_4.$$

Let now (X, G) be arbitrary. By Theorem 7.1 and Corollary 7.4 there exists a form τ with $\sigma(\tau) = 2^{s-1}$ for $\sigma \in C$ and $\sigma(\tau) = -2^{s-1}$ for $\sigma \in X \setminus C$. Choose τ anisotropic. Then by Theorem 6.4 and our information on the finite case we get $\dim(\tau) \leq 4^{s-1}$.

Finally, to bound $l(X)$ consider the form $\rho' = 2^{s-1} \times \langle 1 \rangle + \tau$, as well as the defining form ρ of C. Then $2^{s-w(C)} \times \rho \sim \rho'$, and since ρ is anisotropic, $2^{s-w(C)} l(C) \leq \dim(\rho') \leq 2^{s-1} + 4^{s-1}$. $\qquad\qquad\square$

Corollary 7.8   *Let* $2 \leq s = s(X) < \infty$ *and let* $C \subset X$ *be constructible. Then* C *is a union of no more than*

$$\binom{4^{s-1}}{2 \cdot 4^{s-2} + 2^{s-2}}$$

disjoint basic sets.

Proof.   Choose a form τ as in the proposition. Then $\binom{4^{s-1}}{2 \cdot 4^{s-2} + 2^{s-2}}$ is the number of choices for the signs $\sigma(a_i)$ of the coefficients a_i of τ such that $\sigma(\tau) = 2^{s-1}$.$\square$

Corollary 7.9   *Let* $1 \leq s = s(X) < \infty$ *and let* $C \subset X$ *be contructible.*

   a) If $s \leq 2$, then C is a disjoint union of no more than s basic sets. In particular, $t(X) \leq s$.

   b) If $s > 2$, then C is a union of no more than

$$\tau(s) = \binom{4^{s-1} - 2^{s-1} + 1}{2 \cdot 4^{s-2} - 2^{s-2} + 1}$$

basic sets. In other words, $t(X) \leq \tau(s)$.

Proof.   *a)* We have to consider the case $s = 2$. By Proposition 7.7, we find $\tau = \langle a_1, a_2, a_3, a_4 \rangle$ with $\sigma(\tau) = 2$ for $\sigma \in C$ and $\sigma(\tau) = -2$ for $\sigma \in X \setminus C$. Then $a_1 a_2 = -a_3 a_4$. Now consider

$$\begin{aligned}
\rho &= \langle 1, a_1, a_2, a_1 a_2, 1, a_3, a_4, a_3 a_4 \rangle \\
&= \langle\langle a_1, a_2 \rangle\rangle + \langle\langle a_3, a_4 \rangle\rangle \\
&= \varphi_1 + \varphi_2.
\end{aligned}$$

We have $\sigma(\rho) = 4$ for $\sigma \in C$ and $\sigma(\rho) = 0$ for $\sigma \in X \setminus C$. On the other hand, $\sigma(\varphi_i) = 0$ or 4 for all $\sigma \in X$. Therefore, C is the disjoint union of $\{a_1 > 0, a_2 > 0\}$ and $\{a_3 > 0, a_4 > 0\}$.

   b) Choose a form τ as in Proposition 7.7, and consider the form ω consisting of the first $4^{s-1} - 2^{s-1} + 1$ entries of τ. Then $\sigma \in C$ if and only if $\sigma(\omega) \geq 1$, if and only if at least $2 \cdot 4^{s-2} - 2^{s-2} + 1$ entries of ω are positive. $\square$

Remarks 7.10   From the above results we see that all the invariants s, t, w, l are finite as soon as one of them is; this will be shown for arbitrary spaces of signs in Section V.2.

   For $s \leq 2$ we have found the best possible values for $t(X)$. However, for $s = 3$ we get only $t(X) \leq 1716$, which seems to be a bad bound and for higher s our estimation for $t(X)$ increases with unpleasant velocity ($\tau(4) \geq 2 \cdot 10^{15}$). Thus, it would be desirable to find a better method.

   For finite spaces (X, G) the best bound for $t(X)$, depending on $s(X)$, is known up to a combinatorial function $tf(n)$. This $tf(n)$ is the minimal length

t of a tuple $U_1, \ldots, U_t$ of linear subspaces of $\mathbb{F}_2^{n-1}$ such that for each subset $C \subset \mathbb{F}_2^{n-1}$ there are $a_i \in \mathbb{F}_2^{n-1}$, $1 \leq i \leq t$, with $C = \bigcup_{i=1}^{t}(a_i + U_i)$.

One has the following series of bounds

n	0	1	2	3	4	5	6	7	$\ldots$
$tf(n)$	0	1	2	3	6	≤ 11	≤ 22	≤ 43	$\ldots$

Let (X, G) be a space of orderings which is built up, starting from spaces with stability index $s = 1$, by finitely many additions and extensions with finite groups. Then $t(X) \leq tf(s(X))$. On the other hand, let $(X, G) = (2^n \times E)[\mathbb{Z}_2^{n-1}]$ for $n \geq 1$. Then $s(X) = n$ and $t(X) = tf(n)$, so $t(X) = tf(s(X))$.

Note also that $t(X)$ does not depend solely on $s(X)$. For instance, $t(F_4) = s(F_4) = 4$, whereas $tf(4) = 6$. $\qquad\square$

We conclude this section and the abstract theory of spaces of orderings, with a local-global separation principle. Although separation will be discussed in arbitrary spaces of signs in Section V.3, we advance here the following definition: a function $g \in G$ *separates* two disjoint subsets C, D of X if $C \subset \{g > 0\}$ and $D \subset \{g < 0\}$. Now, let $C \subset X$. Recall that $C^{\perp\perp}$ is the subspace generated by C and that $C^{\perp\perp} = X \cap \mathrm{Adh}(\mathrm{Lin}(C))$. In some cases, even $C^{\perp\perp} = X \cap \mathrm{Lin}(C)$. This happens if C is a finite union of fans (Proposition 1.5 *a)*). One has also:

Theorem 7.11   *Let $C \subset X$ be constructible. Then: $C^{\perp\perp} = X \cap \mathrm{Lin}(C)$. Moreover, there exists a constant $N \geq 1$ such that each $\sigma \in C^{\perp\perp}$ is a product of at most N elements of C.*

Proof.   Let $C = C_1 \cup \cdots \cup C_n$ where C_i is basic, say, $C_i = \{a_{i1} > 0, \ldots, a_{is_i} > 0\}$ and set $\varphi_i = \langle\!\langle a_{i1}, \ldots, a_{is_i} \rangle\!\rangle$. Then $C_i^{\perp} = D(\varphi_i)$ by Proposition III.2.6 *c)*. Thus $C^{\perp} = \bigcap_{i=1}^{n} D(\varphi_i)$. Now let $\sigma \in C^{\perp\perp}$, that is, $\sigma(g) = 1$ for all $g \in \bigcap_{i=1}^{n} D(\varphi_i)$. By Theorem 6.1 there exists a finite subspace Y of X such that $\sigma \in Y$ and $\sigma(g) = 1$ for all $g \in \bigcap_{i=1}^{n} D_Y(\varphi_i)$. Moreover, $\#(Y)$ is bounded by a constant which only depends on the s_i's. But $C \cap Y = \bigcup_{i=1}^{n} Y \cap \{a_{i1} > 0, \ldots, a_{is_i} > 0\}$, hence $(C \cap Y)^{\perp} = \bigcap_{i=1}^{n} D_Y(\varphi_i)$. Now $\sigma \in (C \cap Y)^{\perp\perp} = Y \cap \mathrm{Lin}(C \cap Y)$, since Y is finite (and consequently, a finite union of trivial fans). Thus $\sigma \in X \cap \mathrm{Lin}(C)$ and the second part of the claim is also clear. $\qquad\square$

Theorem 7.12   *Let $C, D \subset X$ be constructible and disjoint. Then there exists $g \in G$ that separates C and D if and only if for all finite subspaces $Y \subset X$ with $cl(Y) \leq 2^{s-1}$, where $s = s(X)$, there exists $h \in G$ (depending on Y) that separates $C \cap Y$ and $D \cap Y$.*

Proof.   First we make the reduction to finite subspaces Y. We may assume that $X = (C \cup D)^{\perp\perp}$, and, by the preceding theorem, $X \subset \mathrm{Lin}(C \cup D)$. We define $C_1 \supset C$, $D_1 \supset D$ such that $X = C_1 \cup D_1$ and $C_1 \cap D_1 = \emptyset$ as follows. Let $\alpha \in X$. Then $\alpha = \sigma_1 \cdots \sigma_m \tau_1 \cdots \tau_n$ with $\sigma_i \in C$, $\tau_j \in D$ for $i = 1, \ldots, m$, $j = 1, \ldots, n$. Note that $m + n$ is odd since $\alpha(-1) = -1$. So define $\alpha \in C_1$ if m is odd and $\alpha \in D_1$ if n is odd. This is well defined under the assumption of the theorem.

Indeed, suppose we also had $\alpha = \sigma_1' \cdots \sigma_{m'}' \tau_1' \cdots \tau_{n'}'$ and $n \not\equiv n'$ mod 2. Consider the subspace Y generated by all the elements σ_i, σ_i', τ_j, τ_j'. We get

$$\sigma_1 \cdots \sigma_m \sigma_1' \cdots \sigma_{m'}' = \tau_1 \cdots \tau_n \tau_1' \cdots \tau_{n'}' = \beta.$$

Since Y is finite, by hypothesis we have $h \in G$ with $\sigma_i, \sigma_i' \in \{h > 0\}$ and $\tau_j, \tau_j' \in \{h < 0\}$. Since $n + n'$ is odd we get $\beta(h) = -1$ on the one hand and $\beta(h) = +1$ on the other hand. Contradiction.

Now, by Theorem 7.11, the elements of C_1 and D_1 are products of a bounded number of elements of $C \cup D$. Thus C_1 and D_1 are closed disjoint and thus constructible in X. Let $F \subset X$ be a fan, $F = \{\alpha_1, \alpha_2, \alpha_3, \alpha_4\}$, such that $\alpha_1, \alpha_2, \alpha_3 \in C_1$. Then by our definition also $\alpha_4 = \alpha_1 \alpha_2 \alpha_3 \in C_1$. This holds correspondingly for D_1. Therefore, by Theorem 7.2 $a)$, C_1 and D_1 are principal, say, $C_1 = \{g > 0\}$ and $D_1 = \{g < 0\}$ and we are done.

Now assume that X is finite. We proceed by induction along the tree of (X, G).

$i)$ Let $s = s(X) = 1$. Then the claim is obvious.

$ii)$ Let $(X, G) = (X_1, G_1) + (X_2, G_2)$ and $s = s(X) = \max\{s(X_1), s(X_2)\}$ such that the claim is true for (X_i, G_i), $i = 1, 2$. We set $C_i = C \cap X_i$ and $D_i = D \cap X_i$ for $i = 1, 2$. By induction we find for each $i = 1, 2$ an element $g_i \in G$ that separates C_i and D_i. Then $g = (g_1, g_2)$ separates C and D.

$iii)$ Let $(X, G) = (X', G')[\mathbb{Z}_2]$ be such that the claim is true for (X', G'). Write $s = s(X)$ and $s' = s(X')$, so that $s = s' + 1$. Let $\mathbb{Z}_2 = \{1, a\}$, $\widehat{\mathbb{Z}}_2 = \{1, \alpha\}$. So $X = X' \cup aX'$ and $G = \mathbb{Z}_2 \times G' = G' \cup aG'$. We write $C = C' \cup \alpha C''$ and $D = D' \cup \alpha D''$. Now, an element $g = (1, g')$, $g' \in G'$, separates C and D if and only if g' separates $C' \cup C''$ and $D' \cup D''$. Also, an element $g = (a, g')$, $g' \in G'$ separates C and D if and only if g' separates $C' \cup D''$ and $D' \cup C''$. Now assume that no $g' \in G'$ separates $C' \cup C''$ and $D' \cup D''$. Then there exists a subspace $Y' \subset X'$ such that no $g' \in G'$ separates $(C' \cup C'') \cap Y'$ and $(D' \cup D'') \cap Y'$, and $cl(Y') \leq 2^{s'-1}$. Take any subspace Z' of X' with $cl(Z') \leq 2^{s'-1}$. It follows directly from the definitions that the subspace U' generated by $Y' \cup Z'$ in X' has chain length $\leq cl(Y') + cl(Z') \leq 2^{s'} = 2^{s-1}$. Now, the extensions of Y', Z' and U' by $\mathbb{Z}_2$ are subspaces Y, Z and U of X with $cl(U) = 2^{s-1}$. So the assumption holds for U, which means that there is $h \in G$ that separates $C \cap U$ and $D \cap U$. Since $Y \subset U$ and $Y' \subset U'$ we see that h is of the form (α, h') with $h' \in G'$, but since also $Z \subset U$ and $Z' \subset U'$, we see that h' separates $(C' \cup D'') \cap Z'$ and $(D' \cup C'') \cap Z'$. This holds for all Z'. Hence, by induction, we find $g' \in G'$ that separates $C' \cup D''$ and $D' \cup C''$. Now $g = (\alpha, g')$ separates C and D. $\square$

Corollary 7.13   *Let C, D be disjoint basic sets in X. If $s(X) \leq 2$ then there exists $g \in G$ that separates C and D.*

Proof. By the theorem and Propositions III.3.11 and III.3.8 $c)$, we may assume that X is a 4-element fan and then the assertion is obvious. $\square$

Examples 7.14   *a)* The separation of basic sets is no longer possible if $s(X) > 2$. Consider, for instance, the space $(X', G') = 3 \times E$ and $(X, G) = (X', G')[\mathbb{Z}_2 \times \mathbb{Z}_2]$. Then $s(X) = 3$ and the sets C, D whose elements are indicated by $\bigcirc$ and $\otimes$ respectively, are basic, but they cannot be separated, as one checks by inspection.

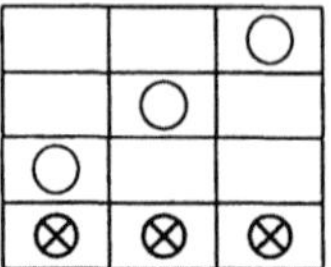

b) It is very likely that the bound $cl(Y) = 2^{s-1}$ in Theorem 7.12 is sharp. This is clear for $s \le 2$ and for $s = 3$ consider the space $(X, G) = ((E + E)[\mathbb{Z}_2] + (E + E)[\mathbb{Z}_2])[\mathbb{Z}_2]$ and the subsets C and D as indicated below

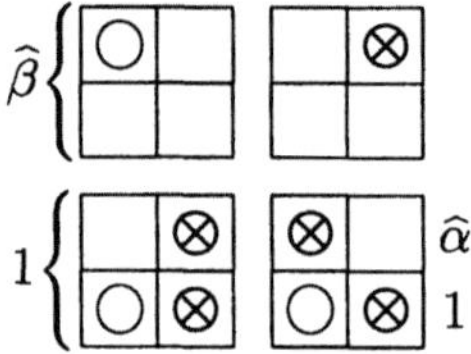

Here C and D cannot be separated but their intersections with any subspace of chain length 3 can be separated. $\square$

Notes

The ideas and results of this chapter are mainly due to Marshall, who introduced abstract spaces of orderings and studied them in five important papers ([Mr1-5]). However, in the special situation of spaces of orderings of fields, most of the constructions and important results had been known long before. They were discovered on the way of studying quadratic forms and Witt rings ([Wt2]) over formally real fields, a theory that became very appealing after Pfister's fundamental result ([Pf], [Le-Lo]). An excellent presentation of this theory can be read in Lam's book [Lm1].

So, let us first make a few comments on what was known in the field situation. Of course, the notion of subspace is classical, and corresponds to the notion of precone ([Se]). As was already mentioned, fans were introduced by Becker and Köpping. The basic constructions, their properties, and spaces of finite type, appear implicitly in [Br3] and more explicitly in [Cr] (see also [Mz]). The theory of extensions of spaces of orderings is rooted to the Baer-Krull theorem ([Ba], [Kr]) and Springer's notion of residue forms ([Sp]). The paper [Br3] contains also the main result of Section 5 in the case of finite spaces of orderings, whose proof is much shorter if one works over fields. The same methods apply to the general case of finite chain length, which corresponds to a field that admits only finitely many real places (see also [Bw-Mr]). Theorem 6.4 comes from [Be-Br]. There, it is deduced from another local-global principle of [Br1] that had been showed by methods due to Prestel ([Pr1]). There is a different proof using the chain length as in the abstract approach, thus reducing the problem to the case of a field with finitely many real places, where the methods of [Br3] can be applied (see [Bw-Mr]). The representation theorem is also due to Becker and Bröcker ([Be-Br]) after it was conjectured by Brown ([Bw]). The first definition of the stability index s of a field appeared in [Br1] as the $\mathbb{F}_2$-dimension of the cokernel of the total signature homomorphism. That this stability index has also the meaning we use here was pointed out by a letter of Knebusch. The stability formula (Corollary 7.4) appeared in [Br2]. Finally, Theorem 7.11 is due to Schwartz [Schw].

Now let us return to the situation of abstract spaces of orderings. Compared with Marshall's original papers, we have made several changes and suplements. Of course, the first three sections are more or less standard. The notion of chain length in Section 4 is one of Marshall's major inventions in order to make the theory work. For the proof of Proposition 4.1 we follow closely the exposition in [Lm2], which uses ideas of Leep. The notions of solid and impervious fans are new; we hope that they clarify a little bit the content of [Mr4]. Section 5 is the only one where we borrow Marshall's presentation [Mr2,4]. The representation theorem can also be shown without going through the more involved local-global principles [Mr3], but these are needed anyway, as for the generation formula (Theorem 7.3). This latter was easily derived from [Mr4] in [Br6], and then used for the first geometric applications. The connection between stability and the other invariants like w, l and t, was discovered in [Br7], where one finds the estimates of Corollaries 7.8 and 7.9 a). That of Corollary 7.9 b) is due to Marshall, while in [Br7] is a bound for disjoint unions, which is even worse. Theorem 7.12 is taken from [Br8], but the remarkable bound for the chain-length is new.

Chapter V. The Main Results

Summary. In an arbitrary space of signs X, generation of basic sets, stability indices, representation of functions by signatures and separation of closed sets reduce to the corresponding problems in the spaces of orderings V^* associated to the subvarieties V of X. This general principle is proved in Sections 1, 2 and 3. Also, in Sections 1 and 2, we obtain criteria for a set to be basic open or to be principal open, and extend to X the inequalities among the invariants s, $\bar{s}$, t, $\bar{t}$, w and l, which were already known for spaces of orderings. Section 4 is devoted to the notions of real divisor and regularity in X. Using them we can bound from below s and $\bar{s}$. Moreover, we compare basicness of an open set and its closure, resp. of a closed set and its interior. Finally, Artin-Lang spaces are introduced in Section 5, jointly with the tilde operator: this is the notion that makes the abstract theory fruitful of geometric applications.

1. Stability Formulae

Let (X, G) be a space of signs. We are going to prove a generation formula, which splits into two parts. Firstly, we consider local spaces of signs, and then we come to the global situation. To start with we show:

Proposition 1.1   *Let (X, G) be local and let F be a non trivial fan in the space of orderings $(X_{\max}, G^*)$ associated to (X, G). Then there exists a subvariety $W \subset X$ such that $\mathrm{Adh}_Z(x) = W$ for all $x \in F$. If, moreover, F is finite, then $(G|F) \setminus \{0\} = G^*|F$, and F is a fan in (X, G).*

The situation here is different from that in Remark III.3.13, since $(X_{\max}, G^*)$ need not be a subspace of (X, G) and so it is not clear whether F is a fan in X.

Proof. Let $x_1, x_2 \in F$ and assume that $\mathrm{Adh}_Z(x_1) \neq \mathrm{Adh}_Z(x_2)$. We can choose $x_3, x_4 \in F$ such that $F' = \{x_1, x_2, x_3, x_4\}$ is a four element fan. Let V be the minimal non–empty Z-closed set of X. There are two cases:

Case 1: $\mathrm{Adh}_Z(x_i) = V$ for some $i = 1, \ldots, 4$. By Proposition III.1.16 we find $g \in G$ with $g(x_1) = g(x_2) = g(x_3) = 1$ and $g(x_4) = -1$. Also $g \in G^*$. Contradiction.

Case 2: $\mathrm{Adh}_Z(x_i) \neq V$ for $i = 1, \ldots, 4$. We choose $x_5 \in X$ with $\mathrm{Adh}_Z(x_5) = V$. Then $x_5 \in X_{\max}$ and again no more than three of $x_1, \ldots, x_5$ have the same support. Thus we find $g \in G$ with $g(x_1) = g(x_2) = g(x_3) = g(x_5) = 1$ and $g(x_4) = -1$. So $g \in G^*$. Contradiction.

Now assume that F is finite. First let $\mathrm{Adh}_Z(x) = V$ for all $x \in F$. If $g \in G$ and $g|F \neq 0$ then automatically $g \in G^*$. So assume that $\mathrm{Adh}_Z(x) = W \supsetneq V$ for all $x \in F$ and let $g \in G$ with $g|F \neq 0$. Using **PE** and **HL** as in the proof of Proposition III.1.16, we find $g' \in G$ such that $g|F = g'|F$ and $g'|V \neq 0$, which means $g' \in G^*$. $\square$

Corollary 1.2 (Local Generation Formula)   *Let (X, G) be a local space of signs and let $C \subset X$ be open and closed. If for all finite fans F in X one has*

$$\#(F) \equiv 0 \bmod \#(C \cap F)$$

and

$$2^k \#(C \cap F) \equiv 0 \bmod \#(F),$$

then there are $f_1, \ldots, f_k \in G^*$ such that $C = \{f_1 > 0, \ldots, f_k > 0\}$.

Proof.   By Proposition III.1.3 *c)*, the set C is contructible and also (see Remark III.4.6) $C \cap X_{\max}$ is constructible in $(X_{\max}, G^*)$. So by the generation formula for spaces of orderings (Theorem IV.7.3) and the assumption we find units $g_1, \ldots, g_k \in G^*$ such that $C \cap X_{\max} = X_{\max} \cap \{g_1 > 0, \ldots, g_k > 0\}$. Let $C' = \{g_1 > 0, \ldots, g_k > 0\} \subset X$. We claim that $C = C'$. In fact, also C' is open and closed since the g_i's are units. Hence $L = (C \cup C') \setminus (C \cap C')$ is open and closed too. Let $x \in L$ with unique closed specialization $\bar{x}$. We get

$$\bar{x} \in ((C \cap X_{\max}) \cup (C' \cap X_{\max})) \setminus ((C \cap X_{\max}) \cap (C' \cap X_{\max})) = \emptyset \, .$$

Contradiction. $\square$

Another way to state this result is:

Corollary 1.3   *Let (X, G) be a local space of signs and let $C \subset X$ be open and closed. If for all subvarieties $W \subset X$ there are $g_1, \ldots, g_k \in G$ (depending on W) such that $C \cap W^* = \{g_1 > 0, \ldots, g_k > 0\} \cap W^*$ then there are $f_1, \ldots, f_k \in G$ such that $C = \{f_1 > 0, \ldots, f_k > 0\}$.*

Proof.   This follows easily from the preceding corollary using Remarks III.3.13.
                                                                      $\square$

Now we turn to the global situation:

Theorem 1.4 (Global Generation Formula)   *Let (X, G) be a space of signs and let $C \subset X$ be constructible such that $C \cap \mathrm{Adh}_Z(\mathrm{Bd}(C)) = \emptyset$. If for all finite fans in X one has*

$$\#(F) \equiv 0 \bmod \#(C \cap F)$$

and

$$2^k \#(C \cap F) \equiv 0 \bmod \#(F),$$

then there are $g_1, \ldots, g_k \in G$ such that $C = \{g_1 > 0, \ldots, g_k > 0\}$.

From this we get:

Corollary 1.5   *Let $C \subset X$ be constructible such that $\mathrm{Adh}_Z(\mathrm{Bd}(C)) \cap C = \emptyset$. Assume that for every subvariety $V \subset X$ there exist $g_1, \ldots, g_k \in G$ (depending on V) such that $C \cap V^* = \{g_1 > 0, \ldots, g_k > 0\} \cap V^*$. Then there exist $f_1, \ldots, f_k \in G$ such that $C = \{f_1 > 0, \ldots, f_k > 0\}$.*

Proof of Theorem 1.4 and Corollary 1.5. Note that, using the generation formula for spaces of orderings and Remark III.3.13, the theorem in turn is a consequence of its corollary. So we show the latter without using the former. First, we may assume that $\mathrm{Bd}(C) = \emptyset$. In fact, by assumption we have

$$C \subset \bigcup_{\substack{g \in G \\ \mathrm{Bd}(C) \subset \{g=0\}}} \{g \neq 0\}.$$

By compactness of C in the constructible topology and **PE** we find $h \in G$ such that $C \subset \{h \neq 0\}$ and $\mathrm{Bd}(C) \subset \{h = 0\}$. Then C has no boundary in the subspace $X' = \{h \neq 0\}$, where we can work instead of X, and multiply the f_i's by h^2 at the end. Thus, we may assume that C is open and closed in the Harrison topology,

Next we proceed by induction on k. It will be enough to find $f \in G$ such that for every V there exist $g_2, \ldots, g_k \in G$ with $C \cap V^* = \{f > 0, g_2 > 0, \ldots, g_k > 0\} \cap V^*$. Then we can switch to the subspace $\{f > 0\}$, where we are done by induction.

Now consider a Z-open constructible set $U \subset X$. By **PE**, $U = \{h \neq 0\}$ for some $h \in G$; in particular, U is a subspace. Such a subspace U will be called *good* if the claim above is true on U, that means, if we find f with the required property on U. We also say that (U, f) is good. By compactness, it suffices to prove the following statements:

a) For every $x \in X$ there exists $U \subset X$, U good, with $x \in U$.
b) If $U_1, U_2 \subset X$ are good, then so is $U_1 \cup U_2$.

To prove a), let $x \in X$ and $V = \mathrm{Adh}_Z(x)$. By Corollary 1.3, we find $f_1, \ldots, f_k \in G$ such that

$$C \cap X_V = \{f_1 > 0, \ldots, f_k > 0\} \cap X_V.$$

Let

$$D = (C \setminus \{f_1 > 0, \ldots, f_k > 0\}) \cup (\{f_1 > 0, \ldots, f_k > 0\} \setminus C).$$

Using the compactness of D in the constructible topology and **PE**, we find $h \in G$ with $D \subset \{h = 0\}$ and $h(x) \neq 0$. So $U = \{h \neq 0\}$ does the job for x.

Now we prove b). Let $U_1, U_2 \subset X$ and $f, g \in G$ be given such that (U_1, f) and (U_2, g) are good. We may assume that $X = U_1 \cup U_2$ and $f = 0$ on $X \setminus U_1$. Applying **HL** to $X \setminus U_2$, f and g we get f' with the following properties:

i) $f'(x) = f(x)$ if $x \notin U_2$, $f'(x) = g(x)$ if $x \notin U_1$.
ii) $\langle f, g \rangle = \langle f', f'fg \rangle$.

We claim that (X, f') is good. so let $V \subset X$ be a subvariety. If $V \subset X \backslash U_2 \subset U_1$ or $V \subset X \setminus U_1 \subset U_2$, there is nothing to show. If $V \not\subset X \setminus U_i$, then $V^* \subset U_i$. Hence, we can suppose $V^* \subset U_1 \cap U_2$. Then both f and g are > 0 on $V^* \cap C$, and since $\langle f, g \rangle = \langle f', f'fg \rangle$, necessarily f' is never zero in $V^* \cap C$. In particular, $f, g, f' \in G(V^*)$. On V^* we have

$$C \cap V^* = \{f > 0, g_2 > 0, \ldots, g_k > 0\} \cap V^* = \{g > 0, h_2 > 0, \ldots, h_k > 0\} \cap V^*$$

for suitable $g_i, h_i \in G$. Also, $\langle f, g \rangle = \langle f', f'fg \rangle$, which means that $f' \in D(\langle f, g \rangle)$ in $(V^*, G(V^*))$. The claim follows from Corollary III.2.8. $\square$

Corollary 1.6 (Global Stability Formulae)   *Let (X, G) be a space of signs. Then*

$$
\begin{aligned}
s(X) \;&=\; \sup\{s(V^*) \,|\, V \subset X \text{ subvariety}\} \;=\; \\
&=\; \sup\{s \,|\, \text{there is a finite fan } F \text{ in } X \text{ with } \#(F) = 2^s\},
\end{aligned}
$$

unless the sup's are zero. In that case, $s(X) \leq 1$.

Corollary 1.7   *Let (X, G) be a space of signs. Then: $s(X) \leq t(X)$, $\bar{s}(X)$, $\bar{t}(X)$; $s(X) = w(X)$ and $2^{s(X)} \leq l(X)$.*

Proof.   Clear from the preceding corollary, since the formulae hold for fans (Proposition III.3.8 c)). Note also that $t = \bar{t}$ and $s = \bar{s}$ for spaces of orderings. $\square$

Corollary 1.8   *Let (X, G) be a space of signs such that $s(X) < \infty$. Then $C \subset X$ is basic open if and only if the following conditions a), b) and one of c) or d) hold true:*

a) C is constructible,
b) $C \cap \mathrm{Adh}_Z(\mathrm{Bd}(C)) = \emptyset$,
c) $C \cap V^*$ is basic in V^* for all subvarieties $V \subset X$,
d) $\#(C \cap F) \neq 3$ for all 4-element fans F in X.

Proof.   For condition c) this follows immediately from Corollary 1.5, and that c) can be replaced by d) follows from Theorem IV.7.2 b). $\square$

   Similarly, we get:

Corollary 1.9   *Let (X, G) be a space of signs. Then $C \subset X$ is principal open (resp. principal closed) if and only if following conditions a), b) and one of c) or d) hold true:*

a) C *is constructible,*

b) $C \cap \mathrm{Adh}_Z(\mathrm{Bd}(C)) = \emptyset$ *(resp.* $\mathrm{Adh}_Z(\mathrm{Bd}(C)) \subset C$*),*

c) $C \cap V^*$ *is principal in* V^* *for all subvarieties* $V \subset X$,

d) $\#(C \cap F) \equiv 0 \bmod 2$ *for all 4-element fans* F *in* X.

Proof. The statement for principal open sets is proved as the preceding one, and the statement for principal closed sets follows by complementation. □

Corollary 1.10   *Let* (X, G) *be a space of signs with* $s(X) \leq 1$. *Then every constructible set* $C \subset X$ *such that* $C \cap \mathrm{Adh}_Z(\mathrm{Bd}(C)) = \emptyset$ *is principal open.*

2. Complexity of Constructible Sets

Let (X, G) be a real space. As usual, we say that X is *noetherian* if every descending chain of Z-closed sets is finite. Recall also that if X is noetherian, every Z-closed set is a union of finitely many maximal subvarieties, called *irreducible components*. If (X, G) is a prespace of signs, X is noetherian if and only if every Z-closed set is constructible (by compactness). If, moreover, **PE** holds for (X, G), that happens if and only if every Z-closed set is principal.

From now on, we fix a noetherian space of signs (X, G). As we know, every form ρ over X defines the signature map $\widehat{\rho} : X \to \mathbb{Z}$; $x \mapsto \rho(x)$, which is continuous with respect to the constructible topology. As for spaces of orderings, we have:

Theorem 2.1 (Representation Theorem)   *For a map* $f : X \to \mathbb{Z}$ *which is continuous with respect to the constructible topology on* X *the following conditions are equivalent:*

a) *There is a form* ρ *over* X *such that* $f = \widehat{\rho}$

b) *For each finite fan* F *in* X *there is a form* ρ *over* F *such that* $f|F = \widehat{\rho}|F$

c) $\sum_{x \in F} f(x) \equiv 0 \bmod \#(F)$ *for each finite fan* F *in* X.

Proof.   When (X, G) is a space of orderings this is Theorem IV.7.1. Then, we also get *a)* $\Rightarrow$ *b)* $\Leftrightarrow$ *c)*. For the implication *b)* $\Rightarrow$ *a)* consider a minimal Z-closed subset $W \subset X$ such that there exists a form ρ over X with $f(x) = \rho(x)$ for all $x \in X \setminus W$. We claim that $W = \emptyset$. Otherwise, we may assume that $\rho(x) = 0$ for all $x \in W$. Also consider an irreducible component V of W and a positive equation p of V. Applying Theorem IV.7.1 to the space of orderings $(V^*, G(V^*))$, and using the trick of Example III.1.20, we find a form τ over X and a Z-open set $U \subset V$ such that $f(x) = \tau(x)$ for all $x \in U$. But then the form $\rho' = \rho + \langle 1, -p \rangle \otimes \tau$ leads to a contradiction to the minimality of W. □

Corollary 2.2   *Let* $C \subset X$ *be basic open, say* $C = \{g_1 > 0, \ldots, g_k > 0\}$, *where* k *is minimal. Then* $w(C) = k$.

Proof. Let ρ be the defining form of C and let $f : X \to \mathbb{Z}$ be the characteristic function of C. Consider also the Pfister form $\phi = \langle\!\langle g_1, \ldots, g_k \rangle\!\rangle$. Then $\hat\phi = 2^k f$ and $\hat\rho = 2^w f$ with $w = w(C) \leq k$. By the global generation formula (Theorem 1.4), we find a finite fan F in X such that $\#(F \cap C) = 2^{-k}\#(F)$. So from the representation theorem we get $w = k$. $\qquad\square$

In Section 1, we have bounded in different ways $s(X)$ by the other invariants $t(X)$, $\bar{s}(X)$, $\bar{t}(X)$, $w(X)$ and $l(X)$, so showing that if any of the latter is finite, so is $s(X)$. Now we shall find bounds for all these invariants depending only on $s(X)$.

Notations 2.3   As usual, the maximal length d of a chain of subvarieties of X

$$X \supset V_d \supsetneq V_{d-1} \supsetneq \cdots \supsetneq V_0$$

is called the *dimension* of X and denoted by $\dim(X)$. Thus, if the chain above is maximal, $\dim(V_i) = i$.

For every i we set

$$s_i = s_i(X) = \sup\{1,\ s(V^*) \,|\, V \text{ is a subvariety of dimension } i\}$$

Lemma 2.4   Let $\dim(X) = d < \infty$ and let $C \subset X$ be constructible. Then *there exist a form ρ_d over X and a Z-closed subset $W \subset X$ such that:*

a) $\dim(\rho_d) = m \leq 2^{w(C)-1}(1 + 2^{s_d - 1})$.
b) $\dim(W) < d$.
c) $\rho_d(x) = 2^{w(C)}$ *for* $x \in C \setminus W$ *and* $\rho_d(x) = 0$ *for* $x \in X \setminus (C \cup W)$.

Proof.   When (X, G) is a space of orderings, ρ_d is the defining form of C (Proposition IV.7.7). Now, let $V_1, \ldots, V_r$ be the irreducible components of X of dimension d. For each V_i we get the result by consideration of the space of orderings $(V_i^*, G(V_i^*))$. So we find a form $\rho_{di} = \langle a_{i1}, \ldots, a_{im_i} \rangle$ doing the job on V_i up to a closed subset $W_i \subset V_i$ with $\dim(W_i) < d$. Then we replace ρ_{di} by $\rho_{di} + (m - m_i) \times \langle p_i \rangle$, where m be the maximum of the m_i's and p_i is a positive equation of V_i. In other words, we assume $m = m_1 = \cdots = m_r$.

Next, we can find $c_j \in G$, $j = 1, \ldots, m$ such that $c_j = a_{ij}$ on $V_j \backslash W$ for $i = 1, \ldots, r$, where W is Z-closed and $\dim(W) < d$. We show how for $r = 2$. We pick a positive equation p (resp. p') of

$$V_2 \cup W_1 \cup W_2 \cup \{a_{i1} = \cdots = a_{im} = 0\}$$

$$(\text{resp. } V_1 \cup V_2 \cup W_1 \cup W_2 \cup \{a_{i1} = \cdots = a_{im} = 0\})$$

and replace a_{1j} by pa_{1j}, a_{2j} by pa_{2j}. Then we apply **HL** to V_1, $f = pa_{1j}$ and $g = p'a_{2j}$.

Finally, $\rho_d = \langle c_1, \ldots, c_m \rangle$ does the job. $\qquad\square$

Theorem 2.5   *Assume that* $\dim(X) = d < \infty$ *and* $1 \le s = s(X) < \infty$, *and write*

$$\lambda(X) = 2^{s_d-1} + \sum_{i=0}^{d-1} 2^{s_i} + 2d + 1.$$

Then, for any constructible set $C \subset X$,

$$l(C) \le 2^{w(C)-1}\lambda(X).$$

In particular, $l(X) \le 2^{s-1}\lambda(X) \le (2d+1)(2^{s-1} + 4^{s-1})$.

For instance, if $s_i \le i$ for all i, we get the estimates:

d	0	1	2	3	4	$\cdots$
$l(X) \le$	1	5	20	72	256	$\cdots$

Proof.   First, we get the form ρ_d and the Z-closed subset $W \subset X$ of Lemma 2.4, with $\dim(W) = e < d$. Replacing X by W and C by $C \cap W$ we do the same, thus getting a form ρ_e and a Z-closed subset $W' \subset W$ with $\dim(W') < e$. Now we patch these forms together, alike we did in the proof of Theorem 2.1. For this, let p be a positive equation for W and assume that ρ_d vanishes on W. Then for $\pi = \rho_d + \langle 1, -p \rangle \otimes \rho_e$ we get

$$\pi(x) = \begin{cases} 2^{w(C)} & \text{for} \quad x \in X \setminus W' \\ 0 & \text{for} \quad x \in X \setminus (C \cup W') \end{cases}$$

and

$$\dim(\pi) \le 2^{w(C)-1}(1 + 2^{s_d-1} + 2(1 + 2^{s_e-1})).$$

Continuing this procedure we get the required estimate. $\square$

Remark 2.6   It seems that there is no bound for $l(C)$ which only depends on $w = w(C)$ and $d = \dim(X)$. Only for $w \le 1$ we have such a bound, namely

a) If $w(C) = 0$, then $l(C) \le 2d - 1$.
b) If $w(C) = 1$, then $l(C) \le 2d + 1$.

Indeed, the argument runs as in the proof above once we note, for any subvariety V of X: a) if $w(C) = 0$, then $C \cap V^*$ is either $\emptyset$ or V^*, hence $C \cap V^*$ can be defined either by $\langle -1 \rangle$ or $\langle 1 \rangle$, and b) if $w(C) = 1$, then either $C \cap V^* = \emptyset$ or $C \cap V^* = V^*$ or $w(C \cap V^*) = 1$, in which case $l(C \cap V^*) = 2$ (see the proof of Proposition III.3.1). $\square$

Let us return to more special constructible sets.

Lemma 2.7   *Let* $\dim(X) = d < \infty$ *and let* $C \subset X$ *be basic closed. Then there is a Z-closed subset $W \subset X$ and* $a_1, \ldots, a_m \in G$ *such that* *a)* $m \le s_d$.

b) $\dim(W) < d$.

c) $C \setminus W = \{a_1 \ge 0, \ldots, a_m \ge 0\} \setminus W$.

Proof. Let V_i, $1 \le i \le r$, be the irreducible components of X of dimension d. The statement is clear for each space of orderings $(V_i^*, G(V_i^*))$ and thus for V_i itself. So

$$(C \cap V_i) \setminus W_i = \{a_{i1} \ge 0, \ldots, a_{im} \ge 0\} \cap V_i \setminus W_i$$

for suitable $a_{ij} \in G$ and a Z-closed subset $W_i \subset V_i$. As in the proof of Lemma 2.4, using **HL** we find $a_j \in G$, $j = 1, \ldots, m$, such that $a_j = a_{ij}$ on $V_i \setminus W$ where W is Z-closed in X and $\dim(W) < d$. $\square$

Lemma 2.8 (Pasting Lemma)   *Let $C \subset X$ be closed constructible and let $W \subset X$ be Z-closed. If*

$$C \setminus W = \{a_1 \ge 0, \ldots, a_k \ge 0\} \setminus W \quad and \quad C \cap W = \{b_1 \ge 0, \ldots, b_l \ge 0\} \cap W$$

for suitable $a_i, b_i \in G$, then there exist $c_1, \ldots, c_m \in G$, $m \le k + l$, such that

$$C = \{c_1 \ge 0, \ldots, c_m \ge 0\}.$$

Proof. After multiplying the a_i's by a positive equation p of W we may assume that $W \subset \{a_1 \ge 0, \ldots, a_k \ge 0\}$. For each $i = 1, \ldots, l$, let $C_i = C \cap \{b_i \le 0\}$. Then, by **HL**, for each $i = 1, \ldots, l$, we find $c_{k+i} \in G$ such that $c_{k+i} = p$ on C_i, $c_{k+i} = b_i$ on W and $c_{k+i} > 0$ on $C \setminus C_i$. Setting $c_i = a_i$ for $i = 1, \ldots, k$, we get $C = \{c_1 \ge 0, \ldots, c_m \ge 0\}$. $\square$

From the preceding lemmas we get immediately

Theorem 2.9   *Let $\dim(X) = d < \infty$. Then every basic closed set $C \subset X$ can be written as*

$$C = \{a_1 \ge 0, \ldots, a_m \ge 0\},$$

where $m \le \sum_{i=0}^{d} s_i$. In other words, $\bar{s}(X) \le \sum_{i=0}^{d} s_i$.

Proposition 2.10   *Let $\dim(X) = 1$ and $s(V^*) \le 1$ for all subvarieties $V \subset X$. Then every closed constructible set $C \subset X$ is principal closed.*

Similarly, we get:

Theorem 2.11   *Let $\dim(X) = d < \infty$. Then, $C \subset X$ is basic closed if and only if the following conditions a), b) and one of c) or d) hold true:*

a) C *is constructible.*

b) C *is closed.*

c) $C \cap V^*$ *is basic in V^* for every subvariety $V \subset X$.*

d) $\#(C \cap F) \ne 3$ *for all 4-element fans F in X.*

For the sequel, we recall the combinatorial function

$$\tau(s) = s \text{ for } 0 \leq s \leq 2, \quad \tau(s) = \binom{4^{s-1} - 2^{s-1} + 1}{2 \cdot 4^{s-2} - 2^{s-2} + 1} \text{ for } s \geq 3$$

introduced in Corollary IV.7.9.

Lemma 2.12   *Let* $\dim(X) = d < \infty$ *and let* $C \subset X$ *be constructible. Then there are basic open sets* $C_1, \ldots, C_m \subset X$ *and a* Z*-closed subset* $W \subset X$ *such that:*

a) $m \leq \tau(s_d)$.
b) $\dim(W) < d$.
c) $C \setminus W = (C_1 \cup \cdots \cup C_m) \setminus W$.

Proof.   For spaces of orderings this is Corollary IV.7.9, and for arbitrary X we argue as in Lemma 2.7. $\Box$

Lemma 2.13 (Pasting Lemma)   *Let* $C \subset X$ *be open constructible and let* $C_1, \ldots, C_m \subset X$ *be basic open such that* $C \setminus W = (C_1 \cup \cdots \cup C_m) \setminus W$ *where* $W \subset X$ *is* Z*-closed. Assume further that* $C \cap W = (C_{m+1} \cup \cdots \cup C_l) \cap W$ *for basic open sets* $C_i \subset X$, $i = m+1, \ldots, l$. *Then* $C = C_1' \cup \cdots \cup C_l'$ *for suitable basic open sets* $C_i' \subset X$.

Proof.   For $i = 1, \ldots, m$, let $C_i' = C_i \setminus W$ which is again basic open. So $C_1' \cup \cdots \cup C_m' = C \setminus W$. For $i = m+1, \ldots, l$, write $C_i = \{g_{i1} > 0, \ldots, g_{ik} > 0\}$, assuming that g_{ij} vanishes on $\mathrm{Adh}_Z(\mathrm{Bd}(C_i))$. Set $D_i = (C_i \cup \mathrm{Adh}_Z(\mathrm{Bd}(C_i))) \setminus C$ and f a positive equation for W. By **HL** we find $g_{ij}' \in G$ with $g_{ij}' = g_{ij}$ on W, $g_{ij}' = -f$ on D_i and $\langle g_{ij}', -g_{ij}'g_{ij}f \rangle = \langle g_{ij}, -f \rangle$. Finally, take $C_i' = C_i$ for $i = 1, \ldots, m$ and $C_i' = \{g_{i1}' > 0, \ldots, g_{ik}' > 0\}$ for $i = l+1, \ldots, m$. $\Box$

From these lemmas we immediately deduce:

Theorem 2.14   *Assume that* $\dim(X) = d < \infty$ *and* $1 \leq s = s(X) < \infty$. *Then* $t(X) \leq \sum_{i=0}^{d} \tau(s_i) \leq (d+1)\tau(s)$, *and, by complementation,* $\bar{t}(X) \leq s^{(d+1)\tau(s)}$.

From all the bounds obtained in this and the preceding section, we deduce:

Corollary 2.15   *For a noetherian space of signs of finite dimension, all the invariants* s, $\bar{s}$, t, $\bar{t}$, w *and* l *are finite if and only if any of them is finite.*

The bound for $\bar{t}$ is even worse than that for t. Only in a very special case we know more:

Proposition 2.16   *Let* $\dim(X) \leq 2$ *and* $s_i \leq i$ *for* $i = 0, 1, 2$. *Then* $\bar{t}(X) \leq 2$.

Proof.   Let $C \subset X$ be constructible and closed. By the global stability formula (Corollary 1.6) and Corollary IV.7.8, $t(V^*) \leq 2$ for all subvarieties V

of X. Hence, we find closed basic sets $C_1, C_2 \subset X$ such that $(C \setminus (C_1 \cup C_2)) \cup ((C_1 \cup C_2) \setminus C) \subset W$, where W is a Z-closed subset of X of dimension ≤ 1. Let $C_i' = (C_i \cup W) \cap C$ for $i = 1, 2$. Then, by Proposition 2.10 and Theorem 2.11, C_i is basic closed, and clearly $C = C_1' \cup C_2'$. $\qquad\square$

Finally, using the same method as in the preceding results, we obtain:

Theorem 2.17   *Let* $\dim(X) = d < \infty$ *and* $1 \leq s = s(X) < \infty$. *Then every constructible set* $C \subset X$ *can be written as*

$$C = \bigcup_{j=1}^{r} \{f_{j1} > 0, \ldots, f_{jk_j} > 0, g_j = 0\},$$

with

$$\text{number of } g_j\text{'s} \leq \textstyle\sum_{i=0}^{d} \tau(s_i) \leq (d+1)\tau(s),$$

$$\text{number of } f_{jk}\text{'s} \leq \textstyle\sum_{i=0}^{d} s_i \tau(s_i) \leq (d+1)s\tau(s).$$

3. Separation

Let (X, G) be a real space. The question of separating subsets of X by elements of G is fundamental in real geometry. To be more precise we introduce the following terminology:

Definition 3.1   *Given two constructible sets* $C, D \subset X$ *and an element* $g \in G$, *we say that* g *separates* C *and* D, *and write* $C \mid_g D$, *if* $g(x) \geq 0$ *for* $x \in C$, $g(x) \leq 0$ *for* $x \in D$ *and* $(C \cup D) \cap \{g = 0\} = (C \cup D) \cap \mathrm{Adh}_Z(C \cap D)$. *If we do not need to specify the element* g, *we just write* $C \mid D$.

Note that if g is ≥ 0 on C and ≤ 0 on D, then g vanishes on $C \cap D$, and consequently, also on $\mathrm{Adh}_Z(C \cap D)$. Hence, the definition asks for the best possible separation. We also see that separation has no meaning if $\mathrm{Adh}_Z(C \cap D) = X$. In case $C \cap D = \emptyset$, $C \mid_g D$ reduces to $g(x) > 0$ for $x \in C$ and $g(x) < 0$ for $x \in D$. This is the only interesting case if G is a group, since then the Zariski topology is trivial.

We henceforth fix a noetherian space of signs (X, G) of finite dimension.

Theorem 3.2   *Let* $C, D \subset X$ *be closed and constructible such that* $\mathrm{Adh}_Z(C \cap D) \neq X$. *Then the following conditions are equivalent:*

    a) $C \mid D$

    b) $C \cap V^* \mid D \cap V^*$ *for all subvarieties* $V \not\subset \mathrm{Adh}_Z(C \cap D)$.

    c) $C \cap Y \mid D \cap Y$ *for all finite subspaces* Y *of the spaces of orderings* $(V^*, G(V^*))$ *associated to all subvarieties* $V \not\subset \mathrm{Adh}_Z(C \cap D)$ *with* $cl(Y) \leq 2^{s(V^*)-1}$.

    d) $C \cap Y \mid D \cap Y$ *for all finite subspaces* $Y \not\subset \mathrm{Adh}_Z(C \cap D)$ *which are spaces of orderings with* $cl(Y) \leq 2^{s(X)-1}$.

Proof.   First we note that $C \cap D \cap V^* = \emptyset$ if $V \not\subset \mathrm{Adh}_Z(C \cap D)$. Now, that *a)* implies all the other conditions is obvious. The equivalence of *b)* and *c)* is Theorem IV.7.12. Then we have *d)* $\Rightarrow$ *c)* by the global stability formula (Corollary 1.6). Finally, we prove *b)* $\Rightarrow$ *a)*. Let $V_1, \ldots, V_r$ be the irreducible components of maximal dimension of X. For the V_i's which are not contained in $\mathrm{Adh}_Z(C \cap D)$ (there must be some) we find an element $f_i \in G$ such that $C \cap V_i^* \subset \{f_i > 0\}$ and $D \cap V_i^* \subset \{f_i < 0\}$. As usual, by compactness and the irreducibility of V_i we find a Z-closed set $W \subset X$ such that $C \cap V_i \setminus W \subset \{f > 0\}$ and $D \cap V_i \setminus W \subset \{f < 0\}$. Clearly, we can suppose that W is the same for all these V_i's, and that $W \supset \mathrm{Adh}_Z(C \cap D)$. For the V_i's which are contained in $\mathrm{Adh}_Z(C \cap D)$, let f be any positive equation of V_i. Next, by means of **HL**, we can patch together these f_i's to get $f \in G$ that separates $C \setminus W$ and $D \setminus W$. If $\mathrm{Adh}_Z(C \cap D \cap W) \neq W$, by induction there is $g \in G$ such that $C \cap W \mid_g D \cap W$. If $\mathrm{Adh}_Z(C \cap D \cap W) = W$, let g be any positive equation for W. Now we apply **HL** to the closed constructible set $(C \setminus \{g > 0\}) \cup (D \setminus \{g < 0\})$, and to f, g. This gives us an $f' \in G$ which separates C and D. $\qquad\square$

Corollary 3.3   *Let $Y = \mathrm{Adh}_Z(C) \cap \mathrm{Adh}_Z(D)$, and assume that either*

i) $s(Y) \leq 1$, or
ii) $s(Y) \leq 2$ and C and D are basic closed.

Then $C \mid D$.

Proof.   Let V be a subvariety not contained in $\mathrm{Adh}_Z(C \cap D)$ such that $C \cap V^* \neq \emptyset$ and $D \cap V^* \neq \emptyset$. Then $V \subset Y$, and $s(V^*) \leq s(Y)$. Now separation follows from the theorem because: *a)* in a space of orderings with stability index ≤ 1 every constructible set is principal (Proposition III.3.1), and *b)* in a space of orderings of stability index ≤ 2 any two disjoint basic sets can be separated (Corollary IV.7.13). $\qquad\square$

4. Real Divisors

Throughout this section (X, G) will be a noetherian space of signs whose dimension is finite.

Proposition and Definition 4.1   *Set $d = \dim(X)$.*

a) Let $Y \subset X$ be a subvariety of dimension $d - 1$ such that:
  i) Every $y \in Y^*$ has exactly two generizations in X.
  ii) For any two elements $y, y' \in Y^*$ their four generizations form a fan.
  Then, all the generizations of the elements $y \in Y$ have the same Zariski closure V in X. This Zariski closure is an irreducible component of dimension d of X. The subvariety Y is called a (real) divisor of X.
b) Let $Y \subset X$ be a divisor and let $u \in G$ be such that $u(x_1)u(x_2) < 0$ for the two generizations x_1, x_2 of some point $y \in Y^*$. Then, this happens for the

two generizations of every point of Y^*; in particular, u vanishes on Y^*. The element u is called a uniformizer *of Y.*

Proof.   *a)* is an immediate consequence of Remarks III.3.13 *a)*. For *b)*, note that since the four generizations x_1, x_2, x_1', x_2' of any two given points $y, y' \in Y^*$ form a fan, we have $u(x_1)u(x_2)u(x_1')u(x_2') > 0$. $\square$

From the definition it follows:

Proposition 4.2   *Let $Y \subset X$ be a divisor and let $F \subset Y^*$ be a fan. Then*

$$F_1 = \{x \in X \mid \text{there exists } y \in F \text{ such that } x \to y\}$$

is a fan in X with $\#(F_1) = 2\#(F)$.

The next result is the key step in order to determine lower bounds for $\bar{s}(X)$.

Proposition 4.3   *Suppose that we are given the following data:*

i) A divisor $Y \subset X$ and a uniformizer $u \in G$ of Y.
ii) Elements $b_1, \ldots, b_m \in G$ such that the basic closed set

$$D = Y \cap \{b_1 \geq 0, \ldots, b_m \geq 0\},$$

cannot be written with fewer inequalities in Y.
iii) A fan $F \subset D \cap Y^*$ with $\#(F) = 2^l$.
iv) Elements $a_1, \ldots, a_l \in G$ such that $\#(\{a_1 > 0, \ldots, a_l > 0\} \cap F) = 1$.

Then the basic closed set of X

$$C = \{u \geq 0, u^2 a_1 \geq 0, \ldots, u^2 a_l \geq 0, b_1 \geq 0, \ldots, b_m \geq 0\}$$

cannot be written with fewer than $l + m + 1$ inequalities. Furthermore, there is $x \in C \cap V^*$ for some irreducible component V of X that specializes to a point $y \in F$.

Proof.   Since m is minimal, the elements b_i do not vanish on Y. For F consider the fan F_1 as in the preceding proposition. Then, since open sets are closed under generization, $F_1 \subset \{b_1 > 0, \ldots, b_m > 0\}$. Thus, by construction, $\#(F_1 \cap C) = 1$. In particular, we get the last assertion of the statement.
Now consider any description $C = \{c_1 \geq 0, \ldots, c_r \geq 0\}$. By Proposition III.3.8 *b)*, we need $l + 1$ of the c_i's, say $c_1, \ldots, c_{l+1}$, to describe $F_1 \cap C$, and none of these c_i's is positive on all F_1. On the other hand, since $F \subset D \subset C$, we have $c_i \geq 0$ on F for $i = 1, \ldots, l + 1$. Hence, $c_i = 0$ on F, which means $c_i = 0$ on Y for $i = 1, \ldots, l+1$. So, $D = C \cap Y = \{c_{l+2} \geq 0, \ldots, c_r \geq 0\}$, and since the description of D requires at least m inequalities, we conclude $r - (l + 1) \geq m$.
    $\square$

Next, we consider the relation between basic open and basic closed sets.

Proposition 4.4   *Let $C \subset X$ be basic closed. Then $C' = C \setminus \mathrm{Adh}_Z(\mathrm{Bd}(C))$ is basic open.*

Proof.   By definition, C' is constructible and $C' \cap \mathrm{Adh}_Z(\mathrm{Bd}(C')) = \emptyset$. So we only have to check condition *c)* of Corollary 1.8. For this let $V \subset X$ be a subvariety. If $V \subset \mathrm{Adh}_Z(\mathrm{Bd}(C))$ then $C' \cap V^* = \emptyset$. Otherwise, $C' \cap V^* = C \cap V^*$. So in any event condition *c)* holds. $\square$

Conversely, one may ask whether the closure of a basic open set is basic closed. First of all, in general it is not clear that the closure is constructible. But even if X is the real space associated to a real R-variety, so that we are dealing with semialgebraic sets, it may happen that the closure of a basic open semialgebraic set fails to be basic closed. We shall present examples of this in Section VI.7. Let us state here a positive result:

Proposition 4.5   *Let $C \subset X$ be basic open such that $\mathrm{Adh}(C)$ is constructible. Then $\mathrm{Adh}(C) \cap Y^*$ is basic for all divisors Y in X.*

Proof.   Let $C' = \mathrm{Adh}(C)$. Of course, $C' \cap Y^*$ is constructible in Y^*, that is, open and closed in the space of orderings $(Y^*, G(Y^*))$. Let $F \subset Y^*$ be a four element fan. According to Theorem IV.7.2 *b)*, we must show that $\#(C' \cap F) \neq 3$. But:

If for some $y \in Y^*$ its two generizations x_1, x_2 belong to C, then this holds for all $y \in Y^*$.

In fact, let $y, y' \in Y^*$ have generizations x_1, x_2 and x_1', x_2' respectively. Then $F' = \{x_1, x_2, x_1', x_2'\}$ is a 4-element fan, hence, C being basic, $\#(F' \cap C) \neq 3$.

For every $y \in C' \cap Y^*$, at least one of its two generizations x_1, x_2 belongs to C.

As $y \in \mathrm{Adh}(C)$, there exists $x \in C$ with $x \to y$. Then $x = x_1$ or $x = x_2$.

Now let F_1 be the 8-element fan consisting off all generizations of F. If $\#(F \cap Y^*) = 3$, by the two remarks above, we get $\#(F_1 \cap C) = 3$ or 6. In any case, $\#(F_1 \cap C) \not\equiv 0 \bmod \#(F_1)$, which contradicts Remarks III.3.13 *b)*. $\square$

The preceding result says that the obstruction for the closure of a basic open set to be basic cannot be supported by a divisor. For principal sets the situation is simpler. Readily from Corollary 1.9 it follows:

Proposition 4.6

   a) Let $C \subset X$ be principal closed. Then $C \setminus \mathrm{Adh}_Z(\mathrm{Bd}(C))$ is principal open.
   b) Let $C \subset X$ be principal open. Then $C \cup \mathrm{Adh}_Z(\mathrm{Bd}(C))$ is principal closed.

As a natural generalization of real divisors, we are also able to define regularity in the abstract setting:

Definition 4.7   *Let $Y \subset X$ be a subvariety. We say that Y is regular in X if there is a sequence $Y = Y_0 \subset Y_1 \subset \cdots \subset Y_m = X$ such that Y_{i-1} is a divisor of Y_i for $i = 1, \ldots, m$.*

From this Definition and Proposition 4.1, we readily deduce:

Proposition 4.8   *Let $Y \subset X$ be a regular subvariety and let $Y = Y_0 \subset Y_1 \subset \cdots \subset Y_m = X$, where Y_{i-1} is a divisor in Y_i for $i = 1, \ldots, m$. Then:*

a) *For every $y \in Y^*$ there are exactly 2^m chains $x_m \to \cdots \to x_1 \to y$, $x_i \in Y_i$ for $i = 1, \ldots, m$. All the x_m's occuring have the same Zariski closure V in X. This Zariski closure is an irreducible component of dimension d of X.*

b) *For every fan $F \subset Y^*$, the set*

$$F_1 = \{x \in X | \text{ there exist a chain } x = x_m \to \cdots \to x_1 \to y, \; x_i \in Y_i, \; y \in F\}$$

is a fan in V^* and $\#(F_1) = 2^m \#(F)$

Corollary 4.9   *With Y and V as in the preceding proposition, let $C \subset X$ be constructible such that $C \cap \mathrm{Adh}_Z(\mathrm{Bd}(C)) = \emptyset$. If*

$$C \cap V^* = \{a_1 > 0, \ldots, a_s > 0\} \cap V^*$$

for $a_1, \ldots, a_s \in G$, then there are also $b_1, \ldots, b_s \in G$ such that

$$C \cap Y^* = \{b_1 > 0, \ldots, b_s > 0\} \cap Y^*.$$

Proof.   Let $F \subset Y^*$ be a finite fan such that $C \cap F \neq \emptyset$. According to Corollary 1.8 we have to show that

$$\#(F) \equiv 0 \bmod \#(C \cap F) \quad \text{and} \quad 2^s \#(C \cap F) \equiv 0 \bmod \#(F).$$

We consider F_1 as in the proposition. Let $y \in F$ and $x \in F_1$ such that $x \to y$. Since C is open, $y \in C$ implies $x \in C$. Conversely, if $x \in C$ and $y \notin C$, it follows $y \in \mathrm{Bd}(C)$, hence $Y = \mathrm{Adh}_Z(y) \subset \mathrm{Adh}_Z(\mathrm{Bd}(C))$, and, $Y \cap C = \emptyset$. This is impossible, since $F \cap C \neq \emptyset$. Consequently, $x \in C$ if and only if $y \in C$. Thus, $2^m \#(C \cap F) = \#(C \cap F_1)$, and since $2^m \#(F) = \#(F_1)$, the fan F verifies the desired congruences as F_1 does. $\square$

In view of this corollary, in the geometric situation one has to check the criterion of the generation formula (Corollary 1.5) only for finitely many subvarieties. Namely, for the irreducible components, then for the irreducible components of the singular locus, then for the irreducible components of the singular locus of the singular locus, and so on.

In the final result of this section we use divisors in order to bound $t(X)$ from below.

Proposition 4.10   *Let Y be a divisor in X and let $C \subset X$ be constructible and open. Assume that there are $y, y', z \in Y^*$ such that*

$$y \notin C, \ x_1 \in C, \ x_2 \notin C; \ y' \notin C, \ x_1' \in C, \ x_2' \in C; \ z \in C,$$

where $x_i \to y$ and $x_i' \to y'$ for $i = 1, 2$. Then C is not a union of two basic open sets.

Proof. Suppose that $C = C_1 \cup C_2$, where C_1 and C_2 are basic open. Let, for instance, $x_1 \in C_1$. As $x_2 \notin C$, $y \in \mathrm{Bd}(C_1)$, hence $Y = \mathrm{Adh}_Z(y) \subset \mathrm{Adh}_Z(\mathrm{Bd}(C_1))$ and, C_1 being basic open, $Y \cap C_1 = \emptyset$. Thus $Y \cap C \subset C_2$. So $z \in C_2$. Now if x_1' or $x_2' \in C_2$, by the same reason $Y \cap C_2 = \emptyset$. Contradiction. So $x_1', x_2' \in C_1$. But $F = \{x_1, x_2, x_1', x_2'\}$ is a fan and $\#(F \cap C_1) = 3$. Contradiction. $\square$

5. The Artin-Lang Property

Let (X, G) be a prespace of signs. Very often, one is not interested in X itself, but in a distinguished subset Y of X: basic open and closed, constructible, Zariski closed subsets of Y. On the other hand, Y will not be a subspace, and the real space $(Y, G|Y)$ will not be a space of signs. A typical example of this appears in the study of semialgebraic subsets of real algebraic varieties, as described in Sections I.3 and I.4. Of course, we want to apply the richer structure of X to deal with the constructible subsets of Y. So we need a transfer from X to Y. The key notion here is patterned upon the classical Artin-Lang homomorphism theorem (Theorem I.1.7):

Definition 5.1 (Artin-Lang Property)   *A set $Y \subset X$ is called an* Artin-Lang *subset if Y is dense in X for the constructible topology, that is, $Y \cap C \neq \emptyset$ for every non-empty constructible set $C \subset X$.*

It is clear from the definition that Artin-Lang subsets remain Artin-Lang after intersection with *constructible* subspaces. It is also clear that any Artin-Lang subset contains all constructible points.

The transfer required from X to Y is now immediate. We use all the notations and terminology introduced in Section I.3 for arbitrary real spaces.

(5.2) The Tilde Operator. Let Y be an Artin-Lang subset of X. Then the map

$$\mathcal{C}(X) \to \mathcal{C}(Y); \ C \mapsto Y \cap C$$

is an isomorphism of Boolean algebras. The inverse map

$$\mathcal{C}(Y) \to \mathcal{C}(X); \ S \mapsto \tilde{S}$$

is called the *tilde operator*. More explicitely, for principal open sets we have

$$S = \{x \in Y \mid f(x) > 0\} \mapsto \tilde{S} = \{x \in X \mid f(x) > 0\},$$

and it is clear how this extends to arbitrary constructible sets. $\square$

This description of the tilde operator shows that it preserves the properties: principal open, principal closed, basic open, basic closed, strictly open, strictly closed, constructible, as well as the corresponding Zariski ones. In fact, we have:

Theorem 5.3 (Abstract Ultrafilter Theorem)   *Let Y be an Artin-Lang subset of X, and let $\tilde{Y}$ be the Stone space of Y. Then the map*

$$\theta : X \to \tilde{Y} : x \mapsto \phi_x = \{S \in \mathcal{C}(Y) \mid x \in \tilde{S}\}$$

is an isomorphism of real spaces.

Proof.   The map is a bijection because the tilde operator is an isomorphism of Boolean algebras and X is its own Stone space. So, it remains to show that the restriction map $G \to G|Y$ is injective. Let $f, g \in G$ be such that $f|Y = g|Y$. Then we have $\{f > 0\} \cap Y = \{g > 0\} \cap Y$ and $\{f = 0\} \cap Y = \{g = 0\} \cap Y$. But this implies $\{f > 0\} = \{g > 0\}$ and $\{f = 0\} = \{g = 0\}$, so that $f = g$. $\square$

Remarks 5.4   Let Y be an Artin-Lang subset of X.

a) The ultrafilter theorem says that X is the Stone space of Y, and the general remarks made in Section I.3 are valid here. Concerning constructibility of closures we recall that **FT** holds for Y if and only if $\tilde{S}$ is open (resp. closed) for every open (resp. closed) constructible set $S \subset Y$ (I.3.5 *b)*), and **AC** and **FT** hold for Y if and only if **AC** holds for X (I.3.5 *c)*). Constructibility of connected components is a more subtle matter. We quote, for instance, that **CC** holds for X in case each constructible set of Y has finitely many connected components (I.3.5 *d)*). $\square$

b) Again by the ultrafilter theorem, forms over Y are the same as forms over X, via the tilde operator. In particular, the defining form of a constructible set $S \subset Y$ is also the defining form of $\tilde{S}$, so that $w(S) = w(\tilde{S})$ and $l(S) = l(\tilde{S})$. Consequently, $w(Y) = w(X)$ and $l(Y) = l(X)$. Analogously, $s(Y) = s(X)$, $\bar{s}(Y) = \bar{s}(X)$, $t(Y) = t(X)$, $\bar{t}(Y) = \bar{t}(X)$. $\square$

c) An element $g \in G$ separates two subsets S and T of Y if and only if it separates $\tilde{S}$ and $\tilde{T}$. However, separation questions refer usually to closed sets, which leads to the extra condition **FT** for Y.

d) X is noetherian if and only if Y is noetherian. In that event, $\dim(Y) = \dim(X)$.

Suppose first that X is noetherian. Since the Z-closed sets of Y are the sets $C \cap Y$ for all Z-closed sets $C \subset X$, we see that Y noetherian. Conversely, suppose that X is not noetherian. Then, there is a Z-closed set $C \subset Z$ which is not constructible, and we find an infinite chain of constructible Z-closed sets C_i containing C. Hence, the sets $S_i = C_i \cap Y$ form an infinite chain of constructible Z-closed sets in Y, and Y is not noetherian. $\square$

e) (Hilbert's 17th Problem) An element $f \in G$ which is ≥ 0 on Y is also ≥ 0 on X. This obvious fact has a special meaning in case X is the space of signs of a commutative ring with unit A (II.1.5). Indeed, then Y is a subset of $\mathrm{Spec}_r(A)$, and we deduce that any $f \in A$ which is ≥ 0 on Y is ≥ 0 on the whole of $\mathrm{Spec}_r(A)$. Hence, by Corollary II.1.15, $f^{2n+1} + fq_1 = q_2$ for suitable $n \geq 0$ and $q_1, q_2 \in \sum A^2$. Of course, there is a corresponding version of all the various abstract Stellensätze. $\qquad\square$

f) We have already remarked that Y contains all constructible points of X, so that the set X_{const} of constructible points of X is the minimal candidate for Artin-Lang subset. More delicate is the role of closed points. One interesting fact is that if Y is compact (with respect to the Harrison topology), then $Y = X_{\max}$.

Indeed, let $x \in X_{\max}$ and let ϕ be the family of all closed constructible sets $C \subset X$ such that $x \in C$. Since Y is compact and $C \cap Y \neq \emptyset$ for all $C \in \phi$, there exists $y \in \bigcap_{C \in \phi} C \cap Y$. In particular, $y \in \mathrm{Adh}(x)$, hence $y = x$. So $X_{\max} \subset Y$. Since Y is Hausdorff, $y \to z$ in Y if and only if $y = z$. Therefore $Y = X_{\max}$. $\qquad\square$

Examples 5.5 *a)* We consider an example already discussed in Sections I.3 and I.4. Let R be a real closed field and let V be a real affine algebraic R-variety. We consider the real space $(V(R), G)$ associated to V as described in Examples I.3.9, and the space of signs (X, G) associated to the ring $R[V]$ of regular functions on V (II.1.5). By the Artin-Lang homomorphism theorem (Theorem I.1.7), the set

$$Y = \{\alpha \in X \mid \kappa(\mathrm{supp}(\alpha)) = R\},$$

is an Artin-Lang subset of X. Also, there is an obvious canonical identification $V(R) \equiv Y$, and we conclude that X is the Stone space of $V(R)$. This gives the standard tilde operator in semialgebraic geometry ([B-C-R 7.2]).

Here, Y is the set X_{const} of all constructible points of X (by the Artin-Lang homomorphism theorem again, Example II.1.8), that is, Y is as small as possible. For arbitrary V, X_{const} is a proper subset of the set $X_{\max}$ of closed points of X; in case $R = \mathbb{R}$ and $V(\mathbb{R})$ is compact, $X_{\mathrm{const}} = X_{\max}$ (Remark 5.4 *f)*). We shall discuss this later on (Sections VI.6-7).

Finally, for both the space of signs X and the Artin-Lang subset Y, the extra conditions **FT, AC, CC** hold true. $\qquad\square$

b) Let us look now at a more subtle example, involving another ring of functions. Namely, let A be the ring of continuous functions $f : \mathbb{R} \to \mathbb{R}$ which are polynomial for $t \leq 0$ and for $t \geq 0$ separately. Then the real spectrum $X = \mathrm{Spec}_r(A)$ contains $\mathbb{R}$, but $\mathbb{R}$ is not an Artin-Lang subset.

Indeed, suppose it were, and consider the function $f : t \to |t|$. As f is ≥ 0 on $\mathbb{R}$, $f(f^{2n} + q_1) = q_2$ with $n \geq 0$ and $q_1, q_2 \in \sum A^2$ (Remark 5.4 *e)*). This equality is polynomial for $t \geq 0$, and we get $q_1', q_2' \in \sum \mathbb{R}[t]^2$ such that

$$1 = \lim_{t \to +\infty} \frac{t(t^{2n} + q_1'(t))}{q_2'(t)},$$

which is impossible, since the degree of the numerator is odd and that of the denominator is even.

The reader can show that in fact

$$\{f < 0\} \equiv \mathrm{Spec}_r(\mathbb{R}(t)), \quad \{f \geq 0\} \equiv \mathrm{Spec}_r(\mathbb{R}[t]),$$

$\mathbb{R} = X_{\mathrm{const}} \subsetneqq X_{\max}$, and $X_{\max}$ is an Artin-Lang subset. Also, that **FT** and **AC** hold for X_{const}, $X_{\max}$ and X, while **CC** holds only for X_{const}. $\quad\square$

Thus, we shall have to consider pairs $X \supset Y$, where X is a space of signs and Y an Artin-Lang subset. The choice of Y depends on the particular application in mind, and the relations among constructible points, closed points and the chosen Artin-Lang subset may vary a lot. We shall see that X_{const} and/or $X_{\max}$ are often Artin-Lang subsets. However, they are not in some important cases, yet there is still a suitable Artin-Lang subset to consider. We shall have examples of this in Chapter VIII. In any case, constructible points and closed points have always a special relevance, and we now finish the section with a few remarks concerning them.

Remarks 5.6   *a)* A point of a noetherian space of signs is constructible if and only if it is isolated in its Zariski closure.

For, in a noetherian space, Z-closed sets are constructible. $\quad\square$

b) Let X be a noetherian space of signs for which **AC** holds. Then specializations of constructible points are constructible. In particular, every constructible point $x \in X$ specializes to a constructible closed point.

Since **AC** holds for X, the set $\mathrm{Adh}(x)$ is constructible. This set is the chain of specializations of x (Corollary III.1.15), and since X is noetherian, it is finite. Hence, every point of the chain is constructible. $\quad\square$

c) A constructible point need not be closed. Let R be a real closed field and let $A = R[[t]]$. Then $\mathrm{Spec}_r(A)$ consists of three constructible points, defined repectively by $t > 0$, $t < 0$ and $t = 0$. Only the latter is closed. $\quad\square$

d) A closed point need not be constructible. Let R be a real closed field and let $A = R[t]$. Consider the prime cone α with support (0) defined by: $f(\alpha) > 0$ if and only if the biggest degree monomial of f has positive coefficient. Then α is closed, but not constructible. $\quad\square$

Notes

The content of the whole chapter is new. We present here the final generalization of the geometric results which are collected, for instance, in [Br10]. Sometimes these generalizations are straightforward, but often new ideas are also involved. In the case of local rings, Proposition 1.1 is due to Knebusch [Kn1], while Corollaries 1.2 and 1.3 appear in [Br6]. Theorem 1.4 has a long history. First of all, Bröcker showed in [Br6] that for a real algebraic variety of dimension d over a real closed field $s = d$ for $d \leq 3$ and $s \leq d(d-2)(d-4)\cdots$ in general. Later this was improved a little bit (see [B-C-R]), until the break through was done in 1988 when Scheiderer observed the usefulness of the elementary Corollary III.2.8. Soon afterwards, Scheiderer ([Sch1]) and Bröcker (unpublished) found proofs that $s = d$ for arbitrary d. The latter proof had the advantage that it could be easily generalized to our Theorem 1.4 for arbitrary noetherian rings ([Br9]). Other approaches were given by Mahé ([Mh]), who presented a direct attack in the geometric case, and Marshall ([Mr6]), who showed Theorem 1.4 for arbitrary rings (commutative with unit). In the geometric setting Corollaries 1.5-1.10 appear all in [Br10]. Theorems 2.1 and 2.5 are new, even in the geometric context. In [Mr7] Marshall uses the minimal length of a separating family for a semialgebraic set, an invariant which is closely related to our ℓ. The rest of Section 2 appeared for the geometric setting in [Br10]. The separation question has a very long history in real geometry, starting with Mostowski's counterexample and separation theorem ([Mw]) (see also [Co], and [Br-St] for a quantitative approach). The content of Section 3, in the geometric situation, is in [Br8]. More direct methods have been used by Acquistapace, Broglia, Fortuna, Galbiati, Tognoli and Vélez ([Ac-Bg-Vz], [Bg-To], [Ft-Gl]). Proposition 4.3 generalizes a result of Scheiderer ([Sch1]), and Propositions 4.5, 4.6 an earlier one by Bröcker ([Br10]). Clearly, Definition 5.1 is the abstract version of the Artin-Lang theorem [Lg1]. The ultrafilter theorem (Theorem 5.3) appeared for the first time explicitely in Brumfiel's book ([Bf1]), although it might have been known before. Finally, Remark 5.4 *c)* has its roots in E. Artin's work on Hilbert's 17th problem ([A]). A slightly different axiomatic approach to real spectra which covers partially the same main results is due to Marshall ([Mr8]).

Chapter VI. Spaces of Signs of Rings

Summary. In view of the global stability formulae (Corollary V.1.6), and the canonical decomposition (III.1.9), the computation of stability indices in the space of signs of a ring reduces to estimations of the size of fans of residue fields of that ring. We obtain in Section 1 such estimations, via real valuations, after proving the so-called trivialization theorem for fans. Then, in Section 2, we deduce upper bounds for the stability index of a field extension in terms of the ground field. These bounds are sharp when the ground field is real closed or the rational numbers field, as follows from the lower bounds discussed in Section 3. In section 4 we generalize the previous upper bounds to algebras. The results are specially good for algebras over a field, which are the matter of Section 5; again, we obtain the best estimations over a real closed field and over the rationals. Section 6 is devoted to totally archimedean rings, which are the abstract counterparts of compact spaces. These rings have two special features: firstly, their complexity bounds are low, and, secondly, generation of basic sets and separation are characterized by multilocal conditions. We end in Section 7 with the translation to concrete semialgebraic geometry of most of the abstract results obtained so far. We as well discuss several examples and counterexamples to questions raised in earlier chapters.

1. Fans and Valuations

Throughout this section, K denotes a formally real field and we set $\Sigma = \sum K^2$. Let X be the space of signs $(\operatorname{Spec}_r(K), G)$ associated to K. As described in Example IV.1.4, $G = K^*/\Sigma^*$, and the subspaces Y of X correspond bijectively to the precones $T \subset K$ by the formulas

$$T = \{t \in K \mid t = 0 \text{ or } t\Sigma^* \subset Y^\perp\},$$

$$Y = \{\sigma \in X \mid \sigma(t) = +1 \text{ for all } t \in T\};$$

we have then $G|Y = K^*/T^*$ and write $Y = X/T$. Concerning fans we have

Proposition 1.1   *Let $T \subset K$ be a precone and $Y = X/T$ the associated subspace. The following assertions are equivalent:*

a) Y is fan.
b) $T + gT = T \cup gT$ for all $g \in K \setminus (-T)$.
c) $\Sigma + g\Sigma \subset T \cup gT$ for all $g \in K \setminus (-T)$.

Proof.   a) $\Rightarrow$ b) Suppose that there are $g \in K \setminus (-T)$, and $t_1, t_2 \in T$ such that $a = t_1 + gt_2 \notin T \cup gT$. Then neither of $-g$, a, ga belong to T, and we find σ_1, σ_2, σ_3 in Y such that $\sigma_1(-g) = \sigma_2(a) = \sigma_3(ag) = -1$. It follows that $\sigma_1(g) = +1$, $\sigma_2(g) = \sigma_3(g) = -1$ and $\sigma_1(a) = \sigma_3(a) = +1$. Consequently, for $\sigma = \sigma_1\sigma_2\sigma_3$ we have $\sigma(g) = +1$ and $\sigma(a) = -1$, which shows that either σ is not an ordering, or, being one, it does not belong to Y.

b) $\Rightarrow$ c) This is trivial, since $\Sigma \subset T$.

c) $\Rightarrow$ a) Let $\sigma_1, \sigma_2, \sigma_3 \in Y$ and $\sigma = \sigma_1\sigma_2\sigma_3$. Suppose that σ is not an ordering. Then, by Proposition III.3.5, there are $a, g \in G$ such that $\sigma(a) = -1$, $\sigma(g) = +1$ and $a \in D(\langle 1, g \rangle)$. By the latter condition, $\tau(g) > 0$ implies $\tau(a) > 0$ for all $\tau \in X$, which by the Positivstellensatz (Theorem II.1.14) means $a \in \Sigma + g\Sigma$. But the other two conditions show that $a \notin T \cup gT$. The implication follows immediately from this. $\square$

Next we want to investigate the valuation theory of fans. We shall use the notations introduced in Section II.3 for a valuation v of K: Γ_v, V, $\mathfrak{m}_v$, k_v and $\lambda_v : a \mapsto \bar{a}$, for its value group, valuation ring, maximal ideal, residue field and place, respectively; often we shall even omit the index refering to v. To start with:

Proposition and Definition 1.2   *Let $T \subset K$ be a precone and v a valuation of K. The following assertions are equivalent:*

a) v is compatible with all the $\sigma \in X/T$.
b) $1 + \mathfrak{m} \subset T$.

If these properties hold, T and v are called compatible

Proof.   Immediate from Proposition and Definition II.3.2. $\square$

Now we restate the Baer-Krull theorem (Proposition II.3.3) in terms of residue spaces and extensions (see IV.2.9):

Theorem 1.3   *Let T be a precone of K, and let v be a valuation of K with residue field k and value group Γ. Suppose that T and v are compatible, and let X (resp. $\overline{X}$) be the space of orderings of K (resp. of k). Then:*

a) $\overline{T} = \{\bar{t} \mid t \in T \cap V\} \subset k$ is a precone.
b) The group $\Gamma/\widehat{v(T^*)}$ is isomorphic to a closed subgroup of the translation group $\mathrm{Tr}(X/T, G/T)$ of the subspace X/T.
c) The residue space with respect to that subgroup is isomorphic to $\overline{X}/\overline{T}$.
d) X/T is isomorphic to the extension $(\overline{X}/\overline{T})[\Gamma/v(T^*)]$.

e) X/T is a fan if and only if $\overline{X}/\overline{T}$ is a fan.

Conversely, given a precone $\overline{T}$ of k, then $\{t \in V \mid \bar{t} \in \overline{T}\}K^2$ is a precone of K, which is compatible with v.

Proof. a) This is a simple verification. The compatibility is used to see that $-1 \notin \overline{T}$.

b) Every homomorphism $\alpha : \Gamma/v(T^*) \to \{+1, -1\}$ can be seen as a homomorphism $\alpha : K^*/T^* \to \{+1, -1\}$ by the formula $\alpha(gT^*) = \alpha(v(g) + v(T^*))$ ($g \in K^*$). It follows from the Baer-Krull theorem that $\alpha\sigma \in X/T$ for all $\sigma \in X/T$.

c) Let D be the closed subgroup provided by the identification above. Then $D^{\perp} = V^*/V^* \cap T^*$, where as usual V^* stands for the units of V, and the place $V \to k$ gives an identification $D^{\perp} \equiv k^*/\overline{T}^*$. Again by the Baer-Krull theorem, the image of X/T in $\widehat{k^*/\overline{T}^*}$ via restriction is $\overline{X}/\overline{T}$.

d) This follows directly from c).

e) This is a consequence of Proposition IV.2.14 c).

The last statement is obvious. $\square$

Corollary 1.4   *Let v be a real valuation of K with maximal ideal $\mathfrak{m}$, residue field k and value group Γ. Let X (resp. $\overline{X}$) denote the space of orderings of K (resp. k). Then*

a) $s(K) \geq s(k) + \dim_{\mathbb{F}_2}(\Gamma/2\Gamma)$.
b) *If $1 + \mathfrak{m} \subset \sum K^2$, then X is isomorphic to $\overline{X}[\Gamma/2\Gamma]$. In particular the equality holds in a).*

Proof. a) Consider the precones $\overline{T} = \sum k^2$ of k and $T = \{t \in V \mid \bar{t} \in \sum k^2\}K^2$ as in Theorem 1.3. Clearly, $T = (1 + \mathfrak{m})\sum K^2$ so that $v(T^*) = 2\Gamma$. Hence, by Theorem 1.3 d), $X/T = \overline{X}[\Gamma/2\Gamma]$. From Proposition IV.2.14 e), we get $s(K) \geq s(X/T) = s(k) + \dim_{\mathbb{F}_2}(\Gamma/2\Gamma)$.

b) If $1 + \mathfrak{m} \subset \sum K^2$, the precone T of the above argument coincides with $\sum K^2$. Hence $X/T = X$, and the conclusion is as wanted. $\square$

Remarks 1.5   a) The most important example in which condition $1 + \mathfrak{m} \subset \sum K^2$ holds in the above corollary is that of a henselian valuation ring. Indeed, in that case $1 + \mathfrak{m} \subset K^2$ (Proposition II.2.4). $\square$

b) Let A be a regular local domain, with residue field k and quotient field K. Then $s(K) \geq s(k) + \dim(A)$.

For, by Lemma II.3.4, there is a discrete valuation ring $V \subset K$ of rank $\dim(A)$ with residue field k. Explicitly, if $u_1, \ldots, u_d$ are a regular system of parameters of A, and the elements $a_1, \ldots, a_s \in A$ are such that the basic set $\{\bar{a}_1 > 0, \ldots, \bar{a}_s > 0\}$ defined by their classes in k cannot be described with less than s inequalities, then the basic set $\{a_1 > 0, \ldots, a_s > 0, u_1 > 0, \ldots, u_d > 0\}$ cannot be described with less than $s + d$ inequalities. $\qquad\square$

The preceding theorem will be used to study a given fan once we can find valuations compatible with that fan. This is our next concern.

For $\sigma \in X$, let $V = V(\sigma)$ denote the convex hull of $\mathbb{Q}$ in K with respect to σ, which is the smallest valuation ring of K compatible with σ (II.3.6). Then, σ induces in the residue field k of V an archimedean ordering, and V is trivial if and only if σ itself is archimedean. Recall also that all the valuation rings compatible with σ form a chain.

Now, let $T \subset K$ be a precone, and $Y = X/T$ the corresponding subspace. Let $V = V(T) \subset K$ denote the ring generated by all the valuation rings $V(\sigma)$, $\sigma \in Y$. Clearly, V is the smallest valuation ring compatible with T; this V can be trivial. In fact, this condition is crucial:

Theorem 1.6 (Fan Trivialization Theorem)   *Let $T \subset K$ be a precone and $Y = X/T$ the corresponding subspace. Let $V(T)$ be the smallest valuation ring of K compatible with T, let k be the residue field of $V(T)$ and $\overline{X}$ the space of orderings of k. Let $\overline{T} \subset k$ be the precone induced by T and $\overline{Y} = \overline{X}/\overline{T}$. The following assertions are equivalent:*

   a) Y is a fan,
   b) $\#(\overline{Y}) \le 2$.

The implication *b) $\Rightarrow$ a)* follows from Theorem 1.3 and Proposition IV.2.14 *h)*. For the converse, we first prove a lemma:

Lemma 1.7   *Let $\sigma_1, \sigma_2, \sigma_3$ be three different elements of X whose product $\sigma = \sigma_1\sigma_2\sigma_3$ is a well defined element of X. Let $V_1 = V(\sigma_1)$, $V_2 = V(\sigma_2)$, $V_3 = V(\sigma_3)$ be the smallest valuation rings compatible with each one. For $i < j$, denote by V_{ij} the valuation ring generated by V_i and V_j, which is the smallest valuation ring compatible with both σ_i and σ_j. Then no V_{ij} is trivial. In particular, no σ_i is archimedean.*

Proof.   To start with, let us suppose there are two archimedean orderings, say $\sigma_1, \sigma_2 : K \hookrightarrow \mathbb{R}$. We choose two elements $a, b \in K$ with

$$\sigma_1(a) < 0, \sigma_2(a) > 0, \sigma_3(a) > 0; \quad \sigma_1(b) > 0, \sigma_2(b) < 0, \sigma_3(b) > 0.$$

If $\sigma_1(a/b) > \sigma_2(a/b)$, then there are positive integers m, n such that $\sigma_1(a/b) > -m/n > \sigma_2(a/b)$. Since σ_1 and σ_2 coincide on $\mathbb{Q} \subset K$, we get $\sigma_1(na + mb) > 0$, $\sigma_2(na + mb) > 0$. As $\sigma_3(na + mb) > 0$ we conclude $\sigma(na + mb) > 0$. But this is imposible, since $\sigma(a) < 0$ and $\sigma(b) < 0$. If $\sigma_1(a/b) < \sigma_2(a/b)$, we replace a and b by $1/a$ and $1/b$; if $\sigma_1(a/b) = \sigma_2(a/b)$, we replace a by $a(1 + a/n)^2$, where n is a big positive integer. Thus we are reduced to the case already settled.

Now suppose one ordering is archimedean, say σ_1. Again, we choose a and b

as above. Since σ_2 is not archimedean, V_2 is not trivial; let $\mathfrak{m}_2$ be the maximal ideal of that valuation ring. If $a \in \mathfrak{m}_2$, we replace a by $1/a$ to have $a \notin \mathfrak{m}_2$. Also, we pick a non-zero element $x \in \mathfrak{m}_2$ and replace b by $x^2 b$ to guarantee that $b \in \mathfrak{m}_2$. Now, we consider the element $a + nb$, where n is a positive integer. Since σ_1 is archimedean, $\sigma_1(a + nb) > 0$ for n big. By the assumptions on a, b and $\mathfrak{m}_2$, $\sigma_2(a + nb)\sigma_2(a) > 0$, so that $\sigma_2(a + nb) > 0$. Finally, $\sigma_3(a + nb) > 0$. Hence, $\sigma(a + nb) > 0$, which is impossible because $\sigma(a) < 0$ and $\sigma(b) < 0$.

Finally, suppose that V_{ij} is trivial, say for $i = 1$, $j = 2$. By what we have already seen, the valuations v_1, v_2 associated to V_1, V_2 are not archimedean, and our assumption means that they are independent. Let $\mathfrak{m}_1$, $\mathfrak{m}_2$ be the maximal ideals of those valuations. Again, we choose $a, b \in K$ such that

$$\sigma_1(a) < 0, \sigma_2(a) > 0, \sigma_3(a) > 0; \ \sigma_1(b) > 0, \sigma_2(b) < 0, \sigma_3(b) > 0.$$

By the approximation theorem ([Bk CA VI.7.2, Cor.1, p.414]), there is $x \in K$ such that

$$v_1(x) > \max\{v_1(a), -v_1(a)\}, \ v_2(x) < \min\{v_2(a), -v_2(a)\}.$$

Replacing a by $x^2 a$ we get $a \in \mathfrak{m}_1$, $a \notin \mathfrak{m}_2$. After a similar modification, we have $b \notin \mathfrak{m}_1$, $b \in \mathfrak{m}_2$. In this situation, $\sigma(a + b) > 0$, but $\sigma(a) < 0$, $\sigma(b) < 0$. Contradiction. $\qquad\square$

After this lemma, we can prove the remaining implication of the trivialization theorem:

Proof of Theorem 1.6. a) $\Rightarrow$ b) Take three different elements $\sigma_1, \sigma_2, \sigma_3 \in Y$ and $\sigma = \sigma_1\sigma_2\sigma_3 \in Y$. Consider the associated valuation rings $V_i = V(\sigma_i)$ and let V be the valuation ring generated by these V_i's. Thus each σ_i is compatible with V and induces an ordering $\bar{\sigma}_i$ in the residue field k of V. Furthermore, the valuation ring $\bar{V}_i$ induced by V_i in k coincides with $V(\bar{\sigma}_i)$. It follows that the valuation ring generated by the $\bar{V}_i$'s is trivial. Now, let $\bar{V}_{ij}$ be the valuation ring of k generated by $\bar{V}_i$ and $\bar{V}_j$, for $i < j$. Since any two $\bar{V}_{ij}$ are compatible with some of the $\bar{\sigma}_l$, they are comparable, and consequently, there is a $\bar{V}_{ij}$ which contains all the three $\bar{V}_l$'s. Hence, it is trivial. Since $\sigma = \sigma_1\sigma_2\sigma_3$ is compatible with V, the product $\bar{\sigma}_1\bar{\sigma}_2\bar{\sigma}_3$ is a well defined ordering of k, and by the lemma, the three $\bar{\sigma}_i$'s cannot be different.

After this preparation we distinguish several cases.

Case 1. There are two orderings $\sigma_1, \sigma_2 \in Y$ such that the valuation rings $V_1 = V(\sigma_1)$, $V_2 = V(\sigma_2)$ are not comparable. We claim that $V(T)$ is the valuation ring W generated by V_1 and V_2. Indeed, first note that the two orderings $\bar{\sigma}_1$ and $\bar{\sigma}_2$ induced by σ_1 and σ_2 in the residue field k_W of W are different. By what has been already seen, for any other $\sigma_3 \in Y$ and for the valuation ring V generated by W and $V(\sigma_3)$, σ_3 induces in k the same ordering as σ_1 or σ_2, which shows that σ_3 is compatible with V_1 or with V_2, hence with W. Thus we get at once that $V(T) = W$ and that Y induces in its residue field a trivial fan.

Case 2. All the valuation rings $V(\sigma)$, $\sigma \in Y$, are comparable. If all of them coincide, they all coincide with $V(T)$, and the previous preparation shows that F induces in the residue field a trivial fan. If there are $\sigma_1, \sigma_2 \in Y$ such that $V(\sigma_1) \subsetneqq V(\sigma_2) = W$, a similar argument as in *Case 1* shows that each ordering $\sigma_3 \in Y$ is compatible with W, and induces in its residue field the same ordering as σ_1 or σ_2. We are done. $\square$

Readily from the trivialization theorem, we obtain the formula for the stability index that will be crucial in all the later computations. For a field k, we set $\varepsilon(k) = 1$ or 0 according to whether or not k admits more than one ordering.

Theorem 1.8   *Let K be a formally real field. Then:*

$$s(K) = \sup_v\{s(k_v) + \dim_{\mathbb{F}_2}(\Gamma_v/2\Gamma_v)\} =$$
$$= \sup_v\{\varepsilon(k_v) + \dim_{\mathbb{F}_2}(\Gamma_v/2\Gamma_v)\} \leq 1 + \sup_v\{\dim_{\mathbb{F}_2}(\Gamma_v/2\Gamma_v)\},$$

where the sup*'s range over all real valuations v of K.*

Proof.   Let $T \subset K$ be a precone such that X/T is a fan. We choose a valuation v of K as given by Theorem 1.6. By Theorem 1.3, $X/T \equiv (\overline{X}/\overline{T})[\Gamma_v/v(T^*)]$, with $\#(\overline{X}/\overline{T}) \leq 2$ and $v(T^*) \supset 2\Gamma_v$. By Proposition IV.2.14 *e)* and Corollary 1.4,

$$s(X/T) = s(\overline{X}/\overline{T}) + \dim_{\mathbb{F}_2}(\Gamma_v/v(T^*)) \leq \varepsilon(k_v) + \dim_{\mathbb{F}_2}(\Gamma_v/2\Gamma_v) \leq$$
$$\leq s(k_v) + \dim_{\mathbb{F}_2}(\Gamma_v/2\Gamma_v) \leq s(K).$$

The result follows clearly from these inequalities. $\square$

2. Field Extensions: Upper Bounds

Here we shall consider a field extension $L \supset K$ and bound $s(L)$ in terms of $s(K)$. We denote by $\deg(L : K)$ the transcendence degree of L over K. We shall need two results from valuation theory. First, we recall that the *rational rank* of a group Γ is the maximum number of elements of the group which are $\mathbb{Z}$-linearly independent; we shall denote it by rat.rk.(Γ).

Proposition 2.1   *Let $(L, w) \supset (K, v)$ be an extension of valued fields of finite transcendence degree. Then*

$$\mathrm{rat.rk.}(\Gamma_w/\Gamma_v) + \deg(k_w : k_v) \leq \deg(L : K).$$

Furthermore, setting $r = \mathrm{rat.rk.}(\Gamma_w/\Gamma_v)$, Γ_w contains, up to isomorphism, the group $\Gamma_v \oplus \mathbb{Z}^r$, and the quotient $\Gamma_w/(\Gamma_v \oplus \mathbb{Z}^r)$ is a torsion group.

Proof.   [Bk CA VI.10.3, Th.1, p.439]. $\square$

Proposition 2.2   *Let (K, v) be a valued field and let $\bar{k}/k_v$ be an algebraic field extension. Then there is an algebraic extension of valued fields $(K', v') \supset (K, v)$ such that $k_{v'} \supset k_v$ is isomorphic to $\bar{k} \supset k_v$ and $\Gamma_{v'} = \Gamma_v$.*

Proof.   [En Th.27.1, p.206]. In fact, much more is proved there, but this is sufficient for us. $\square$

Also, we prove a small remark concerning abelian groups:

Lemma 2.3   *Let H be an abelian group without torsion, Λ a subgroup such that H/Λ is a torsion group, and $p > 0$ a prime number. Then $\#(\Lambda/p\Lambda) \geq \#(H/pH)$, and the equality holds if $\#(H/\Lambda)$ is finite.*

Proof.   We may restrict to the case $q = \#(H/\Lambda)$ is finite, and by induction on q, we may also assume that q is a prime number. Since $(pH + \Lambda)/pH$ is isomorphic to $\Lambda/(pH \cap \Lambda)$ it remains to show that $\#(H/(pH + \Lambda)) = \#((\Lambda \cap pH)/p\Lambda)$. If $q \neq p$, both cardinals are 1. If $q = p$, then $pH + \Lambda = \Lambda$ and $\Lambda \cap pH = pH$, and, since H has no torsion, the map $H/\Lambda \to pH/p\Lambda \, ; a \mapsto pa$ is an isomorphism. $\square$

Now we come back to stability indices. We first consider algebraic extensions:

Proposition 2.4   *Let $L \supset K$ be an algebraic field extension. Then $s(L) \leq s(K) + 1$.*

Proof.   By Theorem 1.8 we have

$$s(L) \leq 1 + \sup_{w}\{\dim_{\mathbb{F}_2}(\Gamma_w/2\Gamma_w)\},$$

where the sup ranges over the real valuations w of L. Let v be the restriction of w to K; since $L \supset K$ is algebraic, Γ_w/Γ_v is a torsion group ([Bk CA VI.8.1, Prop.1, p.417]). Thus, from the preceding lemma we get $\dim_{\mathbb{F}_2}(\Gamma_w/2\Gamma_w) \leq \dim_{\mathbb{F}_2}(\Gamma_v/2\Gamma_v)$. But $\dim_{\mathbb{F}_2}(\Gamma_v/2\Gamma_v) \leq s(K)$ (Corollary 1.4), and we are done.
$\square$

Example 2.5   The inequality in the preceding proposition can be strict. In fact, for every n there is a finite extension $L \supset K$ with $s(L) = 0$ and $s(K) = n$.

We start with the field $K_0 = \mathbb{R}(x_1, \ldots, x_n; t^{1/m} : m \geq 1)$, and two valuations v_1, v_2 with residue field $\mathbb{R}$ and value groups $\Gamma_1 = \mathbb{Z}^n \oplus \mathbb{Q}$, $\Gamma_2 = \mathbb{Q}$ respectively. The definition of v_1 is clear through lexicographically ordering the indeterminates. For v_2 we can pick a transcendental analytic curve $\gamma(t) = (x_1(t), \ldots, x_n(t), t)$, so that for every $f \in K_0$ the substitution $f(\gamma(t))$ is a formal Puiseux series whose order is $v_2(f)$. Next, let $(K_i, \tilde{v}_i)$ be the henselization of (K_0, v_i) inside a fixed algebraic closure of K_0, and put $K = K_1 \cap K_2$. Clearly the restriction to K of v_i, still denoted by v_i, has residue field $\mathbb{R}$ and value group Γ_i; its maximal ideal is denoted by $\mathfrak{m}_i$. We claim that every ordering σ of K is compatible with either v_1 or v_2. Suppose the contrary. Then there are

$c_1, c_2 \in K$ positive in σ such that $1+c_1 \in \mathfrak{m}_1$ and $1+c_2 \in \mathfrak{m}_2$. Since v_2 has rank 1 and both v_1 and v_2 have residue field $\mathbb{R}$, the two valuations are independent. Hence, by the approximation theorem, we find $a_1, a_2 \in K$ such that

$$1 - a_1 \in \mathfrak{m}_1, \ a_1 \in \mathfrak{m}_2, \ a_2 \in \mathfrak{m}_1, \ 1 - a_2 \in \mathfrak{m}_2.$$

Then $1 + c \in \mathfrak{m}_1 \cap \mathfrak{m}_2$ for $c = a_1^2 c_1 + a_2^2 c_2$. By Hensel's lemma we get $-c \in K_1^2 \cap K_2^2 = K^2$, which is impossible since c is positive in σ. Now, the orderings compatible with v_1 form a fan F with 2^n elements, and there is one single ordering σ compatible with v_2. The conclusion is that the space of orderings of K is $X = F \cup \{\sigma\}$, which in the terminology of spaces of finite type (Section IV.3) is $E[\mathbb{Z}_n] + E$ and has stability index n. On the other hand, there is an element $f \in K$ such that $\{\sigma\} = \{f > 0\}$ (pick $a \in \mathfrak{m}_1 \setminus \mathfrak{m}_2$, $b \in \mathfrak{m}_2 \setminus \mathfrak{m}_1$ and set $f = a^2 - b^2$). Hence, the space of orderings of $L = K(\sqrt{f})$ is atomic, and $s(L) = 0$. $\qquad\square$

Notation 2.6 Let K we a field. Then we set

$$s_0(K) = \sup\{s(L) \mid L \supset K \text{ finite field extension }\}.$$

By Proposition 2.4, $s_0(K) \leq s(K) + 1$. $\qquad\square$

Next we consider arbitrary field extensions:

Theorem 2.7 *Let $L \supset K$ be a field extension. Then*

$$s(L) \leq s_0(K) + \deg(L : K).$$

Proof. Clearly, we can assume that $L \supset K$ is finitely generated, so that $d = \deg(L : K)$ is finite. Easily by induction, we are reduced to the case $d = 1$, which we assume henceforth. Thus, let $T \subset L$ be a precone such that $F = X/T$ is a fan. We choose a valuation w of L according to Theorem 1.6, and consider its restriction v to K. By Proposition 2.1 we have:

$$\mathrm{rat.rk.}(\Gamma_w/\Gamma_v) + \deg(k_w : k_v) \leq d = 1.$$

By Theorem 1.3 *d)*,

$$\#(F) \leq \#(\overline{X/T})\#(\Gamma_w/w(T^*)) \leq \#(\overline{X/T})\#(\Gamma_w/2\Gamma_w),$$

where $\#(\overline{X/T}) \leq 2$. Now, since k_v is formally real, we find a fan F' of K with $\#(F') = \#(\Gamma_v/2\Gamma_v)$. As $s(K) \geq s(F')$, we get:

$$\#(\Gamma_v/2\Gamma_v) \leq 2^{s(K)}.$$

Now we consider two cases:

Case 1: rat.rk.$(\Gamma_w/\Gamma_v) = 0$. By Lemma 2.3 $\#(\Gamma_w/2\Gamma_w) = \#(\Gamma_v/2\Gamma_v)$, and by the above inequalities $\#(F) \leq 2^{s_0(K)+1}$.

Case 2: rat.rk.$(\Gamma_w/\Gamma_v) = 1$. Then we have group inclusions $\Gamma_v \subset \Gamma_v \oplus \mathbb{Z} \subset \Gamma_w$ such that $\Gamma_w/(\Gamma_v \oplus \mathbb{Z})$ is a torsion group (Proposition 2.1). Hence, by Lemma 2.3

$$\#(\Gamma_w/2\Gamma_w) \leq 2\#(\Gamma_v/2\Gamma_v).$$

Now, if the trivial fan $\overline{X}/T$ of k_w is a singleton, we get from the preceding bounds

$$\#(F) \leq 2\#(\Gamma_v/2\Gamma_v) \leq 2^{s_0(K)+1}.$$

So, we assume that $\overline{X}/T$ consists of two elements. According to Proposition 2.2, we choose an algebraic extension (K', v') of (K, v) such that $k_{v'} \supset k_v$ is isomorphic to $k_w \supset k_v$ and $\Gamma_{v'} = \Gamma_v$. Then $\#(\Gamma_{v'}/2\Gamma_{v'}) = \#(\Gamma_v/2\Gamma_v)$. In particular, we find a fan F' of K' whose cardinal is $2\#(\Gamma_v/2\Gamma_v) = \#(\Gamma_w/2\Gamma_w)$. Therefore, $\#(\Gamma_w/2\Gamma_w) \leq 2^{s(K')}$, and by the previous bounds $\#(F) \leq 2^{s(K')+1}$.

The proof is thus complete. $\qquad\qquad\qquad\qquad\qquad\qquad\qquad\qquad\qquad$ $\square$

Remark 2.8 The bound above need not be sharp. For instance, if $L \supset K$ is an algebraic extension with $s(L) = 0$ and $s(K) = n > 1$ (Example 2.5), then $s(L(\mathrm{t})) \leq 2 < s_0(K) + 1$. $\qquad\qquad\qquad\qquad\qquad\qquad\qquad\qquad\qquad$ $\square$

We end this section with a variation of the field extension situation. First, some terminology:

Notations 2.9 Let K be a field and σ a fixed ordering of K. Let $V \subset K$ be the convex hull of $\mathbb{Q}$ in K with respect to σ. Let Γ_v be the value group of this valuation ring V, and Γ' the biggest 2-divisible convex subgroup of Γ_v. This convex subgroup defines a valuation ring $U \supset V$ whose value group is $\Gamma_u = \Gamma_v/\Gamma'$. This U is compatible with σ, and σ specializes to an ordering $\bar{\sigma}$ in the residue field k_u of U. We set

$$s_0(\sigma) = \varepsilon + \dim_{\mathbb{F}_2}(\Gamma_u/2\Gamma_u),$$

where $\varepsilon = 1$ or 0 according to whether or not the ordering $\bar{\sigma}$ can be extended in more than one way to some algebraic extension of k_u. $\qquad\qquad\qquad$ $\square$

Now, we are interested in the orderings of a field L that extend a fixed one σ in a subfield $K \subset L$. Those orderings are a subspace of $\mathrm{Spec}_r(L)$ (see Example III.1.7), and consequently they are a space of orderings. The stability index of that space of orderings can be bound as follows:

Proposition 2.10 *Let $L \supset K$ be a field extension, σ an ordering of K, and $\mathrm{Spec}_\sigma(L)$ the space of all orderings of L which extend σ. Then:*

$$s(\mathrm{Spec}_\sigma(L)) \leq s_0(\sigma) + \deg(L : K).$$

Proof. Set $d = \deg(L : K)$. Let F be a fan of L whose orderings restrict all to σ in K. We have to show that $s(F) \leq s_0(\sigma) + d$. To that end, choose a valuation ring $W' \subset L$ according to the trivialization theorem (Theorem 1.6), so that F induces a trivial fan $F_{w'}$ in the residue field $k_{w'}$ of W'. Let $W = W' \cap K$ be the restriction of W' to K. We claim that

$$\dim_{\mathbb{F}_2}(\Gamma_{w'}/2\Gamma_{w'}) \leq \dim_{\mathbb{F}_2}(\Gamma_w/2\Gamma_w) + d - \deg(k_{w'} : k_w).$$

Indeed, by Proposition 2.1, $\Gamma_{w'}$ contains a subgroup Λ isomorphic to $\Gamma_w \oplus \mathbb{Z}^r$ such that $\Gamma_{w'}/\Lambda$ is a torsion group and $r \leq d - \deg(k_{w'} : k_w)$. But, by Lemma 2.3, we have:

$$\dim_{\mathbb{F}_2}(\Gamma_{w'}/2\Gamma_{w'}) \leq \dim_{\mathbb{F}_2}(\Lambda/2\Lambda) = \dim_{\mathbb{F}_2}(\Gamma_w/2\Gamma_w) + r,$$

from which the claim follows.

On the other hand, by Theorem 1.3, $s(F) \leq s(F_{w'}) + \dim_{\mathbb{F}_2}(\Gamma_{w'}/2\Gamma_{w'})$, which combined with the inequality just proved gives:

$$s(F) \leq s(F_{w'}) + \dim_{\mathbb{F}_2}(\Gamma_w/2\Gamma_w) + d - \deg(k_{w'} : k_w).$$

Now, the valuation ring W and the ordering σ are compatible, hence W and U are comparable. Then, easily from the definition of U, we see that

$$\dim_{\mathbb{F}_2}(\Gamma_w/2\Gamma_w) \leq \dim_{\mathbb{F}_2}(\Gamma_u/2\Gamma_u),$$

with strict inequality for $W \supsetneq U$. Note also that $s(F_{w'}) \leq 1$. All of this together, gives the wanted bound except in the case:

$$\deg(k_{w'} : k_w) = 0, \quad W \subset U, \quad s(F_{w'}) = 1.$$

We shall finish by showing that under these hypotheses, $\varepsilon = 1$.

Since $W \subset U$, W induces a valuation $\bar{W}$ in the residue field k_u of U, and $k_{\bar{w}} = k_w$. As σ is compatible with W, it specializes to an ordering τ in k_w. Furthermore, that $s(F_{w'}) = 1$ means that $F_{w'}$ consists of two orderings τ_1 and τ_2, which are extensions of τ to k_w. Finally, since $k_{w'}$ is algebraic over k_w, there is an algebraic extension $(k', \bar{w}') \supset (k_u, \bar{w})$ such that $k_{\bar{w}'} = k_{w'}$ and $\Gamma_{\bar{w}'} = \Gamma_{\bar{w}}$ (Proposition 2.2). In this situation, by the Baer-Krull theorem, the two orderings τ_1, τ_2 of $k_{\bar{w}'}$ have two generizations $\bar{\sigma}_1, \bar{\sigma}_2$ which are extensions to k' of the generization $\bar{\sigma}$ of τ. Thus, $\varepsilon = 1$. We are done. $\square$

3. Field Extensions: Lower Bounds

Let $L \supset K$ be a field extension. In view of Corollary 1.4, lower bounds for $s(L)$ can be obtained by explicit exhibition of real valuations. For instance:

Proposition 3.1 *Let $L \supset K$ be a purely transcendental field extension of transcendence degree $d \geq 1$. Then $s(L) = s_0(K) + d$.*

Proof. After Theorem 2.7, it remains to show $s(L) \geq s(K') + d$ for each finite extension K' of K, say $K' = K[t]/P(t)$. Let $L = K(x_1, \ldots, x_d)$. Then the localization $K[x_1, \ldots, x_d]_{(P(x_1), \ldots, x_d)}$ is a local regular ring of dimension d with residue field K', and by Remark 1.5 *b)*, $s(L) \geq s(K') + d$. $\square$

In general, we have:

Proposition 3.2   *Let $L \supset K$ be a finitely generated extension of formally real fields, and σ an ordering of K. Then $s(L) \geq s(\mathrm{Spec}_\sigma(L)) \geq \deg(L : K)$.*
In particular, if K is real closed, then $s(L) = \deg(L : K)$.

This will follow if we can produce enough real valuations of L. In fact:

Lemma 3.3   *Let $P \in K[x, t]$ be an irreducible polynomial in the indeterminates $x = (x_1, \ldots, x_d)$ and t, monic in t. Let $h_1, \ldots, h_s \in K[x, t]$ be positive in some prime cone which restricts to σ and has support (P). Then there exist a maximal ideal $\mathfrak{m} \subset K[x, t]$ and irreducible polynomials $F_1, \ldots, F_d \in K[x, t]$ such that*

a) $F_1, \ldots, F_d, P \in \mathfrak{m}$ and are a regular system of parameters of the local regular ring $K[x, t]_\mathfrak{m}$,
b) The localization $K[x, t]_{(F_1, \ldots, F_d, P)}/P$ is a local regular ring of dimension d,
c) Each localization $K[x, t]_{(F_i, P)}/P$, $1 \leq i \leq d$, is a rank 1 discrete valuation ring, and
d) $h_1, \ldots, h_s$ are positive in some prime cone which restricts to σ and has support $\mathfrak{m}$.

Proof.   Clearly, *b)* and *c)* are consequences of *a)*. Let β be a prime cone with support (P) such that $h_i(\beta) > 0$ for $i = 1, \ldots, s$, and $\beta | K = \sigma$. Let R be the real closure of K with respect to σ. Then, by the Artin-Lang homomorphism theorem, there is a prime cone $\alpha : K[x, t] \to R$ such that $P(\alpha) = 0$, $(\partial P/\partial t)(\alpha) \neq 0$ and $h_i(\alpha) > 0$ for $i = 1, \ldots, s$; let $\mathfrak{m}$ be the support of α. Then $\mathfrak{m}$ is a maximal ideal, the localization $K[x, t]_\mathfrak{m}$ is a local regular ring of dimension $d + 1$ and $\mathfrak{m}/\mathfrak{m}^2$ has dimension $d + 1$ over the residue field $\kappa(\mathfrak{m})$. But $\partial P/\partial t \notin \mathfrak{m}$ implies that $P \notin \mathfrak{m}^2$, and we can complete P to a basis $F_1, \ldots, F_d, P$ of $\mathfrak{m}/\mathfrak{m}^2$ over $\kappa(\mathfrak{m})$. An immediate application of Nakayama's lemma shows that such a basis generates $\mathfrak{m}K[x, t]_\mathfrak{m}$, hence it is a system of parameters. We have shown *a)*, and *d)* is guaranteed by the construction. $\square$

Once this lemma is proved we turn to Proposition 3.2. By the stability formula for spaces of orderings (Corollary IV.7.4), we have to exhibit a fan F of $\mathrm{Spec}_\sigma(L)$ with 2^d elements, where $d = \deg(L : K)$. In fact:

Proposition 3.4   *Let $L \supset K$ be a finitely generated extension of formally real fields of transcendence degree d. Let σ be an ordering of K and let U be a non-empty open set of $\mathrm{Spec}_\sigma(L)$. Then, there exist a fan $F \subset U$ with $\#(F) = 2^d$.*

Proof.   As is well known, L can be represented as the quotient field of a K-algebra of the form $K[\mathbf{x}, \mathbf{t}]/P$, $\mathbf{x} = (\mathbf{x}_1, \ldots, \mathbf{x}_d)$, where $P \in K[\mathbf{x}, \mathbf{t}]$ is monic in $\mathbf{t}$ and irreducible. Also, we can assume $U = \{h_1 > 0, \ldots, h_s > 0, P = 0\}$ for some $h_i \in K[\mathbf{x}, \mathbf{t}]$. Then, we pick $\mathfrak{m}$ as in the preceding lemma, and there is a prime cone $\alpha \in U$ with support $\mathfrak{m}$. Since $\mathbb{K}[\mathbf{x}, \mathbf{t}]_\mathfrak{m}/P$ is a local regular ring of dimension d with quotient field L, Proposition II.3.4 gives a rank d discrete valuation ring ring W of L with residue field $\kappa(\mathfrak{m})$. Then, the 2^d generizations of α compatible with W form a fan F. Furthermore the h_i's are positive in all those generizations because so is α, and consequently $F \subset U$. $\qquad\square$

We now consider extensions $L \supset \mathbb{Q}$. If the extension is algebraic, and formally real, $s(L) \le s(\mathbb{Q}) + 1 = 1$, and $s(L) = 0$ if and only if L admits a unique ordering. For non-algebraic extensions we have:

Proposition 3.5   *Let $L \supset \mathbb{Q}$ be a formally real finitely generated extension of $\mathbb{Q}$ of transcendence degree $d \ge 1$. Then $s(L) = d + 1$.*

Again, we have to produce real valuations, which we do in the next lemma:

Lemma 3.6   *Let $P \in \mathbb{Q}[\mathbf{x}, \mathbf{t}]$ be an irreducible polynomial in two indeterminates $\mathbf{x}$ and $\mathbf{t}$, monic in $\mathbf{t}$. Let $h_1, \ldots, h_s \in \mathbb{Q}[\mathbf{x}, \mathbf{t}]$ be positive in some prime cone with support (P), and let p be a positive integer. Then there exists an irreducible polynomial $F \in \mathbb{Q}[\mathbf{x}]$ such that:*

a) *The ideal $(F, P) \subset \mathbb{Q}[\mathbf{x}, \mathbf{t}]$ is maximal,*
b) *The residue field $\mathbb{Q}[\mathbf{x}, \mathbf{t}]/(F, P)$ is a number field with at least p orderings which make $h_1, \ldots, h_s$ positive.*
c) *The localization $\mathbb{Q}[\mathbf{x}, \mathbf{t}]_{(F,P)}/P$ is a rank 1 discrete valuation ring.*

Proof.   Let L be the quotient field of $\mathbb{Q}[\mathbf{x}, \mathbf{t}]/P$. By the Artin-Lang homomorphism theorem, there is a point $(a, b) \in \mathbb{R}$ such that $P(a, b) = 0$, $\partial P(a, b)/\partial \mathbf{t} \ne 0$ and $h_i(a, b) > 0$ for all i. By the implicit functions theorem, we find an open interval $I \subset \mathbb{R}$ containing a and a continuous function $\varphi : I \to \mathbb{R}$ such that $\varphi(a) = b$, $P(x, \varphi(x)) = 0$, $\partial P(x, \varphi(x))/\partial \mathbf{t} \ne 0$, and $h_i(x, \varphi(x)) > 0$ for $x \in I$ and $i = 1, \ldots, s$. After a linear translation $\mathbf{x} \mapsto \mathbf{x} - r$ for some rational number $r \in I$, we may assume that $0 \in I$.

Now we need an elementary fact from Galois theory:

> For every prime number p there are infinitely many different totally real cyclic extensions $K \supset \mathbb{Q}$ of degree p.

For instance, if $p = 2$ let $K = \mathbb{Q}(\sqrt{q})$, $q = 2, 3, 5, \ldots$; if $p = 3$, let K be the splitting field of the polynomial $\mathbf{y}^3 - 3q^2\mathbf{y} + q^3$ for $q = 2, 3, 5, \ldots$. In general, one takes a suitable subfield K of the splitting field of $\mathbf{y}^q - 1$, where q is a prime positive integer such that $q \equiv 1 \bmod p$ (and there are infinitely many q's by Dirichlet's theorem).

Now we claim that our extension $K \supset \mathbb{Q}$ can be chosen such that P is irreducible over K. Indeed, any factorization of P over K is also a factorization over $\mathbb{R}$. Hence, if P has a non-trivial monic divisor over K, it is one of its monic divisors over $\mathbb{R}$. These latter are finitely many, and none belongs to $\mathbb{Q}$. Thus, we pick from each a coefficient $a \notin \mathbb{Q}$. Thus, we have to find K such that none of those finitely many a's belongs to K. Since the degree of $K \supset \mathbb{Q}$ is prime, and $a \notin \mathbb{Q}$, the condition $a \in K$ is equivalent to $\mathbb{Q}(a) = K$, and consequently, we have finitely many finite extensions $\mathbb{Q}(a) \supset \mathbb{Q}$ that must be different from K. Since there are infinitely many K's, the claim is proved.

After this choice of K, we pick a primitive element $\theta \in \mathbb{R}$ of K over $\mathbb{Q}$. For each automorphism σ of K, we get a conjugate $\sigma(\theta) \in \mathbb{R}$ of θ, and, the extension being cyclic, these are all the p conjugates of θ. Now, replacing θ by by θ/m for big enough m, we can suppose that all these conjugates belong to I. Since P is irreducible over K, so is $P(\mathbf{x} + \theta, \mathbf{t})$, and by Hilbert's irreducibility theorem ([Lg2 VII.2 Cor.4, pp.148, and VII.3, pp.150-152]) we find a rational number x arbitrarily close to 0 such that $P(\zeta, \mathbf{t})$ is irreducible over K, where $\zeta = x + \theta \in K$. Then, for each automorphism σ (including the identity), $P(\sigma(\zeta), \mathbf{t})$ is irreducible over K. On the other hand, as x is very small, $\sigma(\zeta) = x + \sigma(\theta) \in I$ and we consider $\eta^\sigma = \varphi(\sigma(\zeta))$. The conclusion is that $P(\sigma(\zeta), \mathbf{t})$ is the minimal polynomial of η^σ over K. Finally, since K is cyclic of prime degree p, all the $\sigma(\zeta)$'s have the same minimal polynomial over $\mathbb{Q}$, namely

$$F(\mathbf{x}) = \prod_\sigma (\mathbf{x} - \sigma(\zeta)).$$

After this long preamble, we define p prime cones

$$\alpha^\sigma : \mathbb{Q}[\mathbf{x}, \mathbf{t}] \to \mathbb{R}\, ; \, \mathbf{x} \mapsto \sigma(\zeta),\, \mathbf{t} \mapsto \eta^\sigma.$$

Obviously, all of them are different, and make the h_i's positive, since

$$\alpha^\sigma(h_i) = h_i(\sigma(\zeta), \varphi(\sigma(\zeta))) > 0$$

by the properties of φ. An straightforward computation shows that the support of all of them is the ideal $(F(\mathbf{t}), P(\mathbf{x}, \mathbf{t})) \subset \mathbb{Q}[\mathbf{x}, \mathbf{t}]$. We also see that the domain $\mathbb{Q}[\mathbf{x}, \mathbf{t}]/(F, P)$ is an algebraic extension of $\mathbb{Q}$, hence a field, and deduce that (F, P) is a maximal ideal. Consequently, $\mathbb{Q}[\mathbf{x}, \mathbf{t}]_{(F,P)}$ is a regular local ring of dimension 2, and $\{F, P\}$ is a regular system of parameters. Condition $c)$ follows immediately. $\qquad\qquad\qquad\qquad\qquad\qquad\qquad\qquad\qquad\qquad\qquad\qquad$ $\square$

Thus, we are ready to prove Proposition 3.5. Again, we have to exhibit fans with enough elements, and we do this generically:

Proposition 3.7   *Let $L \supset \mathbb{Q}$ be a formally real finitely generated extension of $\mathbb{Q}$ of transcendence degree $d \geq 1$. Let U be a non-empty open subset of $\mathrm{Spec}_r(L)$. Then there is a fan $F \subset U$ with $\#(F) = 2^{d+1}$.*

Proof.   The field L is the quotient field of a $\mathbb{Q}$-algebra of the form $\mathbb{Q}[\mathbf{x}, \mathbf{t}]/P$, $\mathbf{x} = (\mathbf{x}_1, \ldots, \mathbf{x}_d)$, where $P \in \mathbb{Q}[\mathbf{x}, \mathbf{t}]$ is monic in $\mathbf{t}$ and irreducible. Then, we can suppose $U = \{h_1 > 0, \ldots, h_s > 0, P = 0\}$ for some polynomials $h_i \in \mathbb{Q}[\mathbf{x}, \mathbf{t}]$. Now we choose $\mathfrak{m}$ and $F_1, \ldots, F_d$ as in Lemma 3.3, and consider the prime ideal $\mathfrak{p} = \{F_2, \ldots, F_d, P\}\mathbb{Q}[\mathbf{x}, \mathbf{t}]_{\mathfrak{m}}$. The localization $\mathbb{Q}[\mathbf{x}, \mathbf{t}]_{\mathfrak{p}}/P$ is a local regular ring of dimension $d - 1$ with quotient field L, whose residue field L' is a finitely generated extension of transcendence degree 1 of $\mathbb{Q}$. Moreover, $U' = \{h_1 > 0, \ldots, h_s > 0, F_2 = \cdots = F_d = P = 0\}$ is a non-empty open subset of $\mathrm{Spec}_r(L')$. By Lemma II.3.4 and Theorem 1.3, every fan $F' \subset U'$ lifts to another $F \subset U$ with $\#(F) = 2^{d-1}\#(F')$. We thus have reduced our problem to the case $d = 1$, which we assume henceforth.

We choose an irreducible polynomial $F \in \mathbb{Q}[\mathbf{x}]$ as in Lemma 3.6 for $p = 2$, that is, $V = \mathbb{Q}[\mathbf{x}, \mathbf{t}]_{(F,P)}/P$ is a rank 1 discrete valuation ring whose residue field K has at least 2 orderings $\tau_1, \tau_2 \in U$. By the Baer-Krull theorem, these have four generizations in U that form a fan F. We are done. $\square$

4. Algebras

Here we consider global computations, that is, computations of stability indices of rings, or, more precisely of, algebras. To do that we shall need another important result from valuation theory:

Proposition 4.1   *Let A be a local noetherian domain of dimension d, K its quotient field and k its residue field. Let V a valuation ring of K that dominates A. Then* $\mathrm{rat.rk.}(\Gamma_v) + \deg(k_v : k) \leq d$.

Proof.   [Ab Th.1, p.330]. This is in fact a generalization of Proposition 2.1. $\square$

Back to stability indices, we start with the following:

Theorem 4.2   *Let A be a commutative ring with unit. Let L a field which is an A-algebra and let F be a non-trivial fan of L. Let $\psi : \mathrm{Spec}_r(L) \to \mathrm{Spec}_r(A)$ be the associated map of real spectra, let $\gamma \in F$, $\beta = \psi(\gamma)$ and let α be the unique closed specialization of β in $\mathrm{Spec}_r(A)$. Assume that the localization $A_{\mathrm{supp}(\alpha)}$ is noetherian. Then*

$$s(F) \leq s_0(\kappa(\mathrm{supp}(\alpha))) + \dim(\beta \to \alpha) + \deg(L : \kappa(\mathrm{supp}(\beta))).$$

Proof.   Set $\mathfrak{p} = \mathrm{supp}(\beta)$ and $\mathfrak{m} = \mathrm{supp}(\alpha)$, so that $\dim(\beta \to \alpha) = \dim(A_{\mathfrak{m}}/\mathfrak{p})$. Set $K = \kappa(\mathfrak{p})$. Clearly, we can suppose that $r = \deg(L : K)$ and $d = \dim(A_{\mathfrak{m}}/\mathfrak{p})$ are finite. We argue by induction on d. If $d = 0$, then $\beta = \alpha$, and $\mathfrak{p}$ is the support of α. Hence, the bound follows from the result for fields (Theorem 2.7). Now, let $d > 0$, and consider the convex hull W of $A_{\mathfrak{m}}/\mathfrak{p}$ in L with respect to γ, which is a valuation ring that dominates $A_{\mathfrak{m}}/\mathfrak{p}$. On the other hand, we have a valuation ring W' of L that trivializes F (Theorem 1.6). In particular, W' is compatible with γ, as W is, and so either $W \subset W'$ or $W' \subset W$. Let V denote the bigger of W and W'. Note that this V is compatible with all the orderings of F, and F induces a fan F_v in the residue field L_v of V. By Theorem 1.3, we have:

$$s(F) \leq s(F_v) + \dim_{\mathbb{F}_2}(\Gamma_v/2\Gamma_v).$$

Let V be trivial. If $V = W'$, then W' and so F would be trivial. If $V = W$, then $A_{\mathfrak{m}}/\mathfrak{p} = K$, that is, $\mathfrak{m} = \mathfrak{p}$, so that $d = 0$ Consequently, V is not trivial, which means that V dominates a local ring $A_{\mathfrak{n}}/\mathfrak{p}$, where $\mathfrak{n}$ is a prime ideal of A such that $\mathfrak{p} \subset \mathfrak{n} \subset \mathfrak{m}$. Note that $A_{\mathfrak{n}}$ is noetherian, since $A_{\mathfrak{m}}$ is noetherian by hypothesis.

Now we pick elements $z_1, \ldots, z_r \in L$ algebraically independent over K. After replacing z_i by $1/z_i$, we may assume that $z_i \in V$, so that V contains the domain $D = (A_{\mathfrak{n}}/\mathfrak{p})[z_1, \ldots, z_r]$. Let $\mathfrak{d} = \mathfrak{m}_v \cap D$. Then by the dimension formula for polynomial extensions of noetherian rings ([Mt Th.23, p.84]), we get

$$\mathrm{ht}(\mathfrak{d}) \leq \mathrm{ht}(\mathfrak{n}/\mathfrak{p}) - \deg(\kappa(\mathfrak{d}) : \kappa(\mathfrak{n})) + r.$$

On the other hand, L is an algebraic extension of $\mathrm{qf}(D)$, and consequently V and $V \cap \mathrm{qf}(D)$ have the same rational rank and the residue field of V is algebraic over that of $V \cap \mathrm{qf}(D)$ ([Bk CA, VI.8.1, Prop.1, p.417]). From Proposition 4.1, we get:

$$\dim_{\mathbb{Z}_2}(\Gamma_v/2\Gamma_v) \leq \mathrm{rat.rk.}(V) \leq \mathrm{ht}(\mathfrak{d}) - \deg(L_v : \kappa(\mathfrak{d})).$$

Puting all the above inequalities together, we conclude:

$$s(F) \leq s(F_v) + \mathrm{ht}(\mathfrak{n}/\mathfrak{p}) - \deg(L_v : \kappa(\mathfrak{n})) + r.$$

Finally, F_v is a fan of the field L_v, which is an extension of $\kappa(\mathfrak{n})$, and, for the canonical map $\varphi_v : \mathrm{Spec}_r(L_v) \to \mathrm{Spec}_r(A)$, all closed specializations of $\varphi_v(F_v)$ are closed specializations of $\psi(F)$. In particular, $\gamma \to \gamma_v \in F_v$, and $\beta \to \beta_v \to \alpha$, where β_v is the restriction to $\kappa(\mathfrak{n})$ of γ_v. Since $\mathfrak{n} \supsetneq \mathfrak{p}$, $\dim(A_{\mathfrak{m}}/\mathfrak{n}) < \dim(A_{\mathfrak{m}}/\mathfrak{p})$, and by induction we have:

$$s(F_v) \leq s_0(\kappa(\mathfrak{m})) + \dim(A_{\mathfrak{m}}/\mathfrak{n}) + \deg(L_v : \kappa(\mathfrak{n})).$$

From the two last inequalities we deduce:

$$s(F) \leq s_0(\kappa(\mathfrak{m})) + \dim(A_{\mathfrak{m}}/\mathfrak{n}) + \mathrm{ht}(\mathfrak{n}/\mathfrak{p}) + r \leq s_0(\kappa(\mathfrak{m})) + \dim(A_{\mathfrak{m}}/\mathfrak{p}) + \deg(L : K),$$

as wanted. $\qquad\qquad\qquad\qquad\qquad\qquad\qquad\qquad\qquad\qquad\qquad\qquad\quad\square$

The preceding theorem can be stated in a more general form, using some additional terminology:

Notations 4.3   *a)* Let A be a commutative ring with unit and $X \subset \mathrm{Spec}_r(A)$ any subset. We set

$$s_0(X) = \sup\{s_0(\kappa(\mathrm{supp}(\alpha))) \mid \alpha \text{ closed specialization of } X\}.$$

b) Let A be a commutative ring with unit and B an A-algebra. We set:

$$\deg_r(B : A) = \sup\{\deg(\kappa(\mathfrak{Q}) : \kappa(\mathfrak{P})) \mid \mathfrak{Q} \subset B \text{ minimal real prime, } \mathfrak{P} = \mathfrak{Q} \cap A\}.$$

$\square$

After this, we have:

Theorem 4.4   *Let B be an A-algebra and $\varphi : \mathrm{Spec}_r(B) \to \mathrm{Spec}_r(A)$ the associated map of real spectra. Let Y be a subspace of $\mathrm{Spec}_r(B)$ and let $X \supset \varphi(Y)$. Suppose that for every closed specialization α of X in $\mathrm{Spec}_r(A)$, the localization $A_{\mathrm{supp}(\alpha)}$ is noetherian. Then*

$$s(Y) \leq s_0(X) + \dim_r(A) + \deg_r(B : A).$$

Proof.   By the global stability formula (Corollary V.1.6), we have to show the bound for every fan F of Y. Let $\mathfrak{q} \subset B$ denote the common support of all prime cones of F, and set $L = \kappa(\mathfrak{q})$, $\mathfrak{p} = \mathfrak{q} \cap A$, $K = \kappa(\mathfrak{p})$. Then we can apply the preceding theorem, and with the notations α, $\mathfrak{m}$, etc., used there, it remains to see that

$$\dim(A_{\mathfrak{m}}/\mathfrak{p}) + \deg(L : K) \leq \dim_r(A) + \deg_r(B : A).$$

Since $\mathfrak{q}$ is a real prime ideal, there is a minimal real prime ideal $\mathfrak{Q} \subset B$ contained in $\mathfrak{q}$; we set $\mathfrak{P} = \mathfrak{Q} \cap A$. Then we have the extension of domains $A' = A_{\mathfrak{p}}/\mathfrak{P} \subset B' = B_{\mathfrak{q}}/\mathfrak{Q}$. Let $B^* \subset B'$ be a finitely generated extension of A', and set $\mathfrak{q}^* = \mathfrak{q} \cap B^*$. Since the ring A' is noetherian, we get from the dimension formula:

$$\deg(\kappa(\mathfrak{q}^*) : K) \leq \mathrm{ht}(\mathfrak{p}/\mathfrak{P}) + \deg(\kappa(\mathfrak{Q}) : \kappa(\mathfrak{P})) \leq \mathrm{ht}(\mathfrak{p}/\mathfrak{P}) + \deg_r(B : A).$$

Since $\deg(L : K)$ is the sup of the transcendence degrees $\deg(\kappa(\mathfrak{q}^*) : K)$ for all the B^*'s, we conclude

$$\dim(A_{\mathfrak{m}}/\mathfrak{p}) + \deg(L : K) \leq$$
$$\dim(A_{\mathfrak{m}}/\mathfrak{p}) + \mathrm{ht}(\mathfrak{p}/\mathfrak{P}) + \deg_r(B : A) \leq \mathrm{ht}(\mathfrak{m}/\mathfrak{P}) + \deg_r(B : A),$$

and since $\mathrm{ht}(\mathfrak{m}/\mathfrak{P}) \leq \dim_r(A)$, we are done. $\square$

Remarks 4.5   *a)* We can always take $X = \mathrm{Spec}_r(A)$, and then we write $s_0(A)$ instead of $s_0(\mathrm{Spec}_r(A))$. In particular, for $Y = \mathrm{Spec}_r(B)$ we get the bound

$$s(B) \leq s_0(A) + \dim_r(A) + \deg_r(B : A).$$

Note that when $A = K \subset L = B$ are fields, we recover Theorem 2.7. $\square$

b) We have $\dim_r(B) \leq \dim_r(A) + \deg_r(B : A)$, and $s_0(A) + \dim_r(B)$ would be a better bound. However, this is not a bound in general. Let $A = \mathbb{R}[[x, y]]$, $B = \mathbb{R}[[x, y]][1/x]$. Then $\dim_r(B) = 1$, but $s(B) \geq 2$. $\square$

Next, we can apply the results of Section V.2 to bound the other complexity invariants. Among the many possible statements we can deduce, we single out the following one:

Proposition 4.6   *Let Y be a noetherian subspace of* $\mathrm{Spec}_r(A)$. *Then*

$$s(Y) \le s_0(Y) + d \quad \text{and} \quad \bar{s}(Y) \le s_0(Y)(d+1) + \frac{1}{2}d(d+1),$$

where $d = \dim_r(A)$.

Proof.   In this case $A = B$ and $X = Y$, so that the bound for s comes from Theorem 4.4. But then, Theorem V.2.9 gives the bound for $\bar{s}$. $\square$

These bounds are important to deal with *bounded* subsets in real analytic geometry in Chapter VIII. The abstract notion corresponding to bounded sets is that of archimedian rings, and we discuss it in Section 6. Another important case to which Proposition 4.6 applies is the following:

Proposition 4.7   *Let A be a noetherian henselian local ring with residue field k and $d = \dim_r(A)$. Then*

$$s(A) \le s_0(k) + d \quad \text{and} \quad \bar{s}(A) \le s_0(k)(d+1) + \frac{1}{2}d(d+1).$$

Proof.   Each prime cone of A specializes to a prime cone with support the maximal ideal (Proposition II.2.4), and consequently $s_0(A) = s_0(k)$. $\square$

For instance, let $A = R[[\mathbf{x}_1, \ldots, \mathbf{x}_d]]$ be a ring of formal power series over a real closed field R. Then $s(A) \le d$. Furthermore, A is a local regular ring, hence by Remark 1.5 *b)*, $s(A) \ge d$ and the bound is sharp. Concerning lower bounds for $\bar{s}$ we can do nothing yet, even for a nice ring as this. We shall come back to this problem in Section VII.5.

5. Algebras Finitely Generated over Fields

For algebras over fields the bound of Theorem 4.4 takes a very neat form:

Proposition 5.1   *Let K be a field and let B be a finitely generated K-algebra. Then*

$$\dim_r(B) \le s(B) \le s_0(K) + \dim_r(B)$$

In particular, if K is real closed, $s(B) = \dim_r(B)$.

Proof.   Since B is a finitely generated K-algebra, we know that $\dim_r(B) = \deg_r(B : K)$, and the upper bound follows immediately from Remark 4.5 *a)*.

For the lower bound, let $\mathfrak{p} \subset B$ be a real prime ideal such that $\dim_r(B) =$

$\dim(B/\mathfrak{p})$, so that $\dim_r(B) = \deg(L : K)$, where L is the quotient field of $B/\mathfrak{p}$. By Proposition 3.2, $\dim_r(B) \leq s(L) \leq s(B)$.

Finally, if K is real closed, $s_0(K) = 0$. $\square$

Similarly, from Propositions 3.5, we deduce:

Proposition 5.2   *Let B be a finitely generated $\mathbb{Q}$-algebra with $\dim_r(B) = d \geq 1$. Then $s(B) = 1 + d$.*

Now we turn to the invariant $\bar{s}$:

Proposition 5.3   *Let K be a field and let B be a finitely generated K-algebra with $d = \dim_r(B)$. Then:*

$$\frac{1}{2}d(d+1) \leq \bar{s}(B) \leq s_0(K)(d+1) + \frac{1}{2}d(d+1).$$

In particular, if K is real closed, $\bar{s}(B) = \frac{1}{2}d(d+1)$,

Proof. We can assume that B is a domain and its quotient field L is formally real, so that $d = \dim(B)$. The upper bound follows from Theorem V.2.9 and Proposition 5.1. The lower bound is trivial for $d = 0$, and for $d > 0$ we argue by induction. We choose $x_1, \ldots, x_n \in B$ with $K[x_1, \ldots, x_n] = B$ and after a K-linear change of coordinates we find an irreducible polynomial $P \in K[\mathsf{x}_1, \ldots, \mathsf{x}_d, \mathsf{t}]$, monic in t with non-zero discriminant $\Delta \in K[\mathsf{x}]$, such that:

i) The mapping $A = K[\mathsf{x}_1, \ldots, \mathsf{x}_d, \mathsf{t}]/P \to B$ given by the substitution $\mathsf{x}_i = x_i$ $(1 \leq i \leq d)$, $\mathsf{t} = x_{d+1}$ is finite, injective and the canonical homomorphism $\mathrm{qf}(A) \to L$ is an isomorphism.

ii) For every prime ideal $\mathfrak{q} \subset B$ with $\Delta \notin \mathfrak{q}$ the canonical homomorphism $A_{\mathfrak{q}\cap A} \to B_{\mathfrak{p}}$ is an isomorphism.

(This is the well-known Noether's normalization lemma, plus a standard integral dependence trick as in [Tg I.7.5].) Using this and Lemma 3.3 we find a real prime ideal $\mathfrak{p} \subset B$ such that $V = B_{\mathfrak{p}}$ is a rank 1 discrete valuation ring. Then $Y = \mathrm{Spec}_r(B/\mathfrak{p})$ is a real divisor as defined in Section V.4, and any uniformizer of V is a uniformizer in the sense thereby explained. By induction, $\mathrm{Spec}_r(B/\mathfrak{p})$ contains a basic closed set D that cannot be written with less than $\frac{1}{2}(d-1)d$ inequalities. Moreover, we need the additional induction assumption that D contains some prime cone with support $\mathfrak{p}$. On the other hand, by Proposition 3.4 there is a fan F of the residue field of $\mathfrak{p}$ contained in D with 2^{d-1} elements. Then we can find exactly $d-1$ elements $a_1, \ldots, a_{d-1} \in B$ such that $F \cap \{a_1 > 0, \ldots, a_{d-1} > 0\}$ consists of one single element. Hence, by Proposition V.4.3, we find a basic closed set C that cannot be written with less than

$$(d-1) + \frac{(d-1)d}{2} + 1 = \frac{1}{2}d(d+1).$$

Note that Proposition V.4.3 also guarantees that C contains a total ordering of L, which we need to complete the induction. $\square$

Over the rationals we can compute exactly $\bar{s}$:

Proposition 5.4   *Let B be a finitely generated $\mathbb{Q}$-algebra of dimension $d = \dim_r(B) \geq 1$. Then:*

$$\bar{s}(B) = \frac{1}{2}(d+1)(d+2).$$

Proof.   Suppose first $d = 1$. We may assume that B is a domain with quotient field L, say $B = \mathbb{Q}[y_1, \ldots, y_s]/\mathfrak{p}$. Using Noether's normalization lemma as in the preceding proof, and Lemma 3.6, we find a maximal ideal $\mathfrak{m}$ such that $V = B_\mathfrak{m}$ is a rank 1 discrete valuation ring whose residue field k admits at least three different orderings $\alpha_1, \alpha_2, \alpha_3$. Thus $\mathfrak{m}$ defines a real divisor Y of $X = \mathrm{Spec}_r(B)$ in the sense of Section V.4. In fact, $Y = \mathrm{Spec}_r(k)$, and $s(Y) = 1$. Now, let $b \in A$ be such that $b(\alpha_1) > 0$, $b(\alpha_2) > 0$ and $b(\alpha_3) < 0$. Then $\{b \geq 0\}$ contains the fan $\{\alpha_1, \alpha_2\}$. On the other hand, let $u \in \mathfrak{m}$ be a uniformizer of V, which clearly is a uniformizer in the sense of real divisors. Let finally $a \in B$ be such that $a(\alpha_1) > 0$ and $a(\alpha_2) < 0$. Then, by Proposition V.4.3, the basic closed set $\{u \geq 0, u^2 a \geq 0, b \geq 0\}$ cannot be written with less than $3 = \frac{1}{2}(1+1)(1+2)$ inequalities.

Now, let $d > 1$. Again we can find a real prime ideal $\mathfrak{p} \subset B$ such that $B_\mathfrak{p}$ is a rank 1 discrete valuation ring. Then $Y = \mathrm{Spec}_r(B/\mathfrak{p})$ is a real divisor. By induction, we find a basic closed set $D \subset Y$ that can not be written with less than $\frac{1}{2}d(d+1)$ inequalities and contains some prime cone with support $\mathfrak{p}$. Then, by Proposition 3.7, there is a fan F of the residue field of $\mathfrak{p}$ which is contained in D, with $\#(F) = 2^{(d-1)+1} = 2^d$. Hence by Proposition V.4.3, we get a basic closed set C which cannot be written with less than

$$d + \frac{d(d+1)}{2} + 1 = \frac{1}{2}(d+1)(d+2),$$

and contains some total ordering of L. $\square$

The same arguments as in the preceding proofs, using now Proposition 2.10, can be adapted to get:

Proposition 5.5   *Let K be a field and let B be a finitely generated K-algebra. Let σ be a fixed ordering of K, and $\mathrm{Spec}_\sigma(B)$ the subspace of $\mathrm{Spec}_r(B)$ consisting of all prime cones that restrict to σ; let d denote the dimension of that subspace. Then:*

$$d \leq s(B_\sigma) \leq s_0(\sigma) + d, \quad \frac{1}{2}d(d+1) \leq \bar{s}(B_\sigma) \leq s_0(\sigma)(d+1) + \frac{1}{2}d(d+1).$$

Concerning the other invariants, we also get bounds, but far from sharp. We shall not write them explicitely, except for the following result:

Proposition 5.6   *Let K be a real closed field, and let B be a finitely generated K-algebra with $d = \dim_r(B) \leq 2$. Then:*

  a) If $d = 1$, $t = \bar{t} = 1$.
  b) If $d = 2$, $t = 3$ and $\bar{t} = 2$.

Proof. The only thing that has not been proved yet is that $2 < t$ when $d = 2$. To do it, we shall realize the conditions of Proposition V.4.10. We can assume that B is a domain of dimension 2. Using Lemma 3.3 in the way we know, we find a maximal ideal $\mathfrak{m} \subset B$ such that $B_\mathfrak{m}$ is a local regular ring of dimension 2. Let $u_1, u_2 \in B$ be a regular system of parameters of $B_\mathfrak{m}$, and let $\mathfrak{p} \subset B_\mathfrak{m}$ the prime ideal generated by u_1. Set $C_1 = \{u_1 > 0\}$, $C_2 = \{u_1 < 0, u_2 > 0\}$. Let α be the prime cone with support $\mathfrak{m}$, and β, β' its two generizations in $B_\mathfrak{m}/u_1$, defined by $\beta(u_2) < 0$ and $\beta'(u_2) > 0$. Thus $T = \mathrm{Adh}_Z(\beta) \cap \{u_2 < 0\} \neq \emptyset$, and (Lemma 3.3 once again) there is $\alpha' \in T$ with support a maximal ideal $\mathfrak{m}' \neq \mathfrak{m}$ such that $B_{\mathfrak{m}'}$ is a local regular ring. In fact, $\mathfrak{m}$ and $\mathfrak{m}'$ are the maximal ideals of two different points $a, a' \in K^n$, once we choose a surjective homomorphism $K[\mathbf{x}_1, \ldots, \mathbf{x}_n] \to B$. Since $a \neq a'$, we have, for instance, $a_2 < a_2'$, and the element $h = \mathbf{x}_2 - \frac{1}{2}(a_2 + a_2')$ verifies $h(\alpha) < 0$, $h(\alpha') > 0$. Finally, we set $C_3 = \{h > 0\}$. In this situation, the hypothesis of Proposition V.4.10 hold for $Y = \mathrm{Adh}_Z(T)$, $C = C_1 \cup C_2 \cup C_3$, $y = \beta$, $y' = \beta'$ and $z \to \alpha'$. $\square$

6. Archimedean Rings

Let A be a commutative ring with unit, and $X = \mathrm{Spec}_r(A)$ its real spectrum. We are interested in some special points of X, namely:

Definition 6.1   *A point $\alpha \in X$ is called* archimedean *if it induces an archimedean ordering in the residue field $\kappa(\mathrm{supp}(\alpha))$.*

Since every archimedean ordering in a field comes from an embedding into $\mathbb{R}$, we see that every archimedean point $\alpha \in X$ is in fact a homomorphism $\alpha : A \to \mathbb{R}$. Then, α is closed point: if $\alpha \to \sigma$, $f(\alpha) > 0$ and $f(\sigma) = 0$, the element $\alpha(f) \in \mathbb{R}$ would be infinitesimal! It is clear that, in general, a closed point need not be archimedean. We have the following characterization:

Proposition 6.2   *Let $\alpha \in X$ be a closed point with support $\mathfrak{p}$. The following assertions are equivalent:*

  a) *α is archimedean.*
  b) *For every $f \in A$ there is exists a positive integer n such that $-n \leq f(\alpha) \leq n$.*
  c) *For every generization $\beta \in X$ of α and every $f \in A$ there exists a positive integer n such that $-n \leq f(\beta) \leq n$.*
  d) *For every $f \in A$ there exists a positive integer n such that $-n \leq f(\beta) \leq n$ for all generizations $\beta \in X$ of α.*
  e) *Every valuation ring V of $\kappa(\mathfrak{p})$ compatible with α contains $A/\mathfrak{p}$.*

f) For every generization $\beta \in X$ of α with support $\mathfrak{q}$, every valuation ring of $\kappa(\mathfrak{q})$ compatible with β contains $A/\mathfrak{q}$.

Proof.   *a)* $\Rightarrow$ *b)*, *d)* $\Rightarrow$ *f)* and *f)* $\Rightarrow$ *e)* are trivial, and *b)* $\Rightarrow$ *c)* follows easily from the continuity of specializations.

c) $\Rightarrow$ *d):* Let $f \in A$. The set Y of all generizations of α is proconstructible, hence compact. By *c)*, the constructible sets $U_n = \{-n \leq f \leq n\}$, $n \geq 1$, cover Y, and by compactness, finitely many of these U_n's cover Y too. The maximum of the corresponding n's does the job.

e) $\Rightarrow$ *a):* Suppose that α is not archimedean, the convex hull $V = V(\alpha)$ of $\mathbb{Q}$ in $\kappa(\mathfrak{p})$ with respect to α is a non-trivial valuation ring compatible with α. If $V \supset A/\mathfrak{p}$, the maximal ideal $\mathfrak{m}_v$ of V lies over a non-zero ideal $\mathfrak{m}$ of $A/\mathfrak{p}$. This defines a non-trivial specialization $\alpha \to \sigma$ with $\mathrm{supp}(\sigma) = \mathfrak{m}$, and α is not closed. $\square$

Archimedean points give rise to low bounds for the stability index:

Theorem 6.3   *Let A be a commutative ring with unit.*

a) *Let L be a field which is an A-algebra and let F be a fan of L. Let $\psi : \mathrm{Spec}_r(L) \to \mathrm{Spec}_r(A)$ be the associated map of real spectra, let $\gamma \in F$, $\beta = \psi(\gamma)$ and let α be the unique closed specialization of β in $\mathrm{Spec}_r(A)$. Assume that α is archimedean and the localization $A_{\mathrm{supp}(\alpha)}$ is noetherian. Then*
$$s(F) \leq 1 + \dim(\beta \to \alpha) + \deg(L : \kappa(\mathrm{supp}(\beta))).$$

b) *Let B be an A-algebra and $\varphi : \mathrm{Spec}_r(B) \to \mathrm{Spec}_r(A)$ the associated map of real spectra. Let Y be a subspace of $\mathrm{Spec}_r(B)$ and let $X \supset \varphi(Y)$. Suppose that every α, closed specialization of X, is archimedian and the localization $A_{\mathrm{supp}(\alpha)}$ is noetherian. Then*
$$s(Y) \leq 1 + \dim_r(A) + \deg_r(B : A).$$

Proof.   It is enough to track back the proofs of Theorems 4.2 and 4.4 to see that the valuation ring V compatible with the fan F can be chosen to trivialize F, and so $\#(F_v) \leq 2$, that is, $s(F_v) \leq 1$. By this reason, $s_0(\kappa(\alpha))$ is replaced by 1 in the formulae finally obtained. $\square$

Next we define the rings which are the topic of this section:

Definition 6.4   *A ring A is called* totally archimedean *if all the closed points of its real spectrum are archimedean.*

With this terminology, from Theorem 6.3 and Theorem V.2.9, we get

Proposition 6.5   *Let A be a noetherian totally archimedean ring with $d = \dim_r(A)$. Then*
$$s(A) \leq 1 + d \quad and \quad \bar{s}(A) \leq \frac{1}{2}(d+1)(d+2).$$

We remark that these bounds are not necessarily better than those of Section 4. For instance, if A is henselian with real closed field, then Proposition 4.7 is better.

Now we turn to the most striking feature of archimedean rings, namely, that complexity matters (generation of basic sets, separation) have a multilocal nature. We first consider basicness:

Proposition 6.6   *Let A be a noetherian totally archimedean ring and (X, G), $X = \mathrm{Spec}_r(A)$, its space of signs. Let $C \subset X$ and consider an integer $s \geq 1$. The following assertions are equivalent:*

    a) There are $f_1, \ldots, f_s \in A$ such that $C = \{f_1 > 0, \ldots, f_s > 0\}$.
    b) For any two closed points $x, y \in X$ there are $g_1, \ldots, g_s \in A$ and an open set U containing x, y such that $C \cap U = \{g_1 > 0, \ldots, g_s > 0\} \cap U$.

Proof. For the non-trivial implication, we first remark that any open covering of X_{max} is automatically a covering of X. Thus, *b)* implies that C is locally constructible in X, and by compactness, C is constructible. Next, we shall see that $C \cap \mathrm{Adh}_Z(\mathrm{Bd}(C)) = \emptyset$. Indeed, suppose there is an irreducible component W of $\mathrm{Adh}_Z(\mathrm{Bd}(C))$ which meets C. Then, there are points $\alpha \in W \cap C$ and $\beta \in W \cap \mathrm{Bd}(C)$, and we consider their two closed specializations $\alpha \to x$ and $\beta \to y$. By Proposition II.2.9 we can choose β such that $\mathrm{Adh}_Z(\beta) = W$. Now pick the set U given by *b)*; by continuity $\alpha, \beta \in U$. Furthermore, we have $C \cap U = \{g_1 > 0, \ldots, g_s > 0\} \cap U$ for suitable g_i's. It follows $g_i(\beta) = 0$ for some i, hence g_i vanishes on W and $g_i(\alpha) = 0$. Contradiction.

After this, by the global generation formula (Theorem V.1.4) we must see that no fan F of X gives a numerical obstruction to write C with s inequalities. But F is a fan in the residue field $\kappa = \kappa(\mathfrak{p})$ of a prime ideal $\mathfrak{p}$ of A. Then, there is a valuation ring V of κ compatible with F such that the fan induced by F in the residue field k_v of V is trivial. Since A is totally archimedean, $A/\mathfrak{p} \subset V$, and it follows that the elements of F have two closed specializations, say x, y (possibly $x = y$). Then, an open set U as in *b)* must contain F, and since $C \cap U = \{g_1 > 0, \ldots, g_s > 0\} \cap U$, our F cannot give any numerical obstruction. $\qquad\square$

From the preceding result and Proposition 6.5, we get:

Proposition 6.7   *Let A be a noetherian totally archimedean ring and (X, G), $X = \mathrm{Spec}_r(A)$, its space of signs. Suppose that $\dim_r(A) < \infty$. Then a set $C \subset X$ is basic open if and only if for any two closed points x, y of X there are a basic open set D and an open set U containing x, y such that $C \cap U = D \cap U$.*

Here there is the corresponding statement for basic closed sets, which can be proved in the same way:

Proposition 6.8   *Let A be a noetherian totally archimedean ring and (X, G), $X = \mathrm{Spec}_r(A)$, its space of signs. Suppose that $\dim_r(A) < \infty$. Then a set*

$C \subset X$ *is basic closed if and only if for any two closed points* x, y *of* X *there are a basic closed set* D *and an open set* U *containing* x, y *such that* $C \cap U = D \cap U$.

Now we turn to separation.

Proposition 6.9 *Let* A *be a totally archimedean ring and* (X, G), $X = \mathrm{Spec}_r(A)$, *its space of signs. Then any two disjoint closed subsets of* X *can be separated.*

Proof.   We can assume $\mathbb{Q} \subset A$. First note that every $f \in A$ defines a continuous function $f : X_{\max} = \mathrm{Hom}(A, \mathbb{R}) \to \mathbb{R}$; $\sigma \mapsto \sigma(f)$. Also note that $X_{\max}$ is compact, and by definition A separates points in $X_{\max}$. Let $C, T \subset X$ be closed and disjoint. By Urysohn's theorem ([Dg VII.4.1, p.146]), there is a continuous function $X_{\max} \to \mathbb{R}$ which is $\equiv 1$ on $C \cap X_{\max}$ and $\equiv -1$ on $D \cap X_{\max}$. Then, by the Stone-Weierstrass approximation theorem ([Dg XIII.3.3, p.282]), we find $f \in A$ with $f|C \cap X_{\max} > 0$ and $f|D \cap X_{\max} < 0$. It follows that $f|C > 0$ and $f|D < 0$. Indeed, if, say, $f(\alpha) \leq 0$ for some $\alpha \in C$, we have $f(x) \leq 0$ for the closed specialization x of α. But $x \in C \cap X_{\max}$, contradiction. $\square$

Finally, we turn to the general situation, where we have to separate non-disjoint sets. In that case we have again a multilocal criterion:

Proposition 6.10 *Let* A *be a noetherian totally archimedean ring with* $s = s(A) < \infty$. *Let* (X, G), $X = \mathrm{Spec}_r(A)$, *be the associated space of signs, and let* $C, D \subset X$ *be closed constructible sets. Then* C *and* D *can be separated if and only if any finite set* E *of* 2^{s-1} *closed points of* X *has a neighbourhood* U *such that* $C \cap U$ *and* $D \cap U$ *can be separated.*

Proof. By Theorem V.3.2, it is enough to show that for every real prime ideal $\mathfrak{p} \subset A$ with residue field $\kappa = \kappa(\mathfrak{p})$ and every finite subspace Y of $\mathrm{Spec}_r(\kappa)$ with chain length $cl(Y) \leq 2^{s-1}$, the two sets $C \cap Y$ and $D \cap Y$ can be separated. We proceed as follows. Let E be the set of all closed specializations of elements of Y. We claim that $\#(E) \leq 2^{s-1}$.

In fact, since Y is finite, E is finite too, say $E = \{x_1, \ldots, x_r\}$. Now, by Proposition 6.9, there are $f_1, \ldots, f_r \in A$ such that $f_i(x_i) > 0$ and $f_i(x_j) < 0$ for $i \neq j$. In this situation, the set $Y \cap \{f_i > 0\}$ consists exactly of the generizations of x_i in Y. Thus, we have a covering of Y by r disjoint principal sets, which shows that $r \leq cl(Y) \leq 2^{s-1}$.

Once the claim is proved, let $U \supset E$ be an open set such that $C \cap U$ and $D \cap U$ can be separated. By continuity, U contains Y, and we conclude that $C \cap Y$ and $D \cap Y$ can be separated. We are done. $\square$

7. Coming back to Geometry

Here we descend to a very concrete geometric setting, first discussed in Chapter I. Let K be a formally real field, and $R \supset K$ the real closure of K with respect to a fixed ordering σ. Let V be a (not necessarily irreducible) real affine algebraic R-variety (see Examples I.3.9) and $X = V(R) \subset R^n$ the corresponding algebraic set (for some chosen embedding). We suppose that V is defined over K, or more generally, that X can be defined by equations with coefficients in K. We consider on X three rings of functions, all defined by restriction from R^n:

$\mathcal{P}_K(X) =$ ring of polynomial functions with coefficients in K,
$\mathcal{P}_R(X) =$ ring of polynomial functions with coefficients in R, and
$\mathcal{N}(X) =$ ring of Nash functions (see [B-C-R Ch.8]).

More algebraically, we have:

$$\mathcal{P}_K(X) = K[\mathbf{x}]/I_K, \quad \mathcal{P}_R(X) = R[\mathbf{x}]/I_R, \quad \mathcal{N}(X) = \mathcal{N}(R^n)/I_N,$$

where $\mathbf{x} = (\mathbf{x}_1, \ldots, \mathbf{x}_n)$ are indeterminates, and the I_*'s are the ideals of functions vanishing on X.

We correspondingly equip X with three different structures of real space, namely:

$X_K = (X, G), \quad$ where $G = \{\mathrm{sign}[f] \mid f \in \mathcal{P}_K(X)\},$
$X_R = (X, G), \quad$ where $G = \{\mathrm{sign}[f] \mid f \in \mathcal{P}_R(X)\},$ and
$X_N = (X, G), \quad$ where $G = \{\mathrm{sign}[f] \mid f \in \mathcal{N}(X)\}$

We can identify the Stone space of these three real spaces as follows:

$\widetilde{X_K} = \mathrm{Spec}_\sigma(\mathcal{P}_K(X))$ (see Proposition 5.5),
$\widetilde{X_R} = \mathrm{Spec}_r(\mathcal{P}_R(X)),$
$\widetilde{X_N} = \mathrm{Spec}_r(\mathcal{N}(X)).$

Indeed, by Theorem V.5.3, these equalities follow by identifying each real space X_* with an Artin-Lang subset Y of the right hand side space, which is always $Y = \{\alpha \mid \kappa(\alpha) = R\}$. This was already remarked for X_R in Example V.5.5 a), as a consequence of the Artin-Lang homomorphism theorem. The same works for X_K; note that in this case we get a real spectrum if K admits a unique ordering, and only a subspace of a real spectrum otherwise. For X_N one needs the so-called substitution theorem for Nash functions ([B-C-R 8.5.2, p.159]).

The first important remark is that the boolean algebras of constructible sets for all the three structures are the same, or, in more down to earth words, a set defined by a finite boolean combination of sign conditions on Nash functions is semialgebraic, and can also be described by a finite boolean combination of sign conditions on polynomial functions with coefficients in K. Furthermore, if the initial boolean combination consisted of strict inequalities, the same can be achieved for the latter.

Indeed, this is a consequence of the fact that the canonical mappings

$$\mathrm{Spec}_r(\mathcal{N}(X)) \to \mathrm{Spec}_r(\mathcal{P}_R(X)) \to \mathrm{Spec}_\sigma(\mathcal{P}_K(X))$$

are homeomorphisms for the constructible (resp. Harrison) topologies in each space. To prove this, we can suppose $X = R^n$ and the sequence becomes

$$\mathrm{Spec}_r(\mathcal{N}(R^n)) \to \mathrm{Spec}_r(R[\mathbf{x}]) \to \mathrm{Spec}_\sigma(K[\mathbf{x}]).$$

Then the assertion for the first mapping is another consequence of the substitution theorem for Nash functions ([B-C-R 8.8.1, p.173]). The second mapping is bijective since a prime cone $\alpha : R[\mathbf{x}] \to R_\alpha$ is uniquely determined by $\alpha(\mathbf{x}_1), \ldots, \alpha(\mathbf{x}_n)$. Furthermore, this second mapping is closed with respect to the Harrison topologies by the real going-up for $K[\mathbf{x}] \subset R[\mathbf{x}]$ (Propositions II.4.2 $a)$ and II.4.3). From this the conclusion is immediate.

Consequently, the topological behaviour is the same for the three spaces. In particular, **FT**, **AC**, **CC** hold for all of them (since they hold for X_R). However, concerning complexity this is not true anymore.

Examples 7.1   Let $K = \mathbb{Q}$, $R = \mathbb{R}_{\mathrm{alg}}$.

$a)$ The semialgebraic set $S \subset R^2$ given by the Nash inequality $x + \sqrt{y^2 + 1} > 0$ is not basic for $R[\mathbf{x}, \mathbf{y}]$.

The semialgebraic set $T \subset R^2$ given by the opposite inequality $x + \sqrt{y^2 + 1} < 0$ is also given by $x^2 - y^2 - 1 > 0, x < 0$, but it is not principal for $R[\mathbf{x}, \mathbf{y}]$.

$b)$ The semialgebraic set $S \subset R$ given by $x + \sqrt{2} > 0$ is not basic for $K[\mathbf{x}]$.

The semialgebraic set $T \subset R$ given by $x + \sqrt{2} < 0$ (or equivalently by $x^2 - 2 > 0, x < 0$) is not principal for $K[\mathbf{x}]$.

$c)$ The two disjoint closed semialgebraic sets of R^2, $S = \{x \le 0\} \cup \{y \le 0\}$ and $T = \{x \ge 1, y \ge 1\}$, can be separated by a Nash function, namely $f(x, y) = x + y - \sqrt{(x - y)^2 + 1}$, but cannot by a polynomial.

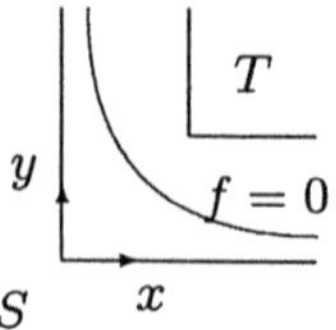

$\square$

We can work out the examples above by finding a suitable fan in the associated Stone space which realizes an obstruction. We only make some drawings and leave the reader the algebraic descriptions. Here they are for Examples 7.1 *a)* and *b)*:

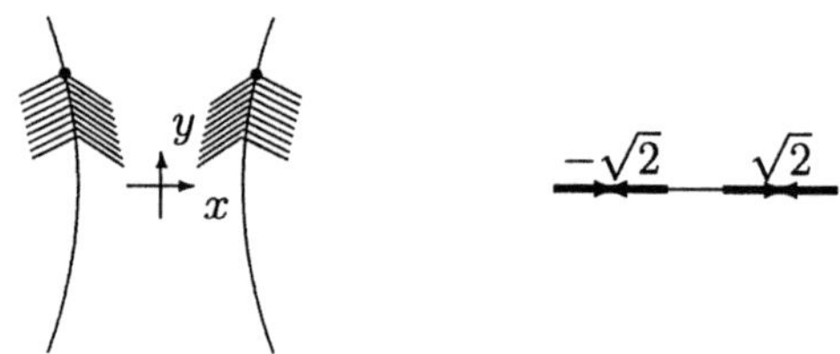

For Example 7.1 *c)* we first embed R^2 in the projective plane, and then draw the fan in a different affine chart:

$$v = 1/y \qquad T = \{v \le u, 1 \le v \le 1\} \qquad u = x/y$$

$$S = \{u \le 0\} \cup \{v \le 0\}$$

$\square$

The fundamental fact now is that we can compute easily the complexity invariants of each of the three spaces. We thus measure the different complexities of the normal forms (I.2.1) and the diagrams (I.2.5) needed to describe the semialgebraic sets of $X = V(R)$ when we use each of the three different types of functions $R[V]$, $K[V]$ and $\mathcal{N}(X)$. We get:

$$s(X, G_R) = s(X, G_N) = \dim(X), \quad s(X, G_K) \le s_0(\sigma) + \dim(X),$$

(see Notations 2.9). Indeed, for G_R this follows directly from Proposition 5.1, for G_N from Theorem 4.4 and for G_K from Proposition 5.5. For instance, in the case $K = \mathbb{Q}$, (X, G_K) is $\mathrm{Spec}_r(\mathbb{Q}[V])$ and $s(X, G_K) = 1 + \dim(V)$. We do not write down more formulae, but instead, we draw some pictures:

Examples 7.2 *a)* Every basic open set of R^d can be written with d inequalities, and here there are some that cannot with less:

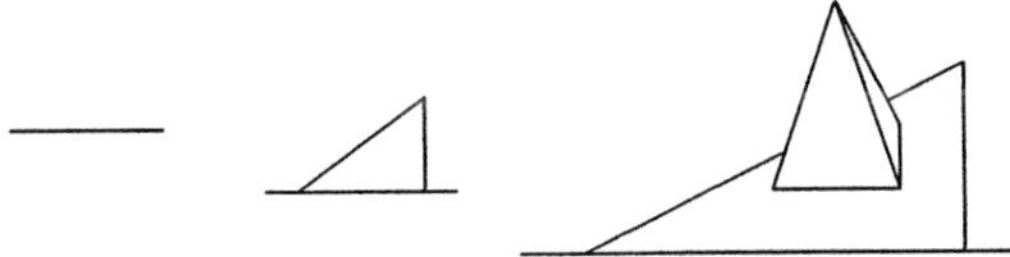

b) Every basic closed set of R^d can be written with $\frac{1}{2}d(d+1)$ inequalities, and here there are some that cannot with less:

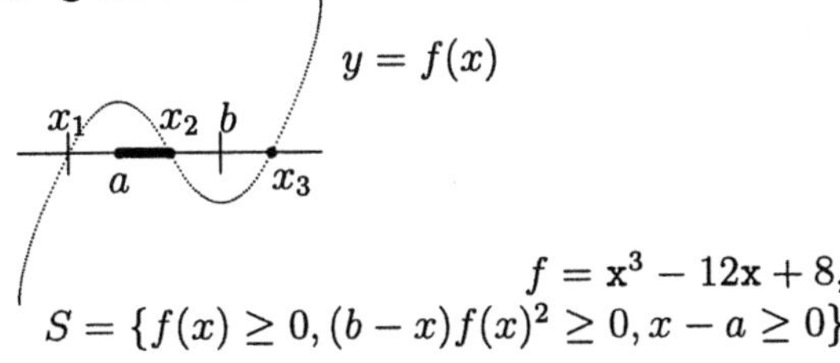

c) Every basic open set of $\mathbb{Q}^d$ can be written with $d+1$ inequalities, and one in $\mathbb{Q}$ that cannot with less was given in Example 7.1 *b)*.

d) Every basic closed set of $\mathbb{Q}^d$ can be written with $\frac{1}{2}(d+1)(d+2)$ inequalities, and here there is one in $\mathbb{Q}$ that cannot with less than 3:

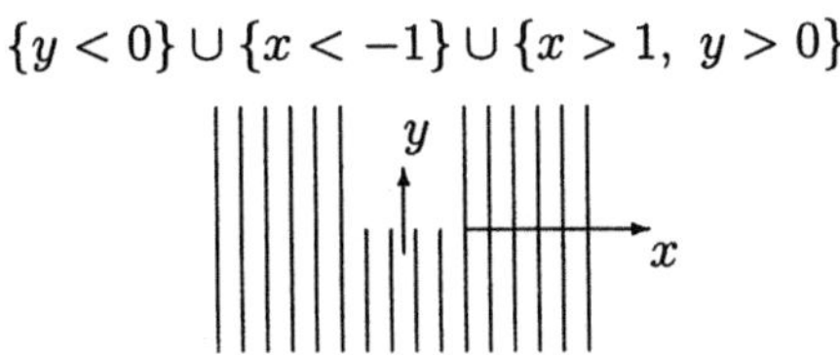

$$f = \mathbf{x}^3 - 12\mathbf{x} + 8,$$
$$S = \{f(x) \geq 0, (b - x)f(x)^2 \geq 0, x - a \geq 0\}$$

e) Every open semialgebraic set of $\mathbb{R}^2$ is the union of 3 basic open sets, and here there is one which is not the union of 2 (see Propositions V.4.10 and VI.5.6):

$$\{y < 0\} \cup \{x < -1\} \cup \{x > 1, \ y > 0\}$$

$\square$

Thus, all the many statements asserted without proof in Chapter I have at last been explained, by systematically giving a geometric meaning to the results concerning algebras finitely generated over a field. We next discuss several counterexamples and concrete results in low dimension. We shall be dealing exclusively with the real space (X, G_R), that is, with semialgebraic subsets of $X = V(R)$ and their descriptions by functions from $R[V]$.

First we consider the question whether the closure of a basic open set is basic closed.

Examples 7.3 *a)* Consider the polynomial

$$f = (x^2t^2 + y^2t^2 + z^2 - t^2)(z^2(x-1)^2 + y^2z^2 - z^2 + t^2) \in R[x, y, z, t],$$

and let V be the corresponding hypersurface. Thus, $R[V] = R[x, y, t, z]/(f)$ and $X = \{f(x, y, z, t) = 0\} \subset R^4$. We see that X has dimension 3 and its singular locus $Y = \{z = t = 0\} \subset X$ has dimension 2. In this situation, $S = X \setminus Y$ is basic open in X, but $\mathrm{Adh}(S)$ is not basic closed. Indeed, we have

$$\mathrm{Adh}(S) \cap Y = \{1 - x^2 - y^2 \geq 0\} \cup \{1 - (x-1)^2 - y^2 \geq 0\} \subset Y \equiv R^2.$$

This set is not basic closed (immediate looking at fans!). $\quad\square$

b) Let now $S \subset R^4$ be the basic open set given by

$$t^2 - (xt - z)^2 - y^2t^2 > 0, \ t > 0, \ y^2 - z^2 > 0.$$

The closure of this set is not basic closed, since

$$\mathrm{Adh}(S) \cap \{z = t = 0\} = \bigcup_{-1 \leq \varepsilon \leq 1} \{(x - \varepsilon)^2 + y^2 \leq 1\}$$

is the sportsfield of Examples I.4.6, which is not even generically basic. $\quad\square$

The examples above are in fact the simplest possible, concerning the dimensions of the ambient space and its singular locus:

Proposition 7.4 *Let Y be the singular locus of X. Suppose that $\dim(X) \leq 3$ and $\dim(Y) \leq 1$. Then the closure of every basic open subset of X is basic closed.*

Proof. This follows immediately from Theorem V.2.11 and Proposition V.4.5, once we notice two facts. First, since the codimension of Y in X is at least 2, all subvarieties of dimension 2 of X are real divisors. Second, since all subvarieties of dimension 1 have stability index 1, everything is basic in them. $\quad\square$

Let us turn to separation. We first translate Theorem V.3.2 into our semi-algebraic setting:

Theorem 7.5 *Let $S, T \subset X$ be two closed semialgebraic sets. The following assertions are equivalent:*

a) *S and T can be separated, that is, there is a function $g \in R[V]$ such that $g|S \geq 0$, $g|T \leq 0$ and $(S \cup T) \cap \{g = 0\} \subset \mathrm{Adh}_Z(S \cap T)$.*

b) *For every subvariety $Y \subset X$ there is a function $f \in R[V]$ not vanishing on Y such that $f|Y \cap S \geq 0$ and $f|Y \cap T \leq 0$.*

In more intuitive words, two closed semialgebraic sets can be separated if and only if they can be separated generically on every subvariety. For instance, "cutting off hairs", that is, separation when $\dim(T) \leq 1$, is always possible.

We next apply the results concerning archimedean rings.

Proposition 7.6   *The ring $R[V]$ is totally archimedean if and only if R is archimedean and X is a bounded affine algebraic set.*

Proof. Suppose that $R[V]$ is totally archimedean. Let α be any closed point of $\mathrm{Spec}_r(R[V])$. Then α is archimedean, hence so is its restriction to R. Thus R is archimedean. Now let $X \subset R^n$, and suppose that X is not bounded. Then, there is $x \in \mathbb{P}^n(R) \setminus X$ which is adherent to X. Hence the sets $U \cap X$, U semialgebraic neighbourhood of x in $\mathbb{P}^n(R)$, form a filter basis of semialgebraic subsets of X. Hence, they are contained in some ultrafilter, which corresponds to a prime cone $\alpha \in \mathrm{Spec}_r(R[V])$. Since $R[V]$ is totally archimedean, $\alpha \to y \in X$. This is impossible, because there are $U \cap X$ as above not adherent to y.

Conversely, suppose that R is archimedean and $X \subset R^n$ is bounded, say $X \subset \{x_1^2 + \cdots + x_n^2 < m\}$ for a given positive integer m. Let H be a real valuation ring of the residue field of some prime ideal $\mathfrak{p}$ of $R[V]$. Since R is archimedean, $R \subset H$, and since $R[V]$ is generated over R by the restrictions of the x_i's, it is enough to see that the classes mod $\mathfrak{p}$ of those restrictions belong to H. But this follows from the convexity of H, since for any prime cone we have $0 < \mathbf{x}_i < \mathbf{x}_1^2 + \cdots + \mathbf{x}_n^2 < m$. $\square$

Let us asume for the sequel that $R \subset \mathbb{R}$ and X is bounded. The set of closed points of the space of signs $\widetilde{X} = \mathrm{Spec}_r(R[V])$ is $\mathrm{Hom}(R[V], \mathbb{R}) = V(\mathbb{R})$, that is, the topological completion of $X = V(R)$. Thus, $\dim(X) = \dim(V(\mathbb{R}))$. On the other hand, since $X \subset V(\mathbb{R})$ and X is an Artin-Lang set of $\widetilde{X}$, also $V(\mathbb{R})$ is a Artin-Lang set. By the ultrafilter theorem (Theorem V.5.3), we conclude that $\widetilde{X}$ is also the Stone space of $V(\mathbb{R})$. Note that here $V(\mathbb{R})$ is endowed with the real space structure corresponding to the ring $R[V]$, that is, the constructible sets of $V(\mathbb{R})$ are the semialgebraic sets defined *over* R. In view of this, Propositions 6.6-6.10 give immediately the following multilocal criteria:

Proposition 7.7   *Let $S \subset V(\mathbb{R})$ and $k \geq 1$. The following assertions are equivalent:*

    a) There are $f_1, \ldots, f_k \in R[V]$ such that $S = \{f_1 > 0, \ldots, f_k > 0\}$.

    b) For any two points $x, y \in V(\mathbb{R})$ there are $g_1 \ldots, g_k \in R[V]$ and an open set U containing x, y such that $S \cap U = \{g_1 > 0, \ldots, g_k > 0\} \cap U$.

In particular, S is basic open over R if and only if for any two points $x, y \in V(\mathbb{R})$ there are D, basic open over R, and an open set U containing both x, y, such that $S \cap U = D \cap U$.

Proposition 7.8   *A subset $S \subset V(\mathbb{R})$ is basic closed over R if and only if for any two points $x, y \in V(\mathbb{R})$ there are a set D, basic closed over R, and an open set U containing both x, y, such that $S \cap U = D \cap U$.*

Proposition 7.9   *Let $S, T \subset V(\mathbb{R})$ be closed semialgebraic over R. Set $d = \dim(X)$.*

    a) If S and T are disjoint, they can be separated over R.

b) The sets S, T can be separated over R if every finite set $E \subset V(\mathbb{R})$ of 2^{d-1} points is contained in an open set U such that $S \cap U$ and $T \cap U$ can be separated over R.

Example 7.10   Separation of closed disjoint sets depends heavily on the archimedean assumption. Let for a moment R be a non-archimedean real closed field. Then the convex hull H of $\mathbb{Q}$ in R is a non-trivial valuation ring, and we can pick an element $\varepsilon > 0$ in the maximal ideal $\mathfrak{m}$ of H. Let D_1, D_2, D_3, D_4 be the four closed squares of R^2 defined by $\varepsilon \leq \pm x, \pm y \leq 1$. Then $S = D_1$ and $T = D_2 \cup D_3 \cup D_4$ are closed disjoint semialgebraic sets that cannot be separated.

The idea to see this is to apply the place associated to H, under which the ε-distance collapses. To be more precise, let k be the residue field of H and Γ its value group, which is divisible. Consider the valuation ring $W_1 = R[\mathbf{x}, \mathbf{y}]_{(\mathbf{y})} \subset R(\mathbf{x}, \mathbf{y})$, whose residue field is $R(\mathbf{x})$, and then the valuation ring $W_2 = H[\mathbf{x}]_{\mathfrak{m}[\mathbf{x}]} \subset R(\mathbf{x})$. Then the composite valuation W of W_1 and W_2 has residue field $k(\mathbf{x})$ and value group $\Gamma \oplus \mathbb{Z}$. Now take two orderings α_1, α_2 of $k(\mathbf{x})$ that specialize to $1/2, -1/2$, respectively, in $\mathrm{Spec}_r(k[\mathbf{x}])$. These α_i have four generizations β_{ij} in $\mathrm{Spec}_r(R[\mathbf{x}, \mathbf{y}])$ which form a fan F of $\mathrm{Spec}_r(R(\mathbf{x}, \mathbf{y}))$ compatible with W. The constructible set of $\mathrm{Spec}_r(R(\mathbf{x}, \mathbf{y}))$ corresponding to D_k contains exactly one of the β_{ij}, and so, to separate S from T we should separate one β_{ij} from the other three, which is impossible.

Note that by Corollary V.3.3, the set T is not basic closed, which it would surely be if ε were not infinitesimal. $\square$

We end now with an striking result in dimension 2. We still assume that R is archimedean, but not that X is bounded.

Proposition 7.11   *Let $\dim(X) = 2$. Then any semialgebraic set $S \subset X$ can be written in the form $S = \{f > 0, g > 0\} \cup \{h > 0\} \cup T$, where $f, g, h \in R[V]$, T is a semialgebraic set of dimension 1 and the union is disjoint.*

Proof.   Since $R \subset \mathbb{R}$, it is enough to prove the result for $R = \mathbb{R}$ (model completeness, Theorem I.1.6), which we assume henceforth. Suppose that we find $f, g, h \in \mathbb{R}[V]$ and a Zariski closed set $Z \subset X$ such that $S \setminus Z = (\{f > 0, g > 0\} \cup \{h > 0\}) \setminus Z$ and the union is disjoint. Then, we set $T = S \cap Z$ and pick a positive equation $p \in \mathbb{R}[V]$ of Z, so that $S = \{pf > 0, g > 0\} \cup \{ph > 0\} \cup T$.

By this remark, replacing S by its interior, we can assume that S is open, and we choose a description by disjunctions/conjunctions of strict inequalities involving $f_1, \ldots, f_s \in \mathbb{R}[V]$. Again by the initial remark, we can move to any birational model of V. Hence, we replace V by a projective model, and then, by resolution of singularities, by a non-singular projective one where all the f_i's are normal crossings. Whence, X is a non-singular projective surface and the sets $\{f_i = 0\} \subset X$ are unions of non-singular curves in general position. Furthermore, any real projective variety is in fact a bounded affine algebraic set. Consequently, by Proposition 7.6, $\mathbb{R}[V]$ is a totally archimedean ring.

Finally, since $\mathrm{Int}(\mathrm{Adh}(S))$ and S differ in a Zariski-closed set of dimension ≤ 1, the initial remark again says that we can assume $S = \mathrm{Int}(\mathrm{Adh}(S))$. In this situation, we easily classify the boundary points x of S. Let $g_1, g_2 \in \mathbb{R}[V]$ a regular system of parameters of the localization $\mathbb{R}[V]_{\mathfrak{m}}$ at the maximal ideal $\mathfrak{m}$ of x, such that $f_i = u_i g_1^{m_{i1}} g_2^{m_{i2}}$, $u_i(x) \neq 0$, in that localization. In a neighbourhood of x there are the following five possibilities:

   i) Only one g_i vanishes at x and $S = \{g_i > 0\}$.
   ii) Both g_i's vanish at x and $S = \{g_1 g_2 > 0\}$.
   iii) Both g_i's vanish at x and $S = \{g_1 > 0, g_2 > 0\}$.
   iv) Both g_i's vanish at x and $S = \{g_i > 0\}$.
   v) Both g_i's vanish at x and $S = \{g_1 > 0\} \cup \{g_2 > 0\}$.

We call the points of types *iv)* and *v) critical.* Clearly, they are finitely many. Let x be one of them. If x is type *iv)*, we choose a very small open ball U in the ambient affine space such that $U \cap X \subset S$, $x \in \mathrm{Bd}(U)$, and the "circle" $\mathrm{Bd}(U) \cap S$ is tangent to $\mathrm{Bd}(S)$ at x. If x is type *v)*, we choose U such that $U \cap X \subset S$, $x \in \mathrm{Bd}(U)$, and the "circle" $\mathrm{Bd}(U) \cap S$ is transversal to $\mathrm{Bd}(S)$ at x. In any case, let u be an equation for U: $U = \{u < 0\}$ and $\mathrm{Bd}(U) = \{u = 0\}$.

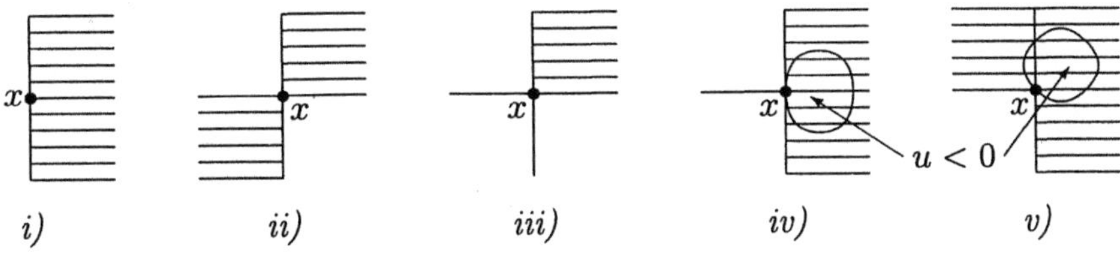

i)   *ii)*   *iii)*   *iv)*   *v)*

Now, let T_1 be the closure of the union of all the $U \cap S$'s and the $\{f_i = 0\} \cap S$'s. Setting $T_2 = X \setminus S$ we get two closed semialgebraic sets whose intersection $T_1 \cap T_2$ is the collection of all the critical points. Furthermore, T_1 and T_2 can be separated. Indeed, up to a 1-dimensional set, $T_1 = \{v < 0\}$ for the product v of the above u's, which means that T_1 and T_2 can be separated generically on every irreducible component of X. Since the other irreducible subvarieties of X have dimension 1, separation is always possible on them. Consequently, by Theorem 7.5, T_1 and T_2 can be separated: there is $h \in \mathbb{R}[V]$ such that

$$h|T_1 \geq 0, \ h|T_2 \leq 0, \ T_1 \cap T_2 = (T_1 \cup T_2) \cap \{h = 0\}.$$

In particular, $\{h > 0\} \subset S$. Finally we consider $T = S \setminus \{h \geq 0\} = S \cap \{h > 0\}$, and considering the different possibilities for all pairs of points $x, y \in X$, Proposition 7.7 shows that T is a basic open set. But $s(X) = 2$, hence $T = \{f > 0, g > 0\}$. We conclude $S = \{f > 0, g > 0\} \cup \{h > 0\}$.   □

Notes

The algebraic study of formally real fields goes back to E. Artin and Schreier ([A],[A-S]). Shortly after that, Baer ([Ba]) and Krull ([Kr]) discovered the connection between orderings and valuations of fields (see Theorem 1.3). As mentioned before, the notion of fan was introduced by Becker and Köpping in [Be-Kö], where one also finds Proposition 1.1. The fan trivialization theorem is due to Bröcker ([Br2]). Here we present a different proof which also works in the p-adic situation. The remaining results of Section 1 appeared in [Br1], as well as most of Section 2. However, the proof of Theorem 2.7 is taken from [An-Be], and Proposition 2.10 is due to Marshall-Walter ([Mr-Wa]). Also Section 3 is partially done in [Br1], but the more delicate Propositions 3.5 and 3.7 are new. The results of Setion 4 are new too. They simplify a lot earlier steps ([An-Br-Rz]) towards the calculation of stability indices of semianalytic sets. Except for finitely generated algebras over the rationals, the applications given in Section 5 were known before. In particular, the equality in Proposition 5.3 is due to Scheiderer ([Sch1]). Again, the study of archimedean rings in Section 6 is new material. However, multilocal properties appeared already in [Br4]. The fact that an open semialgebraic set can always be described by strict inequalities over the field generated by the coefficients of any starting description was proved by Dickmann ([Dk]) answering a question of Bröcker ([Br5]); his original proof belongs to model theory. Examples 7.1-7.3 and 7.10 are well-known, except perhaps for the ones over the rationals. In Example 7.1 we recover the pictures of 'geometric fans' described in detail in Section I.4.4. Much work is done towards proving that obstructions coming from fans show up already in these geometric fans ([An-Rz1,2,4]). This can be extended to finite subspaces, in order to treat separation questions ([Ac-An-Bg]). Thus one can take a step to obtaining constructible results ([Ac-Bg-Vz], [Ac-An-Bg]). The effort to visualize valuative obstructions can be traced back to Schülting ([Schü]; see also [Br-Schü] and [Al-Gm-Rz]). Proposition 7.4 is taken from [Br10], and "cutting off hairs" was first proved over the reals in [Rz2]. The striking Propositions 7.9-7.11 seem to be new.

Chapter VII. Real Algebra of Excellent Rings

Summary. This chapter is devoted to excellent rings, and contains the results that allow to extend what we have already seen for semialgebraic sets to semianalytic sets. In Sections 1 and 2 we collect the commutative algebra needed later. Very few proofs are given, since almost everything can be found in our general references [Mt] and the more elementary [At-Mc], [Bs-Is-Vg]; an important exception is our proof that local-ind-etale limits of excellent rings are again excellent. In addition, we state without proof the fundamental Rotthaus's theorem on M. Artin's approximation property. In Section 3 we characterize the extension of prime cones under completion, a crucial result for all that follows. The curve selection lemma which is proved in Section 4 has many important applications: existence theorems for valuations and fans (Section 5), and constructibility of closures (Section 6) are some. It is also needed in Section 7 for the proof of another key theorem: the real going-down for regular homomorphisms. After this, we characterize local constructibility of connected components in Section 8.

1. Regular Homomorphisms

We start by setting the terminology and reviewing quickly some notions and facts from commutative algebra which are essential for our purposes. Many results are given without proof, but we include references where they can be found. We also present some proofs that are surely well known to specialists but for which we have not a precise or accesible reference. Remember that all rings are assumed to be commutative with 1.

Definition 1.1 *Let A be a ring. An A-module M is called* flat *if for any exact sequence of A-modules*

$$N' \to N \to N''$$

the corresponding tensorized sequence

$$M \otimes N' \to M \otimes N \to M \otimes N''$$

is exact.

Definition 1.2   *A ring homomorphism $\varphi : A \to B$ is called* flat *if B is a flat A-module with the structure induced by φ.*

Concerning notations, whenever we have a homomorphism $\varphi : A \to B$ we shall use the same symbols as for inclusions; for instance, ax instead of $\varphi(a)x$ for $a \in A$ and $x \in B$, or $I \cap A$ instead $\varphi^{-1}(I)$ for $I \subset B$. Also, we shall not mention φ explicitely, unless there is some risk of confusion.

There is a characterization of flat homomorphims in terms of linear equations which is very close to the spirit of this book:

Proposition 1.3   *A ring homomorphism $A \to B$ is flat if and only if for any relation*

$$\sum_{i=1}^{r} a_i x_i = 0$$

where $a_i \in A$ and $x_i \in B$, there exist an integer s and elements $b_{ij} \in A$ and $y_j \in B$ such that

$$\sum_{i=1}^{r} a_i b_{ij} = 0 \ (1 \le j \le s), \qquad and \qquad x_i = \sum_{j=1}^{s} b_{ij} y_j \ (1 \le i \le r).$$

Proof.   [Mt Th.1(6), p.18].   □

Examples 1.4   *a)* Any vector space M over a field k is a flat k-module. Consequently, any homomorphism $k \to B$ is flat.   □

b) Let A be a ring and B a local-ind-etale A-algebra (Definition II.7.5). Then the structure homomorphism $A \to B$ is flat ([Rn Ch.VIII, Th.3, p.94]). In particular, the canonical monomorphisms of a local ring A into its henselization and its real strict localizations, are flat.   □

The following are some basic properties of flatness which will be often used.

Proposition 1.5   *It holds:*

a) (Localization) *Let A be a ring and Σ a multiplicative subset of A. Then $\Sigma^{-1}A$ is a flat A-module.*

b) (Base Change) *Let $A \to B$ be any ring homomorphism and let M be a flat A-module. Then $M \otimes_A B$ is a flat B-module.*

c) (Transitivity) *Let $A \to B$ be a flat homomorphism. Then a flat B-module is also flat over A.*

d) *Let $A \to B$ be a flat homomorphism, let $\mathfrak{q} \subset B$ be a prime ideal and let $\mathfrak{p} = \mathfrak{q} \cap A$. Then*

$$\mathrm{ht}(\mathfrak{q}) = \mathrm{ht}(\mathfrak{p}) + \mathrm{ht}(\mathfrak{q}/\mathfrak{p}B).$$

Proof.   See [Mt 3.B,C and D, p.19] for *a),b)* and *c)*, and [Mt 13.B, Th.19(2), p.79] for *d)*.   □

Definition 1.6   *An A-module M is called* faithfully flat *if it holds that a sequence of A-modules*

$$N' \to N \to N''$$

is exact if and only if the corresponding tensorized sequence

$$M \otimes N' \to M \otimes N \to M \otimes N''$$

is exact. Correspondingly, a homomorphism $\varphi : A \to B$ is called faithfully flat *if B is a faithfully flat A-module with the structure induced by φ.*

As flatness, faithful flatness is also transitive and is preserved by base change. Moreover:

Proposition 1.7   *It holds:*

   a) (Descent) *Let $\varphi : A \to B$ be a homomorphism of rings, and assume that M is a faithfully flat B-module which is also faithfully flat over A. Then φ is a faithfully flat homomorphism.*
   b) Let $A \to B$ be a faithfully flat homomorphism. Then for any ideal I of A, we have $A \cap IB = I$ and $\mathrm{ht}(I) = \mathrm{ht}(IB)$.
   c) Let $\varphi : A \to B$ be a local homomorphism of local rings (that is, it sends the maximal ideal of A into the maximal ideal of B). Then φ is flat if and only if it is faithfully flat.

Proof.   For *a)* see [Mt 4.B, p.27]. For *b)* see [Mt 4.C, p.28] and [Mt 13.B, Th.19(3), p.79]. Finally, *c)* is [Mt 4.D, Th.3, p.28]. $\square$

We now remember that a ring is called *regular* if all its localizations at prime ideals are regular.

Definition 1.8   *Let $\varphi : A \to B$ be a flat homomorphism of rings. Then, φ is called* regular *if for every prime ideal $\mathbf{p} \subset A$ and every finite extension L of its residue field $\kappa(\mathbf{p})$, the ring $B \otimes_A L$ is regular.*

Remarks 1.9   *a)* Let $K = k_1 \times \cdots \times k_r$ be a product of fields and let $k \subset K$ be a subfield of K of characteristic zero. Then the extension $k \subset K$ is regular.

Indeed, let L be a finite extension of k. Then $L = k[\mathsf{t}]/P$ for some polynomial P irreducible over k. Then

$$K \otimes_k L = k_1[\mathsf{t}]/P \times \cdots \times k_r[\mathsf{t}]/P.$$

Hence, if P factorizes over k_i as $P_{i1} \cdots P_{is_i}$, by the chinese remainder theorem,

$$K \otimes_k L = (k_1[\mathsf{t}]/P_{11} \times \cdots \times k_1[\mathsf{t}]/P_{1s_1}) \times \cdots \times (k_r[\mathsf{t}]/P_{r1} \times \cdots \times k_r[\mathsf{t}]/P_{rs_r}),$$

where each factor is now a field, and therefore $K \otimes_k L$ is regular. $\square$

b) Let A be a local ring and assume that $\mathbb{Q} \subset A$. Let $\varphi : A \to B$ be a local-ind-etale homomorphism. Then φ is regular. In particular the homomorphisms of A into its henselization and its real strict localizations are regular.

Indeed, Example 1.4 *b)* says that ϕ is flat. Now let $\mathfrak{p}$ be a prime ideal of A and let L be a finite extension of the residue field $\kappa(\mathfrak{p})$. Then $B \otimes_A L = B \otimes_A \kappa(\mathfrak{p}) \otimes_{\kappa(\mathfrak{p})} L$. By Proposition II.7.6 *c)*, $B \otimes_A \kappa(\mathfrak{p})$ is a product of fields, and by remark *a)* we get that $B \otimes_A L$ is regular, as wanted. $\qquad\square$

The next proposition collects some basic facts of regular homomorphisms.

Proposition 1.10   *It holds:*

   a) The composition of regular homomorphisms is a regular homomorphism.
   b) (Descent) Let $A\overset{\varphi}{\longrightarrow}B\overset{\psi}{\longrightarrow}C$ be ring homomorphisms and assume that $\psi\varphi$ is regular and ψ is faithfully flat. Then φ is regular.
   c) (Base change) Let $A \to B$ be a regular homomorphism and let A' be a finitely generated A-algebra. Then $A' \to A' \otimes_A B$ is regular.
   d) (Polynomial extensions) Let $A \to B$ be a regular homomorphism and let t be an indeterminate. Then the induced homomorphism $A[\mathsf{t}] \to B[\mathsf{t}]$ defined by $\mathsf{t} \mapsto \mathsf{t}$ is regular.
   e) Let $A \to B$ be a regular faithfully flat homomorphism. If A is reduced (resp. normal, regular), then so is B.

Proof.   *a)* and *b)* are [Mt 33.B, Lemma 1, p.250]. For *c)* see [Mt 33.C, Lemma 4, p.253]. Part *d)* follows from *c)* by the base change $- \otimes_A A[\mathsf{t}]$. Finally, *e)* is [Mt 33.B, Lemma 2, p.251]. $\qquad\square$

Corollary 1.11   *Let $A \to B$ be a regular homomorphism. Let $\mathfrak{q}$ be a prime ideal of B and set $\mathfrak{p} = \mathfrak{q} \cap A$. Then $A_\mathfrak{p}$ is regular if and only if $B_\mathfrak{q}$ is regular. Moreover, if $t_1,\ldots,t_d$ is a regular system of parameters of $A_\mathfrak{p}$, then $t_1,\ldots,t_d$ are part of a regular system of parameters of $B_\mathfrak{q}$.*

Proof.   Consider the local homomorphism $A_\mathfrak{p} \to B_\mathfrak{q}$ induced by the given one $A \to B$. Clearly it is regular, and therefore, by Proposition 1.7 *c)*, it is faithfully flat. Thus, by Proposition 1.10 *e)*, if $A_\mathfrak{p}$ is regular so is $B_\mathfrak{q}$. On the other hand, if $B_\mathfrak{q}$ is regular, it follows from [Mt 21.D, Th.51, p.155] that $A_\mathfrak{p}$ is regular.

Going further, assume that $t_1,\ldots,t_d$ is a regular system of parameters of $A_\mathfrak{p}$. Since the induced morphism $\kappa(\mathfrak{p}) \to B_\mathfrak{q}/\mathfrak{p}B_\mathfrak{q}$ is also regular, Proposition 1.10 *e)* implies that both $B_\mathfrak{q}$ and $B_\mathfrak{q}/\mathfrak{p}B_\mathfrak{q}$ are regular rings, of dimensions, say, s and e respectively. This implies that $\mathfrak{p}B_\mathfrak{q}$ is a prime ideal and

$$\mathrm{ht}(\mathfrak{p}B_\mathfrak{q}) = \dim(B_\mathfrak{q}) - \dim(B_\mathfrak{q}/\mathfrak{p}B_\mathfrak{q}).$$

On the other hand, by faithful flatness, Proposition 1.7 *b)*, $\mathrm{ht}(\mathfrak{p}B_\mathfrak{q}) = \mathrm{ht}(\mathfrak{p}) = d$, whence $s = d + e$. Now let $t_{d+1},\ldots,t_{d+e} \in \mathfrak{q}$ be such that their classes in $B_\mathfrak{q}/\mathfrak{p}B_\mathfrak{q}$ form a regular system of parameters. Then $t_1,\ldots,t_{d+e}$ generate $\mathfrak{q}B_\mathfrak{q}$ and therefore they are a regular system of parameters of $B_\mathfrak{q}$. $\qquad\square$

We finish the section with a technical result on power series rings which will be used to prove that local-ind-etale limits of excellent rings are excellent too. Let k be a field of characteristic zero and $\mathbf{x} = (\mathbf{x}_1, \ldots, \mathbf{x}_n)$ indeterminates. A power series $f \in k[[\mathbf{x}]]$ is called *regular of order p in $\mathbf{x}_n$* if $f(0, \ldots, 0, \mathbf{x}_n) = \mathbf{x}_n^p u(\mathbf{x}_n)$, with $u(0) \neq 0$. An extremely useful fact is that this can always be achieved by a generic linear change of the $\mathbf{x}_i$'s. A polynomial $P \in k[[\mathbf{x}']][\mathbf{x}_n]$, $\mathbf{x}' = (\mathbf{x}_1, \ldots, \mathbf{x}_{n-1})$, of degree p is called *distinguished* if it is regular of order p in $\mathbf{x}_n$. The two fundamental results concerning power series are:

Theorem 1.12   *Let $f \in k[[\mathbf{x}]]$ be regular of order p in $\mathbf{x}_n$. Then:*

a) (Weierstrass's preparation theorem) *There are a distinguished polynomial of degree p, $P \in k[[\mathbf{x}']][\mathbf{x}_n]$, and $u \in k[[\mathbf{x}]]$ such that $u(0) \neq 0$ and $f = uP$. Such u and P are unique.*

b) (Rückert's division theorem) *For every $g \in k[[\mathbf{x}]]$ there are $Q \in k[[\mathbf{x}]]$ and $R \in k[[\mathbf{x}']][\mathbf{x}_n]$ of degree $< p$ such that $g = Qf + R$. Such Q and P are unique*

For the proof of these theorems as well as their many important consequences, we refer to [Rz12]. Now, we shall use the preparation theorem in the next proposition.

Proposition 1.13   *Let $I \subset k[[\mathbf{x}]]$ be any ideal and let $K \supset k$ be a field extension. Then, the canonical homomorphism $k[[\mathbf{x}]]/I \to K[[\mathbf{x}]]/IK[[\mathbf{x}]]$ is regular.*

Proof.   By the base change $- \otimes_{k[[\mathbf{x}]]} k[[\mathbf{x}]]/I$, it is enough to prove the result for $I = (0)$. We argue by induction on n, the case $n = 0$ being trivial. So we assume that $n > 0$. We show first that the homomorphism is flat. By Proposition 1.3 we have to see that for any equation $\sum_i a_i h_i = 0$, with $a_i \in k[[\mathbf{x}]]$, $h_i \in K[[\mathbf{x}]]$, there are equations

$$h_i = \sum_j c_{ij} g_j, \qquad \sum_i a_i c_{ij} = 0, \qquad (1.12.1)$$

where $c_{ij} \in k[[\mathbf{x}]]]$, $g_j \in K[[\mathbf{x}]]$.

Consider the commutative diagram

$$
\begin{array}{ccc}
K[[\mathbf{x}']][\mathbf{x}_n]_{(\mathbf{x}', \mathbf{x}_n)} & \xrightarrow{\varphi'} & K[[\mathbf{x}]] \\
\uparrow{\scriptstyle \psi'} & & \uparrow{\scriptstyle \psi} \\
k[[\mathbf{x}']][\mathbf{x}_n]_{(\mathbf{x}', \mathbf{x}_n)} & \xrightarrow{\varphi} & k[[\mathbf{x}]]
\end{array}
$$

where $\mathbf{x}'$ stands for $(\mathbf{x}_1, \ldots, \mathbf{x}_{n-1})$. By induction hypothesis the homomorphism ψ' is flat, and so are also the completion homomorphisms φ and φ'. Now, after a linear change of the $\mathbf{x}_i$'s and applying the Weierstrass preparation theorem, we may assume that $a_i = u_i P_i$ where P_i is a polynomial in $k[[\mathbf{x}']][\mathbf{x}_n]$, and $u_i \in k[[\mathbf{x}]]$ is a unit. We set $f_i = h_i u_i$. Then we have

$$\sum_i P_i f_i = 0,$$

which can be seen as an equation with coefficients in $k[[\mathbf{x}']][\mathbf{x}_n]_{(\mathbf{x}',\mathbf{x}_n)}$. Since the homomorphism $\psi'\varphi'$ is flat, we get $b_{ij} \in k[[\mathbf{x}']][\mathbf{x}_n]_{(\mathbf{x}',\mathbf{x}_n)}$ and $g_j \in K[[\mathbf{x}]]$ such that

$$ f_i = \sum_j b_{ij} g_j, \qquad \sum_i P_i b_{ij} = 0. $$

Thus it is enough to take $c_{ij} = u_i^{-1} b_{ij}$ in 1.12.1.

Next, let $\mathfrak{p} \in k[[\mathbf{x}]]$ be a prime ideal and let L be a finite extension of $\kappa(\mathfrak{p})$. Again, by a linear change of the $\mathbf{x}_i$'s and the Weierstrass preparation theorem, we may assume that $\mathfrak{p}$ is generated by monic polynomials $f_1, \ldots, f_r \in k[[\mathbf{x}']][\mathbf{x}_n]$. Then, by Proposition 1.7 $b)$ these polynomials generate also the ideal $\mathfrak{q} = \mathfrak{p} \cap k[[\mathbf{x}']][\mathbf{x}_n]_{(\mathbf{x}',\mathbf{x}_n)}$ and the canonical mapping

$$ k[[\mathbf{x}']][\mathbf{x}_n]_{(\mathbf{x}',\mathbf{x}_n)}/\mathfrak{q} \to k[[\mathbf{x}]]/\mathfrak{p}, $$

is an isomorphism. Thus

$$ K[[\mathbf{x}]] \otimes_{k[[\mathbf{x}]]} L = K[[\mathbf{x}]] \otimes_{k[[\mathbf{x}']][\mathbf{x}_n]_{(\mathbf{x}',\mathbf{x}_n)}} L. $$

Now, by induction hypothesis and Proposition 1.10 $d)$, ψ' is regular. On the other hand, it is well known that the homomorphism φ' is regular, see [Mt 33.G, Th.77, p.254]. Therefore the homomorphism $\varphi'\psi'$ is regular, which shows that the tensor product $K[[\mathbf{x}]] \otimes_{k[[\mathbf{x}']][\mathbf{x}_n]_{(\mathbf{x}',\mathbf{x}_n)}} L$ is regular and completes the proof. $\square$

2. Excellent Rings

In this section we recall the notion and elementary properties of excellent rings. Roughly speaking, an excellent ring has good dimension properties, good behaviour under completion and its singular locus is Zariski closed. More precisely:

Definition 2.1   *A noetherian ring A is called* excellent *if it verifies the following conditions:*

  i) For any finitely generated A-algebra B and any prime ideals $\mathfrak{q} \subset \mathfrak{p}$ of B, it holds

  $$ \mathrm{ht}(\mathfrak{p}) = \mathrm{ht}(\mathfrak{p}/\mathfrak{q}) + \mathrm{ht}(\mathfrak{q}). $$

  (that is, A is universally catenary*).*
  ii) For any localization $B = A_{\mathfrak{p}}$ at a prime $\mathfrak{p} \subset A$, the homomorphism $B \to \widehat{B}$ from B into its completion is regular.
  iii) For any finitely generated A-algebra B, there exist elements $h_1, \ldots, h_r$ in B such that $B_{\mathfrak{p}}$ is regular if and only if $h_i \notin \mathfrak{p}$ for some i (that is, the singular locus is the zero set of the h_i's).

The following proposition collects the basic properties of excellent rings.

Proposition 2.2   *It holds:*

a) *Let A be an excellent ring and let Σ be a multiplicative subset of A. Then $\Sigma^{-1}A$ is excellent.*

b) *Let A be an excellent ring and let B be a finitely generated A-algebra. Then B is excellent.*

c) (Dimension Formula) *Let A be an excellent domain and let B be a finitely generated A-algebra which is a domain. Let K, L be their respective fields of fractions, let $\mathfrak{q} \in B$ be a prime ideal and set $\mathfrak{p} = \mathfrak{q} \cap A$. Then*

$$\mathrm{ht}(\mathfrak{q}) = \mathrm{ht}(\mathfrak{p}) + \deg(L : K) - \deg(\kappa(\mathfrak{q}) : \kappa(\mathfrak{p})).$$

d) *Let A be a local excellent ring. Then its adic completion $\widehat{A}$ is reduced (resp. normal, regular) if and only if A is reduced (resp. normal, regular).*

e) *Let A be a local excellent ring and let $\widehat{A}$ be its adic completion. Let $\mathfrak{q} \subset \widehat{A}$ be a prime ideal and set $\mathfrak{p} = \mathfrak{q} \cap A$. Then, if $A_{\mathfrak{p}}$ is regular, $\widehat{A}_{\mathfrak{q}}$ is regular too.*

f) *Let A be an excellent domain. Then there is $h \in A$, $h \neq 0$, such that $A_{\mathfrak{p}}$ is regular whenever $h \notin \mathfrak{p}$.*

g) *Let A be an excellent domain with quotient field K, and let L be a finite extension of K. Let A' be the integral closure of A in L. Then A' is a finite A-module.*

h) *Let A a local noetherian ring that verifies condition i) of Definition 2.1 and such that the homomorphism $A \to \widehat{A}$ is regular. Then A is excellent.*

Proof.   *a)* is immediate.

b) That B verifies condition 2.1 *ii)* follows from [Mt 33.G, Th.77, p.254]. That it satisfies condition 2.1 *iii)* appears in [Mt 32.B, Th.73, p.246].

c) This is [Mt 14.C, Th.23, p.84].

d) follows from Proposition 1.10. *e)*.

e) follows from Corollary 1.11.

f) Since (0) is a prime ideal we have that some $h_i \neq 0$. Thus take $h = h_i$.

g) See [Mt 33.H, Th.78, p.257].

h) A verifies condition 2.1 *ii)* by [Mt 33.C, Th.75, p.251]. Now by [Mt 32.B, Th.73, p.246], to check Definition 2.1 *iii)* it is enough to consider the case when B a finite A-algebra. Then B is semilocal and the result follows from [Mt 33.D, Th.76, p.252] and [Mt 33.G, Th.77, p.254].   □

Examples 2.3   *a)* Finitely generated algebras over a field k are excellent.

b) Let k be a field of characteristic 0. *Formal* (resp. *formal-algebraic*) algebras over k, that is, homomorphic images of the ring of formal (resp. algebraic) power series over k, are excellent.   □

c) Analytic algebras over $\mathbb{R}$ or $\mathbb{C}$, that is, homomorphic images of the ring of convergent power series over $\mathbb{R}$ or $\mathbb{C}$, are excellent.   □

The proof in the four cases can be done along the same pattern, by combining Proposition 2.2 *h)* above with the Zariski-Nagata Jacobian criteria, see [Rz12 II.4, p.31ff, V.5, p.106ff]. In fact the following result can be seen as an abstract formulation of this method, and will be applied in Chapter VIII to rings of analytic functions.

Theorem 2.4   *Let k be a field of characteristic zero and let A be a regular ring containing k, such that*

  i) *For any maximal ideal $\mathfrak{m} \in A$ the extension $k \to A/\mathfrak{m}$ is algebraic.*
  ii) *All maximal ideals of A have the same height, say n.*
  iii) *There exist derivations $D_1, \ldots, D_n$ of A over k and elements $x_1, \ldots, x_n \in A$ such that $D_i x_j = \delta_{ij}$.*

Then A is an excellent ring.

Proof.   [Mt 40.F, Th.102, p.291].   □

Of special importance in our setting is the following result:

Theorem 2.5   *Let A be a local excellent ring containing $\mathbb{Q}$ and let B be a local-ind-etale A-algebra. Then B is excellent. In particular, the henselization and the real strict localizations of A are excellent rings.*

Proof.   Let $\varphi : A \to B$ denote the structure homomorphism from A into B. First of all, by Proposition II.7.6, we know that B is noetherian. Now we check that B verifies Definition 2.1 *i)*. We only need to prove the catenarity formula for polynomial rings over B, say $C = B[\mathsf{x}_1, \ldots, \mathsf{x}_n]$. Let $\{B_\lambda\}_{\lambda \in \Lambda}$ be a family of local-etale A-algebras such that $B = \varinjlim\{B_\lambda\}$ and, for each λ, let $\varphi_\lambda : B_\lambda \to B$ be the canonical homomorphism. By Example I.4 *b)*, φ_λ is faithfully flat. Set $C_\lambda = B_\lambda[\mathsf{x}_1, \ldots, \mathsf{x}_n]$. Clearly the family $\{C_\lambda\}_{\lambda \in \Lambda}$ is also a direct system, where the homomorphism $\phi_{\alpha\beta} : C_\alpha \to C_\beta$ is the extension of $\varphi_{\alpha\beta} : B_\alpha \to B_\beta$ sending $\mathsf{x}_i \mapsto \mathsf{x}_i$ for $i = 1, \ldots, n$. Then $C = \varinjlim\{C_\lambda\}$, and the canonical homomorphisms $C_\lambda \to C$ are flat.

Let $\mathfrak{q} \subset \mathfrak{p}$ be prime ideals of C, and set $\mathfrak{q}_\lambda = \mathfrak{q} \cap B_\lambda$ and $\mathfrak{p}_\lambda = \mathfrak{p} \cap B_\lambda$. Since B is noetherian, $\mathfrak{q}$ and $\mathfrak{p}$ are finitely generated, and there exists λ such that for every $\alpha \geq \lambda$ we have $\mathfrak{q} = \mathfrak{q}_\lambda C$ and $\mathfrak{p} = \mathfrak{p}_\lambda C$. By Proposition 1.5 *d)*, we have

$$\mathrm{ht}(\mathfrak{q}) = \mathrm{ht}(\mathfrak{q}_\lambda), \ \ \mathrm{ht}(\mathfrak{p}) = \mathrm{ht}(\mathfrak{p}_\lambda), \ \text{ and } \ \mathrm{ht}(\mathfrak{p}/\mathfrak{q}) = \mathrm{ht}(\mathfrak{p}_\lambda/\mathfrak{q}_\lambda).$$

On the other hand, since each B_λ is of the form $B_\lambda = A[\mathsf{t}]_I/(f)$, it is excellent (Proposition 2.2), and therefore it verifies the catenarity formula. Hence we get:

$$\mathrm{ht}(\mathfrak{p}) = \mathrm{ht}(\mathfrak{q}) + \mathrm{ht}(\mathfrak{p}/\mathfrak{q}),$$

as required.

To finish, by Proposition 2.2 *h)*, we only have to check that the homomorphism $\mu : B \to \widehat{B}$ is regular. Let $\mathfrak{p}$ be a prime ideal of B and let L be a finite extension of $\kappa(\mathfrak{p})$. We have to show that $\widehat{B} \otimes_B L$ is regular. Let $\lambda \in \Lambda$ be such that $\mathfrak{p} = \mathfrak{p}_\lambda B$, where $\mathfrak{p}_\lambda = \mathfrak{p} \cap B_\lambda$. Then $\widehat{B} \otimes_B L$ is a localization of $\widehat{B} \otimes_{B_\lambda} L$, and we claim that it is enough to see that the homomorphism $B_\lambda \to B$ is regular. Indeed, suppose this true. Since $\kappa(\mathfrak{p})$ is algebraic over $\kappa(\mathfrak{p}_\lambda)$ (Proposition II.7.6 c)), so is L. Let $\mathfrak{q}$ be a prime ideal of $\widehat{B} \otimes_{B_\lambda} L$ and let $L' \subset L$ be a finite extension of $\kappa(\mathfrak{p})$ such that $\mathfrak{q}' = \mathfrak{q} \cap (\widehat{B}) \otimes_{B_\lambda} L'$ generates $\mathfrak{q}$. Now, $\widehat{B} \otimes_{B_\lambda} L' \to \widehat{B} \otimes_{B_\lambda} L$ is faithfully flat (since so is $L' \to L$) and therefore $\mathrm{ht}(\mathfrak{q}') = \mathrm{ht}(\mathfrak{q})$. Since by assumption $(\widehat{B} \otimes_{B_\lambda} L')_{\mathfrak{q}'}$ is regular this means that the ideal $\mathfrak{q}'$ is generated in this local ring by $\mathrm{ht}(\mathfrak{q}')$ elements. Since they also generate $\mathfrak{q}$ in $(\widehat{B} \otimes_{B_\lambda} L)_{\mathfrak{q}}$ we get that this ring is regular too, as wanted.

Thus, let us see that the homomorphism $B_\lambda \to \widehat{B}$ is regular. Since $B_\lambda = A[\mathfrak{t}]_I/(f)$ is excellent, and obviously B is a local-ind-etale B_λ-algebra, replacing A by B_λ we are reduced to show that the homomorphism $\mu\varphi : A \to \widehat{B}$ is regular. By hypothesis the homomorphism $j : A \to \widehat{A}$ of A into its completion is regular. Let K and k respectively denote the residue fields of B and A. Then K is an algebraic extension of k, and therefore it is a local-ind-etale limit of local-etale k-algebras, namely, the finitely generated subfields of K. Therefore $K \otimes_k \widehat{A}$ is a local-ind-etale $\widehat{A}$-algebra. In particular it is henselian, and its residue field is K. Let $\psi : \widehat{A} \to K \otimes_k \widehat{A}$ denote the canonical homomorphism. Let $\widehat{\varphi} : \widehat{A} \to \widehat{B}$ be the unique extension of φ to the completions. Then by the universal property of local-ind-etale algebras (Proposition II.7.6 d)), there exists a unique local homomorphism $\sigma : K \otimes_k \widehat{A} \to \widehat{B}$ such that $\widehat{\varphi} = \sigma\psi$, and whose restriction to K is the identity. Let $\vartheta : K \otimes_k \widehat{A} \to \widehat{K \otimes_k \widehat{A}}$ be the canonical homomorphism, and let $\widehat{\sigma} : \widehat{K \otimes_k \widehat{A}} \to \widehat{B}$ be the unique extension of σ. Finally, again by the universal property II.7.6 d), there is a unique homomorphism $u : B \to \widehat{K \otimes_k \widehat{A}}$ such that $u\varphi = \vartheta\psi j$ and whose restriction to K is the identity. Let $\widehat{u} : \widehat{B} \to \widehat{K \otimes_k \widehat{A}}$ be the corresponding extension of u.

Summarizing, we have a commutative diagram of local homomorphisms:

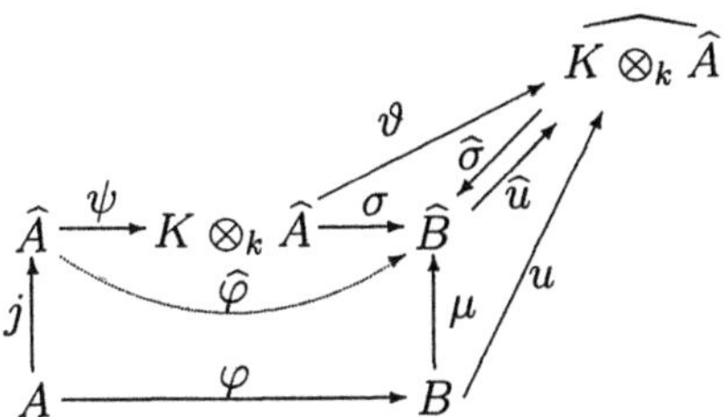

in which a straightforward computation shows that $\widehat{\sigma}$ and $\widehat{u}$ are mutually inverse isomorphisms. Therefore $\mu\varphi$ is regular if and only if so is $u\varphi$. On the other hand, by Cohen's structure theorem we can write $\widehat{A} = k[[\mathbf{x}]]/I$, where $\mathbf{x} = (\mathbf{x}_1, \ldots, \mathbf{x}_n)$, and it follows easily that $\vartheta\psi : \widehat{A} \to \widehat{K \otimes_k \widehat{A}}$ is the canonical homomorphism $k[[\mathbf{x}]]/I \to K[[\mathbf{x}]]/IK[[\mathbf{x}]]$. Hence, by Proposition 1.13, $\vartheta\psi$ is regular, and $u\varphi = (\vartheta\psi)j$ is a composition of regular homomorphisms, therefore regular. This completes the proof. $\qquad\square$

We next analyze the behaviour of normalization under regular homomorphisms of excellent rings. Recall that the *normalization* A^ν of a ring A is the integral closure in its total ring of fractions K; A is called *normal* if $A = A^\nu$ ([Mt 17, pp.115-127]). If A is a reduced noetherian ring, K is canonically isomorphic to the product $K_1 \times \cdots \times K_s$ of the residue fields K_i of the minimal primes $\mathfrak{p}_i$. Moreover, via this isomorphism, A^ν is the product of the normalizations of the rings $A/\mathfrak{p}_i$ (this is an easy consequence of the chinese remainder theorem). If, furthermore, A is excellent, then A^ν is a finite A-module. (by Proposition 2.2 g)).

Proposition 2.6   *Let $A \to B$ be a regular homomorphism of reduced excellent rings, and let K and L be the total rings of fractions of A and B, respectively. Then $B^\nu = B \otimes_A A^\nu$.*

Proof.   By flatness, we have inclusions $B \subset B \otimes_A A^\nu \subset B \otimes_A K \subset L$. Since A^ν is a finite A-module, $B \otimes_A A^\nu$ is a finite B-module, and consequently, is contained in B^ν. But A^ν is normal, and $A^\nu \to B \otimes_A A^\nu$ is regular (by finite base change, Proposition 1.10 c)), hence $B \otimes_A A^\nu$ is normal too, thus $B^\nu = B \otimes_A A^\nu$. $\qquad\square$

Corollary 2.7 (Equidimensionality)   *Let $A \to B$ be a regular homomorphism of excellent local rings, and assume that A is a domain. Then, $\dim(B/\mathfrak{q}) = \dim(B)$ for every minimal prime $\mathfrak{q}$ of B.*

Proof.   By the properties of normalizations, there is a maximal ideal $\mathfrak{n}$ of B^ν such that $(B/\mathfrak{q})^\nu = B^\nu_\mathfrak{n}$, and consequently, $\dim(B/\mathfrak{q}) = \dim(B^\nu_\mathfrak{n})$. Now, by the preceding proposition, we have a regular homomorphism $A^\nu \to B^\nu$, and $\mathfrak{n}$ lies over a maximal ideal $\mathfrak{m}$ of A^ν. Thus

$$\dim(B/\mathfrak{q}) = \dim(B^\nu_\mathfrak{n}) \;=\; \dim(A^\nu_\mathfrak{m}) + \dim(B^\nu_\mathfrak{n}/\mathfrak{m}B^\nu) =$$
$$= \; \dim(A) + \dim(B^\nu_\mathfrak{n}/\mathfrak{m}B^\nu).$$

Now note that the extension $B/\mathfrak{m}_A B \to B^\nu/\mathfrak{m}B^\nu$ is integral, and, consequently,

$$\dim(B^\nu_\mathfrak{n}/\mathfrak{m}B^\nu) = \dim(B/\mathfrak{m}_A B) = \dim(B) - \dim(A).$$

We are done. $\qquad\square$

We finish this section by stating a deep result due to C. Rotthaus which will be used at several key steps along the chapter. Given a local ring A, we denote by $\mathfrak{m}$ its maximal ideal and by $\widehat{\mathfrak{m}}$ the maximal ideal of its completion $\widehat{A}$.

Theorem 2.8   *Let A be a local henselian excellent ring containing $\mathbb{Q}$. Then A has the* approximation property, *that is, if a system of polynomial equations in $A[\mathbf{x}_1 \ldots, \mathbf{x}_n]$,*

$$F_1(\mathbf{x}_1, \ldots, \mathbf{x}_n) = \cdots = F_r(\mathbf{x}_1, \ldots, \mathbf{x}_n) = 0,$$

has a solution $(\widehat{a}_1, \ldots, \widehat{a}_n)$ *in* $\widehat{A}$, *then, for any integer* $\nu \geq 1$, *it has a solution* $(a_1, \ldots, a_n)$ *in* A, *such that*

$$a_i \equiv \widehat{a}_i \mod \widehat{\mathfrak{m}}^\nu, \qquad i = 1, \ldots, n.$$

Proof. [Rt 4.2]. □

3. Extension of Orderings Under Completion

We now work out what can be considered as a key question of the whole chapter, namely, when a prime cone extends under completion. We start by defining the notions of henselian and formal branches and studying the relationship between them. All rings are assumed to contain $\mathbb{Q}$.

Let A be a local excellent domain with quotient field K and maximal ideal $\mathfrak{m}$. Let A^h be the henselization of A, and let $\widehat{A}$ be the $\mathfrak{m}$-adic completion of A. Let I be a prime ideal of A. We have $\widehat{A} = \widehat{A^h}$ and $\dim(A) = \dim(A^h) = \dim(\widehat{A})$. It follows from Proposition 2.2 that the ideals IA^h and $I\widehat{A}$ are radical. Let us denote by $\mathfrak{q}_1, \ldots, \mathfrak{q}_s$ the minimal prime ideals of IA^h, and by $\mathfrak{p}_1, \ldots, \mathfrak{p}_t$ the minimal prime ideals of $I\widehat{A}$. The ideals $\mathfrak{q}_i$ (resp. $\mathfrak{p}_i$) are called the *henselian* (resp. *formal*) *branches of* I. Sometimes we shall refer also to the rings $A^h/\mathfrak{q}_i$ (resp. $\widehat{A}/\mathfrak{p}_i$) as the *henselian* (resp. *formal*) *branches of* A/I. We have $\mathrm{ht}(\mathfrak{q}_i) = \mathrm{ht}(\mathfrak{p}_i) = \mathrm{ht}(I)$ and $\dim(A^h/\mathfrak{q}_i) = \dim(\widehat{A}/\mathfrak{p}_i) = \dim(A/I)$ for all i (by Corollary 2.7, after the base change $- \otimes_A A/I$). In particular, for $I = (0)$, the ideals $\mathfrak{q}_1, \ldots, \mathfrak{q}_s$ are the zero divisors of A^h, the ideals $\mathfrak{p}_1, \ldots, \mathfrak{p}_t$ are the zero divisors of $\widehat{A}$, and the rings $A^h/\mathfrak{q}_i$ (respectively $\widehat{A}/\mathfrak{p}_i$) are called the *henselian* (resp. *formal*) *branches of* A.

We shall show that there exists a one to one correspondence between these two sets. In order to make this statement more precise, let us denote by A^ν the integral closure of A in K. We know by Proposition 2.2 *g)*, that A^ν is a finite A-module and a semilocal domain. Let $\mathfrak{n}_1, \ldots, \mathfrak{n}_r$ be its maximal ideals. In this situation we have:

Proposition 3.1   *The three numbers* r, s, t *coincide, and after reordering* $\widehat{A^\nu_{\mathfrak{n}_i}} = (\widehat{A}/\mathfrak{p}_i)^\nu$, *and* $(A^\nu_{\mathfrak{n}_i})^h = (A^h/\mathfrak{p}_i)^\nu$ *for each* $i = 1, \ldots, r$. *In particular there is a one to one correspondence between the henselian and formal branches, given by* $\widehat{A^h/\mathfrak{q}_i} = \widehat{A}/\mathfrak{p}_i$.

Proof. We show the statement for the completion. The proof for the henselization is the same. Let B be the completion of A^ν with respect to its radical. It is well known, [Mt 24.C, p.174],

$$B = \widehat{A^\nu_{\mathfrak{n}_1}} \times \cdots \times \widehat{A^\nu_{\mathfrak{n}_r}}.$$

Now, let $\widehat{K}$ be the total ring of fractions of $\widehat{A}$, and let K_i be the field of quotients of $\widehat{A}/\mathfrak{p}_i$. Finally let C_i be the integral closure of $\widehat{A}/\mathfrak{p}_i$ in K_i, and C the integral closure of $\widehat{A}$ in $\widehat{K}$. By Proposition 2.6, we have

$$\widehat{A^{\nu}_{\mathfrak{n}_1}} \times \cdots \times \widehat{A^{\nu}_{\mathfrak{n}_r}} = C_1 \times \cdots \times C_t.$$

Next, every $A^{\nu}_{\mathfrak{n}_i}$ is an excellent normal domain by Proposition 2.2 a), g), b). Therefore by Proposition 2.2 d), the local ring $\widehat{A^{\nu}_{\mathfrak{n}_i}}$ is also normal and so it is a domain too. Thus, we conclude that $r \geq t$. On the other hand, C_i is a domain by construction, and therefore we have $t \geq r$. Whence, after renumbering if necessary, we get $C_i = \widehat{A^{\nu}_{\mathfrak{n}_i}}$ and we are done. $\square$

Notice that from the above proposition it follows that $\mathfrak{q}_i \widehat{A} = \mathfrak{p}_i$. This shows in particular that if $\mathfrak{q}$ is a prime ideal of A^h its extension $\mathfrak{q}\widehat{A}$ is also prime. In fact, it is an easy exercise using Rotthaus's Theorem 2.8 to show that if $\mathfrak{p}$ is a prime ideal of an excellent henselian ring A, then $\mathfrak{p}\widehat{A}$ is prime.

We come next to the central result of this section: every total ordering of a local excellent ring making convex the maximal ideal extends to a total ordering of some formal branch.

Theorem 3.2   *Let A be a local excellent domain with maximal ideal* $\mathfrak{m}$, *and* $\widehat{A}$ *its adic completion. Let* $\beta \to \alpha$ *be a specialization in* $\mathrm{Spec}_r(A)$ *with* $\mathrm{supp}(\alpha) = \mathfrak{m}$. *Then there exists* $\widehat{\beta} \to \widehat{\alpha}$ *in* $\mathrm{Spec}_r(\widehat{A})$ *lying over* $\beta \to \alpha$ *such that* $\mathrm{ht}(\mathrm{supp}(\widehat{\alpha}))$ $= \mathrm{ht}(\mathrm{supp}(\alpha))$ *and* $\mathrm{ht}(\mathrm{supp}(\widehat{\beta})) = \mathrm{ht}(\mathrm{supp}(\beta))$. *Moreover, if* $\widehat{\beta}_1 \to \widehat{\alpha}_1$ *and* $\widehat{\beta}_2 \to \widehat{\alpha}_2$ *are two such chains we have* $\widehat{\alpha}_1 = \widehat{\alpha}_2$ *and* $\mathrm{supp}(\widehat{\beta}_1) = \mathrm{supp}(\widehat{\beta}_2)$.

Proof.   First of all we may identify α with the order that it defines in the residue field k of A. Since k is also the residue field of $\widehat{A}$, we identify α with a point $\widehat{\alpha}$ of $\mathrm{Spec}_r(\widehat{A})$. This $\widehat{\alpha} = \alpha$ is the only point of $\mathrm{Spec}_r(\widehat{A})$ with support the maximal ideal of $\widehat{A}$ and lying over α. Now suppose that $\widehat{\beta} \in \mathrm{Spec}_r(\widehat{A})$ lies over β and $\dim(\widehat{\beta}) = \dim(\beta)$. Then we have

$$\mathrm{ht}(\mathrm{supp}(\widehat{\beta})) = \mathrm{ht}(\mathrm{supp}(\beta)) = \mathrm{ht}(\mathrm{supp}(\beta)\widehat{A})$$

and therefore $\mathrm{supp}(\widehat{\beta})$ is an associated prime of the extension $\mathrm{supp}(\beta)\widehat{A}$, that is, a formal branch of $\mathrm{supp}(\beta)$. On the other hand, since $\widehat{A}$ is henselian, the point $\widehat{\beta}$ specializes to some point of $\mathrm{Spec}_r(\widehat{A})$ whose support is the maximal ideal of $\widehat{A}$ and which lies over α. Hence it must be $\widehat{\alpha}$. Therefore, after replacing A by $A/\mathrm{supp}(\beta)$ we may assume that A is a domain with quotient field K and we are reduced to show that the total ordering defined by β in K extends to a total ordering in the quotient field of some formal branch of A.

Next, let A^{ν} be the integral closure of A in its quotient field. By Proposition II.4.3, α extends to a point α' in $\mathrm{Spec}_r(A^{\nu})$ whose support is one of the maximal ideals $\mathfrak{n}$ of A^{ν}. By the preceding proposition, there is a formal branch $\widehat{A}/\mathfrak{p}$ of A such that $\widehat{A^{\nu}_{\mathfrak{n}}} = (\widehat{A}/\mathfrak{p})^{\nu}$. Therefore replacing A by $A^{\nu}_{\mathfrak{n}}$ we may assume that A is normal and consequently that $\widehat{A}$ is a domain. Let us denote by $\widehat{K}$ its quotient field. Finally, let A^h be the henselization of A and let K^h stand for its quotient field. It follows from Theorem II.7.11, that there is a prime cone β^h of A^h lying over β. Since $\operatorname{supp}(\beta^h)$ lies over $\operatorname{supp}(\beta) = (0)$, by the properties of henselizations (Proposition II.7.6 c)) we conclude that $\operatorname{supp}(\beta^h) = (0)$ and β^h is a total ordering of K^h. We are going to show that β^h extends to a total ordering of $\widehat{K}$.

This is the situation where Rotthaus's Theorem 2.8 helps. Indeed, we have to see that for any finite family $f_1, \ldots, f_r$ of elements of A^h positive in β^h, the equation

$$f_1 \mathbf{x}_1^2 + \cdots + f_r \mathbf{x}_r^2 = 0$$

has only the trivial solution in $\widehat{A}$ (this is *Serre's criterion*, a particular case of the Positivstellensatz). Suppose that it has a non-trivial solution in $\widehat{A}$. Then by Theorem 2.8, it has also a non-trivial solution in A^h, which is a contradiction.

It remains to show uniqueness. Assume that $\widehat{\beta}_1$ and $\widehat{\beta}_2$ are two extensions of β to different formal branches of A. Then, by Proposition 3.1, they restrict to two extensions $\beta_1{}^h$ and $\beta_2{}^h$ of β to different henselian branches of A. In particular $\beta_1{}^h \neq \beta_2{}^h$, against Theorem II.7.11. $\square$

Corollary 3.3   *Let A be a henselian local excellent ring and let $\widehat{A}$ be its completion. Then the canonical mapping $i^* : \operatorname{Spec}_r(\widehat{A}) \to \operatorname{Spec}_r(A)$ is surjective.*

Proof.   By Proposition II.2.4, any point $\beta \in \operatorname{Spec}_r(A)$ is a generization of some point α supported on the maximal ideal $\mathfrak{m}$ of A. Then the result follows immediately from the theorem. $\square$

4. Curve Selection Lemma

In this section we obtain the curve selection lemma for excellent rings, which roughly asserts that every constructible set adherent to a point contains some 1-dimensional generization of that point.

Let A be a local ring with residue field k, α an ordering of k and $\kappa(\alpha)$ the corresponding real closure. Then any non-trivial local homomorphism $\gamma : A \to \kappa(\alpha)[[\mathbf{t}]]$ induces two generizations γ_+ and γ_- with the same support $\mathfrak{p} = \operatorname{Ker}(\gamma)$, by restriction to the residue field $\kappa(\mathfrak{p})$ of the two orderings $\mathbf{t} > 0$ and $\mathbf{t} < 0$ of $\kappa(\alpha)((\mathbf{t}))$. These two generizations may well be equal, and have very big dimension. In case the dimension is 1, that is, in case $\dim(A/\mathfrak{p}) = 1$, γ is called a *curve germ at α*, and γ_+, γ_- are called the *half-branches of γ*. In fact, this is the situation for any generization $\beta \to \alpha$ with $\dim(\beta \to \alpha) = 1$, at least for A excellent, as shown by consideration of the formal branches of $\mathfrak{p}$ (Section 3). After this preamble, we start with a lemma:

Lemma 4.1   *Let k be a field of characteristic zero and $\mathbf{x} = (\mathbf{x}_1, \ldots, \mathbf{x}_d)$ indeterminates. Let $\beta \to \alpha$ be a specialization in $\operatorname{Spec}_r(k[[\mathbf{x}]])$, where β is an ordering of $k((\mathbf{x}))$ and α one of k. Let $g_1, \ldots, g_r \in k[[\mathbf{x}]]$ be positive at β. Then,* *there are formal power series $x(t) = (x_1(t), \ldots, x_d(t))$ with coefficients in $\kappa(\alpha)$ such that $x(0) = 0$ and $g_i(x(t)) = c_i t^{q_i} + \cdots$, $c_i > 0$, for all $i = 1, \ldots, r$.*

Proof.   We can suppose $d \geq 2$. After a permutation and replacing if needed x_i by $-x_i$, we may assume $0 < x_d(\beta) < x_1(\beta)$. Also, after linear change of the type

$$x_i = \tilde{x}_i + c_i \tilde{x}_d, \ 1 \leq i < d; \ x_d = \tilde{x}_d$$

with $c_{d-1}(\alpha) < 0$ the g_i's are regular in $\tilde{x}_d$. (In fact, any $c = (c_1, \ldots, c_{d-1}, 1)$ which is not a zero of any initial homogeneous form of any g_i does the job.) Then, by the preparation theorem, we have $f_i = u_i P_i$, where P_i is a distinguished polynomial in $\tilde{x}_d$ and $u_i(0) \neq 0$. Changing u_i and P_i by $-u_i$ and $-P_i$ if necessary, we may assume that $u_{i0} = u_i(\alpha) > 0$; then, for any substitution $\tilde{x} = x(t)$ we have $u_i(x(t)) = u_{i0} + u_{i1} t + \cdots$. Also, since $c_{d-1}(\alpha) < 0$, we still have $0 < \tilde{x}_d(\beta) < \tilde{x}_1(\beta)$. Therefore, we may assume that $g_1, \ldots, g_s$ are distinguished polynomials in $k[[x']][x_d]$, where $x' = (x_1, \ldots, x_{d-1})$.

Next we consider the open constructible set

$$\{g_1 > 0, \ldots, g_r > 0, x_1 > x_d > 0\} \subset \mathrm{Spec}_r(k[[x']][x_d]).$$

By Propositions II.1.9 and II.4.2, the canonical map $\pi : \mathrm{Spec}_r(k[[x']][x_d]) \to \mathrm{Spec}_r(k[[x']])$ sends the set above onto an open constructible set of $\mathrm{Spec}_r(k[[x']])$. Then, by cylindrical decomposition (Proposition II.6.3), there are $h_1, \ldots, h_t \in k[[x']]$ and a semialgebraic section s of π on $\{h_1 > 0, \ldots, h_t > 0\}$ whose image is contained in $\{g_1 > 0, \ldots, g_r > 0, x_1 > x_d > 0\}$.

We argue by induction. Suppose we are given power series with coefficients in $\kappa(\alpha)$, $x'(t) = (x_1(t), \ldots, x_{d-1}(t))$, such that $x'(0) = 0$ and every substitution $h_j(x'(t))$ is a power series whose first non-zero coefficient is positive. Then the substitution $x' = x'(t)$ gives a local homomorphism $\gamma' : k[[x']] \to \kappa(\alpha)[[t]]$ such that γ'_+ makes all the h_j's positive. Thus, we can consider $s(\gamma'_+)$. By the properties of semialgebraic functions (Examples II.5.4), $\kappa(s(\gamma'_+)) = \kappa(\gamma'_+)$, and the latter is the field R of Puiseux series with coefficients in $\kappa(\alpha)$. Thus, $s(\gamma'_+)$ is in fact a homomorphism $k[[x']][x_d] \to R$ that extends the substitution $x' = x'(t)$. Moreover, by construction the image $x_d(t)$ of x_d is positive and smaller than $x_1(t) \in \kappa(\alpha)[[t]]$. Since the ring W of Puiseux series is convex in its quotient field R, we conclude that $x_d(t)$ has only positive exponents, and for some positive integer p, we get $x_d(t^p) \in \kappa(\alpha)[[t]]$, $x_d(0) = 0$. Since $s(\gamma'_+) \in \{g_1 > 0, \ldots, g_r > 0\}$, the power series $x(t) = (x'(t^p), x_d(t^p))$ verify the conditions of the statement. $\square$

After this, we prove:

Theorem 4.2 (Curve Selection Lemma)   *Let A be a ring and let $\alpha \in \mathrm{Spec}_r(A)$ be such that the localization $A_{\mathrm{supp}(\alpha)}$ is excellent. Let $C \subset \mathrm{Spec}_r(A)$ be a constructible set with $\dim_\alpha(C) \geq 2$. Then there is a curve germ at α, with two different half-branches, which belong both to C.*

Proof. By induction, it is enough to show that there is a local homomorphism $\gamma : A \to \kappa(\alpha)[[\mathbf{t}]]$ which is not injective, such that the two prime cones γ_+ and γ_- are different and belong to C. On the other hand, by definition of $\dim_\alpha$, α has a generization $\beta \in C$ with $\dim(\beta \to \alpha) \geq 2$. Working modulo $\mathrm{supp}(\beta)$ we may assume that β has support (0), and then $\beta \in \{f_1 > 0, \ldots, f_s > 0\} \subset C$ for some $f_i \in A$. Localizing at $\mathrm{supp}(\alpha)$, we can suppose that A is a local ring and α is an ordering of the residue field k. Then, A contains $\mathbb{Q}$, and by Zorn's lemma we can choose a maximal subfield k_0 of A such that k is algebraic over k_0. Finally, let $y = (y_1, \ldots, y_n)$ be generators of the maximal ideal $\mathfrak{m}$ of A. After this preparation, we shall prove the theorem by finding a local homomorphism $\gamma : A \to \kappa(\alpha)$ such that its kernel $\mathfrak{p}$ and the two associated generizations $\gamma_+ \to \alpha$ and $\gamma_- \to \alpha$ verify the following conditions:

a) $f_1(\gamma) > 0, \ldots, f_s(\gamma) > 0$
b) $k_0[y] \cap \mathfrak{p} \neq \{0\}$.
c) $\gamma_+ \mid k_0[y] \neq \gamma_- \mid k_0[y]$.

Moreover, by extension under completion (Theorem 3.3), we can replace A by the formal branch to which β extends, or in other words, we can suppose that A is a complete local domain of dimension $d \geq 2$.

Next, we use local parametrization for A. By Cohen's structure theorem ([Mt 28.J, Th.60, p.205]), $A = k[[\mathbf{y}]]/\mathfrak{P}$, where $\mathbf{y} = (\mathbf{y}_1, \ldots, \mathbf{y}_m)$ are indeterminates and $\mathbf{y}_i + \mathfrak{P} = y_i$. By Noether's normalization theorem (which is a consequence of the preparation and division theorems [Rz12 II.2.6, p.24]), there exist k_0-linear combinations $\mathbf{x}_1, \ldots, \mathbf{x}_d, \mathbf{z}$ of $\mathbf{y}_1, \ldots, \mathbf{y}_m$, and an irreducible distinguished polynomial $P \in k[[\mathbf{x}]][\mathbf{z}]$, $\mathbf{x} = (\mathbf{x}_1, \ldots \mathbf{x}_d)$, with non-zero discriminant $\Delta \in k[[\mathbf{x}]]$, such that

i) the canonical homomorphism $B = k[[\mathbf{x}, \mathbf{z}]]/P \to A$ is injective, finite, and induces an isomorphism $B[1/\Delta] \to A[1/\Delta]$,
ii) the canonical homomorphism $k[[\mathbf{x}]][\mathbf{z}]/P \to B$ is an isomorphism.

By *i)*, $Y = \{f_1 > 0, \ldots, f_s > 0, \Delta \neq 0\} \subset \mathrm{Spec}_r(A)$ can be seen as a constructible set of $\mathrm{Spec}_r(B)$, which we do henceforth. By *ii)* and Proposition II.1.9, the canonical map $\mathrm{Spec}_r(B) \to \mathrm{Spec}_r(k[[\mathbf{x}]])$ sends Y onto a constructible set $Z \subset \mathrm{Spec}_r(k[[\mathbf{x}]])$. Moreover, since $\beta \mid k[[\mathbf{x}]] \in Z$ has support (0), there are power series $g_1, \ldots, g_r \in k[[\mathbf{x}]]$ such that $\{g_1 > 0, \ldots, g_r > 0\} \subset Z$ and $g_1(\beta) > 0, \ldots, g_r(\beta) > 0$. Consequently, we can apply Lemma 4.1, and find power series $x(\mathbf{t}) = (x_1(\mathbf{t}), \ldots, x_d(\mathbf{t}))$ with coefficients in $\kappa(\alpha)$ such that $x(0) = 0$ and

iii) $g_i(x(\mathbf{t})) = c_i \mathbf{t}^{q_i} + \cdots,\ c_i > 0$ for all $i = 1, \ldots, r$.

After substitution by a truncation big enough, we can suppose that the $x_i(\mathbf{t})$'s are in fact polynomials. Thus, they define by substitution a homomorphism $k_0[\mathbf{x}] \to \kappa(\alpha)[\mathbf{t}]$, which cannot be injective, since the transcendence degree over k_0 of the source is $d \geq 2$, and that of the target is 1. Thus, there is a polynomial

$Q(\mathbf{x}) \neq 0$ such that $Q(x(\mathbf{t})) = 0$; we can suppose that $\frac{\partial Q}{\partial \mathbf{x}_d}(x(\mathbf{t})) \neq 0$.

Next, we consider the local homomorphism $k[[\mathbf{x}]] \to \kappa(\alpha)[[\mathbf{t}]]$ given by the substitution $\mathbf{x}_i = x_i(\mathbf{t})$. This homomorphism defines a generization $\rho \to \alpha$, which by *iii)* belongs to Z. Hence, there is a prime cone $\eta \in Y \subset \mathrm{Spec}_r(B)$ lying over ρ such that $f_j(\eta) > 0$ for all j, and $\Delta(\eta) \neq 0$. In view of condition *ii)* above, this η is completely determined by $x(\mathbf{t})$ and a root ζ of the polynomial $P(x(\mathbf{t}), \mathbf{z})$ in the field $\kappa(\rho)$, and the latter is the field of Puiseux series with coefficients in $\kappa(\alpha)$. But, since $\rho \to \alpha$, that root must be infinitesimal with respect to $\kappa(\alpha)$. Consequently, $\zeta = z(\mathbf{t})$ is a Puiseux series with positive exponents, and after a substitution $\mathbf{t} = \mathbf{t}^p$ we can suppose that they are all integers.

Summing up, we have found polynomials $x_1(\mathbf{t}), \ldots, x_d(\mathbf{t}) \in \kappa(\alpha)[\mathbf{t}]$ and a power series $z(\mathbf{t}) \in \kappa(\alpha)[[\mathbf{t}]]$ such that

iv) $P(x(\mathbf{t}), z(\mathbf{t})) = 0$.

v) $\Delta(x(\mathbf{t})) \neq 0$, hence $\frac{\partial P}{\partial \mathbf{z}}(x(\mathbf{t}), z(\mathbf{t})) \neq 0$,

vi) $f_i(x(\mathbf{t}), z(\mathbf{t})) = a_i\mathbf{t}^{m_i} + \cdots$, $a_i > 0$ $(1 \leq i \leq s)$.

From all of it, we get:

vii) $P(x(\mathbf{t}^2), z(\mathbf{t}^2)) = 0$,

viii) $\Delta(x(\mathbf{t}^2)) \neq 0$ and $\frac{\partial P}{\partial \mathbf{z}}(x(\mathbf{t}^2), z(\mathbf{t}^2)) = a\mathbf{t}^e + \cdots$, $a \neq 0$,

ix) $f_i(x(\mathbf{t}^2), z(\mathbf{t}^2)) = a_i\mathbf{t}^{2m_i} + \cdots$, $a_i > 0$ $(1 \leq i \leq s)$,

x) $Q(x(\mathbf{t}^2)) = 0$ and $\frac{\partial Q}{\partial \mathbf{x}_d}(x(\mathbf{t}^2)) = c\mathbf{t}^{2l} + \cdots$, $c \neq 0$.

Now, we choose integers

$$\ell_1 > e, 2m_1, \ldots, 2m_s; \quad \ell_2 > 2l,$$

and consider

$$x^*(\mathbf{t}) = (x_1(\mathbf{t}^2), \ldots, x_{d-1}(\mathbf{t}^2), x_d(\mathbf{t}^2) + \mathbf{t}^{\ell_2}).$$

Clearly, we can choose ℓ_2 odd and big enough such that

xi) $P(x^*(\mathbf{t}), z(\mathbf{t}^2)) = a^*\mathbf{t}^{\ell_1 + e} + \cdots$,

xii) $\Delta(x^*(\mathbf{t})) \neq 0$,

and, for any $h(\mathbf{t}) \in \kappa(\alpha)[[\mathbf{t}]]$,

xiii) $\frac{\partial P}{\partial \mathbf{z}}(x^*(\mathbf{t}), z(\mathbf{t}^2) + \mathbf{t}^{\ell_1}h(\mathbf{t})) = a\mathbf{t}^e + \cdots$, $a \neq 0$,

xiv) $f_i(x^*(\mathbf{t}), z(\mathbf{t}^2) + \mathbf{t}^{\ell_1}h(\mathbf{t})) = a_i\mathbf{t}^{2m_i} + \cdots$, $a_i > 0$ $(1 \leq i \leq s)$.

We have

$$P(x^*(\mathbf{t}), z(\mathbf{t}^2) + \mathbf{t}^{\ell_1}h) = P(x^*(\mathbf{t}), z(\mathbf{t}^2)) + \frac{\partial P}{\partial \mathbf{z}}(x^*(\mathbf{t}), z(\mathbf{t}^2))\mathbf{t}^{\ell_1}h + G(t, h)h^2,$$

and by *xi)* and *xiii)*:

$$P(x^*(\mathbf{t}), z(\mathbf{t}^2) + \mathbf{t}^{\ell_1}h) =$$
$$(a^*\mathbf{t}^{\ell_1 + e} + \cdots) + (a\mathbf{t}^{\ell_1 + e} + \cdots)h + G(t, h)h^2 = \mathbf{t}^{\ell_1 + e}H(t, h),$$

where $\frac{\partial H}{\partial h}(0,0) = a \neq 0$. Hence, by the implicit functions theorem, there is $h(t) \in \kappa(\alpha)[[t]]$ such that $h(0) = 0$ and $H(t, h(t)) = 0$. Then, setting

$$z^*(t) = z(t^2) + t^{\ell_1} h(t)$$

we get:

xv) $P(x^*(t), z^*(t)) = 0,$
xvi) $\Delta(x^*(t)) \neq 0,$
xvii) $f_i(x^*(t), z^*(t)) = a_i t^{2m_i} + \cdots, \ a_i > 0, \ (1 \leq i \leq s),$
xviii) $Q(x^*(t)) = ct^{\ell_2 + 2l} + \cdots \ c \neq 0.$

Indeed, only the last equality has not been explained yet, but

$$Q(x^*(t)) = Q(x(t^2)) + \frac{\partial Q}{\partial x_d}(x(t^2))t^{\ell_2} + \cdots,$$

and *xviii)* follows from *x)* and the fact that $\ell_2 > 2l$. Finally, note that, as for $x(t)$, there exists a non-zero polynomial $Q^*(x) \in k_0[x]$ such that

xix) $Q^*(x^*(t)) = 0.$

By *xv)* and *ii)*, the local homomorphism $\gamma : B \to \kappa(\alpha)[[t]]$ defined by the substitutions $x_i = x_i^*(t)$ $(1 \leq i \leq d)$, $z = z^*(t)$, is well defined, and we claim that it extends to A. Indeed, by *xvi)*, it extends to $\gamma : B[1/\Delta] \to \kappa(\alpha)((t))$. Now, by *i)*, $A \subset B[1/\Delta]$ is integral over B, hence $\gamma(A)$ is integral over $\gamma(B)$. Since $\gamma(B) \subset \kappa(\alpha)[[t]]$ and the latter ring is normal, we conclude $\gamma(A) \subset \kappa(\alpha)[[t]]$. This extension $\gamma : A \to \kappa(\alpha)[[t]]$ is the local homomorphism we sought.

Indeed, its kernel $\mathfrak{p}$ contains $Q^*(x) \in k_0[x] \subset k_0[y]$, so that $\mathfrak{p} \cap k_0[y] \neq (0)$. The two generizations γ_+ and γ_- make all the f_i's positive by *xvii)*, and they give different signs to $Q(x) \in k_0[x] \subset k_0[y]$ by *xviii)*, since $\ell_2 + 2l$ is odd. The proof is thus complete. $\qquad \square$

Remarks 4.3 To make clear the geometric meaning of the curve selection lemma, let us look at some particular cases.

a) (Algebraic Curve Selection Lemma) Let R be a real closed field. Let V be a real affine algebraic R-variety and set $A = R[V]$. Let $S \subset V(R)$ be a semialgebraic subset, and consider a point $a \in \mathrm{Adh}(S) \setminus S$. Then, via the tilde operator (Sections V.5 and VI.7) we have a constructible set $C = \tilde{S} \subset \mathrm{Spec}_r(A)$ and $a \equiv \alpha \in \mathrm{Adh}(C) \setminus C$. Then, by Theorem 4.2, there is a prime cone $\gamma \in C$ such that $\dim(A/\mathrm{supp}(\gamma)) = 1$ and $\gamma \to \alpha$. As explained in Section I.4, such a γ is a half-branch at a of an algebraic curve contained in $V(R)$. We have thus obtained the classical curve selection lemma ([B-C-R 2.5.5]). $\qquad \square$

b) (Analytic Curve Selection Lemma) Let A be an an analytic algebra over $\mathbb{R}$, that is, a homomorphic image of a ring $\mathbb{R}\{x_1, \ldots, x_n\}$ of convergent power series. Let $C \subset \mathrm{Spec}_r(A)$ be a constructible set of dimension ≥ 1. Since every prime cone of A specializes to the unique closed point with support the maximal ideal, we can apply Theorem 4.2 to find a prime cone $\gamma \in C$ of dimension 1. Hence, $B = A/\mathrm{supp}(\gamma)$ is an analytic algebra of dimension 1, and in fact a domain. It follows that the normalization of B is another analytic algebra of dimension 1, and being normal, $\mathbb{R}$-isomorphic to $\mathbb{R}\{t\}$ ([Rz12 III.1.2 b)]). Thus we get a homomorphism $\varphi : A \to \mathbb{R}\{t\}$ such that for every $f \in A$, $f(\gamma) > 0$ if and only if $\varphi(f) > 0$ (in the unique ordering of $\mathbb{R}\{t\}$ with $t > 0$).

This description of γ gives a very clear geometric picture. Suppose we are given an isomorphism $A = \mathbb{R}\{x_1, \ldots, x_n\}/I$, where the ideal I is generated by, say, $h_1, \ldots, h_m$. Then, we consider the zero set

$$\{x \in U \mid h_1(x) = \cdots = h_m(x) = 0\},$$

which is defined in some small enough neighbourhood U of the origin where the h_i's converge. In addition, suppose C is defined by $f_1 > 0, \ldots, f_r > 0, g = 0$. Then again, in a maybe smaller U, we can consider the set

$$S = \{x \in U \mid h_1(x) = \cdots = h_m(x) = 0, f_1(x) > 0, \ldots, f_r(x) > 0, g(x) = 0\}$$

Now the point γ can be seen as the half branch $t > 0$ of the analytic curve

$$(-\varepsilon, \varepsilon) \to \mathbb{R}^n \,;\, t \mapsto (x_1(t), \ldots, x_n(t)),$$

where $x_i(t) = \varphi(x_i)$ for $1 \leq i \leq n$. More precisely, γ consists of all power series which are ≥ 0 over that half branch. Furthermore, the fact that γ belongs to C means that the half branch is contained in S. $\square$

c) (Formal Curve Selection Lemma) Let again R be a real closed field, and A a formal algebra over R, that is, a homomorphic image of a ring of formal power series $R[[x_1, \ldots, x_n]]$. Let $C \subset \mathrm{Spec}_r(A)$ be a constructible set of dimension ≥ 1. Then there is a prime cone $\gamma \in C$, which is defined by a local homomorphism $\varphi : A \to R[[t]]$.

The argument is the same as for *b)*. $\square$

d) (Algebraic Curve Selection Lemma, Second Form) Let R be a real closed field, and A a finitely generated R-algebra. Let $\mathfrak{m}$ be maximal ideal of A with residue field R, and let α be the unique prime cone of A with support $\mathfrak{m}$. The real strict localization B of $A_\mathfrak{m}$ at α is a homomorphic image of a ring of algebraic power series $R[[x_1, \ldots, x_n]]_{\mathrm{alg}}$ (by Example II.7.12). Let $C \subset \mathrm{Spec}_r(A)$ be a constructible set such that $\alpha \in \mathrm{Adh}(C) \setminus C$. Then C defines a non-empty constructible set in $\mathrm{Spec}_r(B)$, and arguing as in the preceding remarks, we find $\gamma \in C$, which is defined by a local homomorphism $\varphi : A \to R[[t]]_{\mathrm{alg}}$.

This is in fact a more precise formulation of the algebraic curve selection lemma ([B-C-R 8.1.17]). $\square$

We remark finally that our version of the curve selection lemma differs from the standard ones in the fact that we get the two half-branches of the same curve germ inside the given constructible set. This makes the proof longer, but will be essential for the construction of big fans.

5. Dimension, Valuations and Fans

In this section we shall apply extension of orderings under completion (Theorem 3.2) and the curve selection lemma (Theorem 4.2) to obtain several important existence theorems. We start with specialization chains:

Proposition 5.1 (Real Dimension) *Let A be a ring and let $\alpha \in \operatorname{Spec}_r(A)$ be such that the localization $A_{\operatorname{supp}(\alpha)}$ is excellent. Let $C \subset \operatorname{Spec}_r(A)$ be a constructible set with $d = \dim_\alpha(C)$. Then there is a chain $\alpha_d \to \cdots \to \alpha_1 \to \alpha$, with $\alpha_i \in C$ for $i = 1, \ldots, d$.*

Proof.   Let $\beta \in C$ be such that $d = \dim(\beta \to \alpha)$. If $d = 1$ we are done. Let $d > 1$. Replacing A by $A/\operatorname{supp}(\beta)$ we may assume $\operatorname{supp}(\beta) = (0)$. By hypothesis, there is $h \in A \setminus (0)$ such that for every prime ideal $\mathfrak{p} \subset \operatorname{supp}(\alpha)$, the localization $A_\mathfrak{p}$ is regular. By the curve selection lemma (Theorem 4.2), we find a generization $\alpha_1 \to \alpha$ with $\dim(\alpha_1 \to \alpha) = 1$, and $\alpha_1 \in C \cap \{h \neq 0\}$; let $\mathfrak{p}$ be the support of α_1. Then $A_\mathfrak{p}$ is regular, and α_1 has a generization β_1 with support (0) (Proposition II.3.4). Thus, $\dim(\beta_1 \to \alpha_1) = d - 1$, and the proof ends by induction. $\square$

Proposition 5.2 (Existence of Real Valuations) *Let A be a local noetherian domain, with residue field k and quotient field K. Let α be an ordering of k and suppose that there is an ordering β of K which is a generization of α. Then there is a real valuation ring V of K centered at α, whose residue field is a finite extension of k.*
    If A is excellent, then V can be chosen discrete of rank equal to $\dim(A)$.

Proof.   We first suppose that A is excellent. By the preceding proposition we may take $\beta = \alpha_d \to \cdots \to \alpha_1 \to \alpha$, with $d = \dim(\beta \to \alpha) = \dim(A)$. Setting $V_i = V_{\alpha_d \alpha_{d-i}}$, $i = 0, \ldots, d - 1$, we have the following chain of valuation rings

$$V = V_d \subsetneq V_{d-1} \subsetneq \cdots \subsetneq V_1 \subset K.$$

Thus, $\operatorname{rank}(V) \geq d$. But $V = V_{\beta\alpha}$ dominates the local ring A, which has dimension d, and it is a general fact from valuation theory, [Ab Th.1, p.330], that in this situation V is a discrete valuation ring of rank d and its residue field k_V is a finite extension of k.

Now, let A be an arbitrary noetherian local domain of dimension d and $\beta \to \alpha$ with $\operatorname{supp}(\beta) = (0)$. Consider the valuation ring $V = V_{\beta\alpha}$, which is centered at α. If $\dim(A) = 1$, again from [Ab Th.1, p.330] we deduce that V is a discrete rank 1 valuation ring whose residue field is a finite extension of k, and we are done. Now, let $\dim(A) > 1$. We replace V by the rank 1 valuation ring containing V, and let $\mathfrak{n} = \mathfrak{m}_V \cap A$ denote the center of V in A. We have $\beta \to \gamma_V$ in $\operatorname{Spec}_r(V)$ and $\beta \to \gamma \to \alpha$ in $\operatorname{Spec}_r(A)$, where γ_V is an ordering of k_V and γ its restriction to $\kappa(\mathfrak{n})$.

If $\beta \neq \gamma \neq \alpha$, by induction we have a real valuation ring W of $K = \operatorname{qf}(A_\mathfrak{n})$

centered at γ whose residue field k_W is a finite extension of $\kappa(\mathbf{n})$. Pick a primitive element θ of k_W over $\kappa(\mathbf{n})$. We can choose θ integral over $A/\mathbf{n}$, and set $B = (A/\mathbf{n})[\theta]$. By the real going-up for integral extensions (Proposition II.4.3), we have $\gamma_W \to \alpha_W$ in B, where α_W lies over α. Now, $\dim(B_{\mathrm{supp}(\alpha_W)}) \le \dim(A/\mathbf{n}) < \dim(A)$, and by induction there is a real valuation W' of $k_W = \mathrm{qf}(B)$ centered at α_W whose residue field is a finite extension of $\kappa(\mathrm{supp}(\alpha_W))$. The composite of W' and W is the valuation we sought.

Finally, suppose that $\mathbf{n} = \mathbf{m}$. Then, we have a local homomorphism $A \to V$, and since V has rank 1, this homomorphism is continuous with respect to the adic topology in the source and the valuation topology in the target. Consequently, it extends to the respective completions $\hat{A} \to \hat{V}$. As is well-known, [Bk CA VI.5.3, Prop. 5, p.402], $\hat{V}$ has the same value group and residue field as V. By the Baer-Krull theorem (Proposition II.3.3), there is a total ordering $\hat{\beta}$ of $\hat{V}$ lying over β, which restricts, via the homomorphism $\hat{A}_{\mathbf{n}} \to \hat{V}$, to a prime cone of $\hat{A}$, still denoted by $\hat{\beta}$. Thus, $\hat{\beta} \to \gamma$ in $\hat{A}$. Let $B = \hat{A}/\mathrm{supp}(\hat{\beta})$. This ring B is a complete local domain with residue field k, and dominates A. Furthermore, it is excellent, and by the case already settled, there is a real valuation W of its quotient field L centered at α whose residue field k_W is a finite extension of k. The restriction of W to $K \subset L$ is the valuation we sought. $\qquad\qquad\square$

Next, we shall obtain lower bounds for the complexity invariants s and $\bar{s}$. Let A be a ring and let $\alpha \in \mathrm{Spec}_r(A)$ be a prime cone such that $A_{\mathrm{supp}(\alpha)}$ is excellent. The real strict localization A_α of A at α is then an excellent ring (Theorem 2.5); recall that it is a local henselian ring whose residue field is $\kappa(\alpha)$. The real spectrum of this ring is canonically homeomorphic to the set $U_\alpha \subset \mathrm{Spec}_r(A)$ of all generizations of α (Theorem II.7.11). Furthermore, the dimension of a specialization $\beta \to \alpha$ is the same in A and in A_α (Corollary 2.7). Although the preceding homeomorphism is not an isomorphism of spaces of signs, it gives the inequalities:

$$s(A) \ge s(U_\alpha) \ge s(A_\alpha), \qquad \bar{s}(A) \ge \bar{s}(U_\alpha) \ge \bar{s}(A_\alpha).$$

Let us first look at the stability index:

Proposition 5.3   *In the situation above, it holds:*

$$s(A_\alpha) \ge \sup\{\dim(\beta \to \alpha) \mid \beta \in A_\alpha\}.$$

Proof.   Let $\beta \in \mathrm{Spec}_r(A_\alpha)$ be such that $\dim(\beta \to \alpha) = d$. Then, by Proposition 5.2, there is a rank d discrete valuation ring of $\kappa(\mathrm{supp}(\beta))$ whose residue field is $\kappa(\alpha)$. Hence, α has 2^d generizations with support $\mathrm{supp}(\beta)$, which form a fan F. By Corollary VI.1.4 *a)*, $s(A_\alpha) \ge d$. $\qquad\qquad\square$

In order to bound $\bar{s}$, we need a more careful construction of fans, which in fact motivated our version of the curve selection lemma.

Proposition 5.4 (Existence of Fans)   *Let $C \subset A_\alpha$ be a constructible set whose dimension $d = \dim_\alpha(C)$ is at least 2. Then, there is a fan $F \subset C$ with $\#(F) = 2^d$.*

Proof. Let $\beta \in C$ be a generization of α with $\dim(\beta \to \alpha) = d$; Replacing A by $A/\mathrm{supp}(\beta)$, we can assume that $\mathrm{supp}(\beta) = (0)$, and that $C = \{f_1 > 0, \ldots, f_s > 0\}$. Pick $0 \neq h \in A_\alpha$ such that $(A_\alpha)_{\mathfrak{p}}$ is regular if $h \notin \mathfrak{p}$. By the curve selection lemma, there is a curve germ γ at α such that the two half-branches γ_+ and γ_- are different and belong to $C \cap \{h \neq 0\}$; let $\mathfrak{p}$ be the common support of these two half-branches. Then, $h \notin \mathfrak{p}$ and $(A_\alpha)_{\mathfrak{p}}$ is regular. By Proposition II.3.4, there is a rank $d-1$ discrete valuation ring of $K = \mathrm{qf}(A)$ with residue field $\kappa(\mathfrak{p})$. Hence each half-branch has 2^{d-1} generizations compatible with that valuation ring, which all together form a fan F with $\#(F) = 2^d$. By construction and continuity, $F \subset \{f_1 > 0, \ldots, f_s > 0\}$. $\qquad\qquad\Box$

After this result, we can bound $\bar{s}$ from below as follows:

Proposition 5.5   *Let $\beta \in A_\alpha$ be a generization of α with $d = \dim(\beta \to \alpha)$. Then:*

$$\bar{s}(A_\alpha) \geq \frac{1}{2}d(d+1) - 1.$$

Proof. We shall mimic the argument of Propositions VI.5.3 and VI.5.4 based on Proposition V.4.3. We can suppose $d \geq 2$. We construct by induction on $i \geq 2$ a basic closed set C_i that cannot be written with less than $m_i = \frac{1}{2}i(i+1) - 1$ inequalities in its Zariski closure Z_i, and such that $\dim_\alpha(C_i) \geq i$.

Construction for $i = 2$: By Proposition 5.1, we may suppose $\dim(\beta \to \alpha) = 2$. Then, let $h \in A_\alpha \setminus \mathrm{supp}(\beta)$ be such that $(A_\alpha)_{\mathfrak{p}}/\mathrm{supp}(\beta)$ is regular if $h \notin \mathfrak{p}$. By the curve selection lemma, we find such a $\mathfrak{p}$ which is the common support of two different generizations γ_+ and γ_- of α. Let $f \in A_\alpha$ be positive at γ_+ and negative at γ_-. Let $C_1 = \{f \geq 0\} \cap \mathcal{Z}(\mathfrak{p})$; note that $Z_1 = \mathcal{Z}(\mathfrak{p})$ is the Zariski closure of C_1. Since $(A_\alpha)_{\mathfrak{p}}/\mathrm{supp}(\beta)$ is regular of dimension 1, Z_1 is a real divisor of $Y_2 = \mathcal{Z}(\mathrm{supp}(\beta))$ in the sense of Section V.4. Thus, by Proposition V.4.3, using the fan $F = \{\gamma_+\} \subset C_1$, we obtain a basic closed set $C_2 \subset Y_2$ which cannot be written with less than $1 + 0 + 1 = m_2$ inequalities in Y_2, and which contains a generization of γ_+ with the same support as β. This latter condition implies $\dim_\alpha(C_2) \geq 2$ and $Y_2 = \mathrm{Adh}_Z(C_2)$.

Induction step: Let $i > 2$. By Proposition 5.1, we can change β to find β' such that $\beta \to \beta' \to \alpha$, $\dim(\beta \to \alpha) = i$ and $(A_\alpha)_{\mathrm{supp}(\beta')}/\mathrm{supp}(\beta)$ is regular of dimension 1. By induction, there is a basic closed set C_{i-1} such that $C_{i-1} \cap \mathcal{Z}(\mathrm{supp}(\beta'))$ cannot be written with less than m_{i-1} inequalities in its Zariski closure $Z_{i-1} = \mathcal{Z}(\mathrm{supp}(\beta'))$, and $\dim_\alpha(C_{i-1}) \geq i - 1$. Then, by Proposition 5.4, there is a fan $F \subset C_{i-1}$ with $\#(F) = 2^{i-1}$. Again, Z_{i-1} is a real divisor of $Y_i = \mathcal{Z}(\mathrm{supp}(\beta))$, and from Proposition V.4.3 we get a basic closed set C_i which cannot be written with less than $1 + (i-1) + m_{i-1} = \frac{1}{2}i(i+1) - 1$ inequalities in Y_i, and such that $\dim_\alpha(C_i) \geq i$. We are done. $\qquad\Box$

Remarks 5.6 *a)* Let A be an excellent henselian local ring with residue field k and $d = \dim_r(A)$. From the lower bounds above and Proposition VI.4.7 we get:

$$d \le s(A) \le s_0(k) + d, \qquad \frac{1}{2}d(d+1) - 1 \le \bar{s}(A) \le s_0(k)(d+1) + \frac{1}{2}d(d+1).$$

If k is real closed, these inequalities reduce to:

$$s(A) = d, \qquad \frac{1}{2}d(d+1) - 1 \le \bar{s}(A) \le \frac{1}{2}d(d+1).$$

b) The above proof shows why the -1 appears in the lower bound for $\bar{s}$: it comes from the construction of C_2. Consequently, the bound can be improved to $\frac{1}{2}d(d+1)$ when C_1 is not a singleton.

c) The preceding results have a strong local nature, since they deal with A_α, or equivalently, with U_α. If α varies, then the bound is usually better. This happens for algebras finitely generated over fields, as was shown in Section VI.5. A similar improvement will be found for global analytic sets in Chapter VIII.

□

6. Closures of Constructible Sets

Here, we discuss when the closure of a constructible set is also constructible. First, we obtain a quite general positive result:

Proposition 6.1 (Constructibility of Closures) *Let A be a ring and let $Y \subset \mathrm{Spec}_r(A)$ be a subspace such that*

a) For every $\alpha \in Y$ the localization $A_{\mathrm{supp}(\alpha)}$ is noetherian.
b) Y is a noetherian space, that is, every Z-closed subset of Y is constructible.
c) For every $\alpha \in Y$ there exists $h \in A \setminus \mathrm{supp}(\alpha)$ such that for every $\beta \in \{h \ne 0\} \cap Y \cap \mathrm{Adh}_Z(\alpha)$ the ring $A_{\mathrm{supp}(\beta)}/\mathrm{supp}(\alpha)$ is regular.

Then $\mathbf{AC}$ holds for Y, that is, the closure in Y of a constructible subset of Y is a constructible subset of Y.
In particular, $\mathbf{AC}$ holds for $Y = \mathrm{Spec}_r(A)$ if A is excellent.

Proof. By definition of subspace (Definition III.1.6), $Y = \mathrm{Adh}_Z(Y) \cap \bigcap_{i \in I} \{f_{0i} > 0\}$, where $f_{0i} \in A$. We can replace A by A/I, where I is the ideal of Y, or, equivalently, we can assume that $Y = \bigcap_{i \in I} \{f_{0i} > 0\}$. Let $C \subset Y$ be constructible and let $\mathrm{Adh}(C)$ denote its closure in Y. We must see that $\mathrm{Adh}(C)$ is a constructible subset of Y. As remarked above, Y is a proconstructible subset of $\mathrm{Spec}_r(A)$, and by Proposition II.1.11 we have to show that $\mathrm{Adh}(C)$ is open and closed in the constructible topology. The latter is obvious, since the constructible topology is finer than the Harrison topology. Consequently, we have to prove that for every $\alpha \in \mathrm{Adh}(C)$ there is a constructible subset T of Y with $\alpha \in T \subset \mathrm{Adh}(C)$.

To that end, we first pick some generization $\beta \in C$ of α, which exists by Proposition II.2.3. Now notice that, by Proposition II.1.11 again, $C = D \cap Y$ for some constructible subset D of $\mathrm{Spec}_r(A)$. Hence, from the boolean description of D we get $f_1, \ldots, f_r, g \in A$ with

$$\beta \in \{f_1 > 0, \ldots, f_r > 0, g = 0\} \cap Y \subset C.$$

Consider $A_1 = A_{\mathrm{supp}(\alpha)}/\mathrm{supp}(\beta)$, and $B = A_1[\sqrt{f_1}, \ldots, \sqrt{f_r}] \subset k(\beta)$. The extension $A_1 \subset B$ is integral, we have $\beta \to \alpha$ in $\mathrm{Spec}_r(A)$, and β is also a point of B. Thus, by the real going-up (Proposition II.4.3), β has a specialization $\alpha' \in \mathrm{Spec}_r(B)$ which lies over α. Now, by Proposition 5.2, we find a real valuation ring V of $\mathrm{qf}(B)$ centered at α' whose residue field k_V is a finite extension of $k(\mathrm{supp}(\alpha'))$. Consequently, k_V is a finite extension of $k(\mathrm{supp}(\alpha))$. By Proposition II.1.9, the induced mapping $\mathrm{Spec}_r(k_V) \to \mathrm{Spec}_r(k(\mathrm{supp}(\alpha)))$ sends constructible sets onto constructible sets. In particular, there are $h_1, \ldots, h_s \in A$ such that any total ordering of $k(\mathrm{supp}(\alpha))$ in $\{h_1 > 0, \ldots, h_s > 0\}$ extends to a total ordering of k_V, and $h_1(\alpha) > 0, \ldots, h_s(\alpha) > 0$.

Now, by condition $b)$, there are $g_1, \ldots, g_t \in \mathrm{supp}(\alpha)$ such that

$$\{g_1 = 0, \ldots, g_t = 0\} = Y \cap \mathrm{Adh}_Z(\alpha).$$

Finally, we choose $h \notin \mathrm{supp}(\alpha)$ according to condition $c)$, and put

$$T = \{h_1 > 0, \ldots, h_s > 0, g_1 = 0, \ldots, g_t = 0, h \neq 0\}.$$

By construction $\alpha \in T$, and we claim that $T \cap Y \subset \mathrm{Adh}(C)$. Indeed, let $\gamma \in T$. Then $\mathrm{Adh}_Z(\gamma) \subset \mathrm{Adh}_Z(\alpha)$, and by the real Nullstellensatz (Theorem II.2.8), $\mathfrak{p} = \mathrm{supp}(\gamma) \supset \mathrm{supp}(\alpha)$. We also have $h \notin \mathfrak{p}$, so that $A_{\mathfrak{p}}/\mathrm{supp}(\alpha)$ is regular by the choice of h. Hence, by Lemma II.3.4, there is a generization α' of γ which is a total ordering of $A_{\mathfrak{p}}/\mathrm{supp}(\alpha) \subset k(\mathrm{supp}(\alpha))$. But $h_i(\gamma) > 0$ implies $h_i(\alpha') > 0$, for $i = 1, \ldots, s$, and so α' extends to some total ordering α_V in k_V. Then, using again Lemma II.3.4, there is a generization β_V of α_V which is a total ordering of V. As $\sqrt{f_i} \in \mathrm{qf}(V)$, we have $f_i(\beta_V) > 0$ for all i, and as $g \in \mathrm{supp}(\beta)$, we have $g(\beta_V) = 0$. Hence, denoting by β' the restriction of β_V to $A/\mathrm{supp}(\beta) \subset V$, we get

$$\beta' \in \{f_1 > 0, \ldots, f_r > 0\} \subset C,$$

and $\beta' \to \alpha' \to \gamma$. In addition, $\beta' \in Y$. Indeed, $f_{0i}(\gamma) > 0$ and by continuity $f_{0i}(\beta') > 0$ for all i. Thus $\gamma \in \mathrm{Adh}(C)$ and the proof is complete. $\square$

Example 6.2 Here there is a real spectrum for which **AC** fails. Let A be the ring of continuous functions $f : \mathbb{R} \to \mathbb{R}$ and let $X = \mathrm{Spec}_r(A)$ be its real spectrum. Then the closure of the set $C = \{f > 0\} \subset X$, where $f(t) = t$, $t \in \mathbb{R}$, is not constructible.

By way of contradiction, suppose that $\mathrm{Adh}(C)$ is constructible. By Corollary II.1.13, there are $h_{ij} \in A$ such that

$$\mathrm{Adh}(C) = \bigcup_{1 \leq i \leq r} \{h_{i1} \geq 0, \dots, h_{is} \geq 0\},$$

and then

$$\mathrm{Adh}(C) = \{g = 0\} \quad \text{for} \quad g = \prod_{1 \leq i \leq r} \sum_{1 \leq j \leq s} (|h_{ij}| - h_{ij})^2.$$

This follows from the fact that for any $h \in A$ and $\alpha \in X$, it holds $|h|(\alpha) = |h(\alpha)|$. Indeed, since $\sqrt{|h|} \in A$, we have $|h|(\alpha) \geq 0$. But $(|h| - h)(|h| + h) = 0$, hence either $|h|(\alpha) = h(\alpha)$ or $|h|(\alpha) = -h(\alpha)$. We are done.

On the other hand, we embed $\mathbb{R} \subset X$ by identifying each $t \in \mathbb{R}$ with the prime cone $A \to \mathbb{R}$ defined by evaluation at t. It is clear then that $\mathrm{Adh}(C) \cap \mathbb{R}$ is the infinite interval $I = \{t \in \mathbb{R} \mid t \geq 0\}$, and we deduce that g vanishes on I and is > 0 on $\mathbb{R} \setminus I$. Thus, $h = -1/\ln(g) \in A$, and for all $n \geq 1$ it holds

$$\lim_{t \to 0, t < 0} \frac{h^n(t)}{g(t)} = +\infty.$$

This implies that $h \notin \sqrt[r]{gA}$, and by the real Nullstellensatz (Proposition II.2.8), there is a prime cone α such that $g(\alpha) = 0$ and $h(\alpha) \neq 0$. Finally, by construction, $h(|f| + f) = 0$ and so h vanishes on C. By continuity, h must vanish on $\mathrm{Adh}(C) = \{g = 0\}$. This is the contradiction. $\square$

We end the section with some consequences concerning the constructible points of an excellent ring A. Since **AC** holds for $\mathrm{Spec}_r(A)$, every constructible point specializes to a closed constructible point (Remark V.5.6 $b)$). We have also seen an example of a constructible non-closed point (Remark V.5.6 $c)$), and, in fact, there is no other example:

Proposition 6.3   *Let A be an excellent ring and let $\alpha \in \mathrm{Spec}_r(A)$ be a closed point. Let β be a constructible point which is a generization of α. Then $\dim(\beta \to \alpha) \leq 1$.*

Proof.   Let $d = \dim(\beta \to \alpha)$, so that $\dim_\alpha(\beta) = d$. By Proposition 5.1, the constructible set $\{\beta\}$ must contain a specialization chain consisting of d different prime cones, hence $d \leq 1$. $\square$

Thus, the constructible points of A are either closed or half-branches of curve germs at closed points. As a complement to this, we can characterize closed constructible points as follows:

Proposition 6.4   *Let A be an excellent ring and let $\alpha \in \mathrm{Spec}_r(A)$ be a closed point. Let $\mathrm{Adh}_Z(\alpha)^*$ be the proconstructible set $\mathrm{Spec}_r(\kappa(\mathrm{supp}(\alpha))$, which is space of orderings and a subspace of $\mathrm{Spec}_r(A)$. Then, α is constructible if and only if it is isolated in $\mathrm{Adh}_Z(\alpha)^*$.*

Proof. As was seen in Remark V.5.6 *a)*, α is constructible if and only if

$$\{\alpha\} = \{f_1 > 0, \ldots, f_s > 0\} \cap \mathrm{Adh}_Z(\alpha)$$

for some $f_1, \ldots, f_s \in A$. Hence, the "only if part" of the statement is clear. For the "if part", suppose that α is isolated in $\mathrm{Adh}_Z(\alpha)^*$, that is, there are $g_1, \ldots, g_s \in A$ be such that $\{\alpha\} = \{g_1 > 0, \ldots, g_s > 0\} \cap \mathrm{Adh}_Z(\alpha)^*$. We choose $h \in A \setminus \mathrm{supp}(\alpha)$ such that $A_{\mathfrak{p}}/\mathrm{supp}(\alpha)$ is regular whenever $h \notin \mathfrak{p}$. We claim that $\{\alpha\} = \{g_1 > 0, \ldots, g_s > 0, h \neq 0\} \cap \mathrm{Adh}_Z(\alpha)$, which concludes the proof. To prove this claim, pick $\beta \in \mathrm{Adh}_Z(\alpha)$ such that $h(\beta) \neq 0$. Then $A_{\mathrm{supp}(\beta)}/\mathrm{supp}(\alpha)$ is regular, and by Lemma II.3.4, β has a generization α' whose support is $\mathrm{supp}(\alpha)$, that is, $\alpha' \in \mathrm{Adh}_Z(\alpha)^*$. If, in addition, for all i we have $f_i(\beta) > 0$, continuity gives $f_i(\alpha') > 0$. Thus, $\alpha' \in \{g_1 > 0, \ldots, g_s > 0\} \cap \mathrm{Adh}_Z(\alpha)^*$, and we conclude $\alpha' = \alpha$. Consequently, $\alpha \to \beta$, and α being closed, $\alpha = \beta$. We are done. $\qquad\square$

The reader can realize the geometric meaning of these remarks in the global situation of semialgebraic sets, and in the local one of Nash germs. We shall later find the same behaviour in the analytic context.

7. Real Going-down for Regular Homomorphisms

We saw in Example II.4.4 that the real going-down requires some regularity conditions to hold. The goal of this section is to prove the following:

Theorem 7.1   *Let A be a ring and $\varphi : A \to B$ a homomorphism. Let $\beta \to \alpha$ in $\mathrm{Spec}_r(A)$, and let $\alpha' \in \mathrm{Spec}_r(B)$ be such that $\varphi^*(\alpha') = \alpha$. Suppose that the local domain $A_{\beta\alpha}$ is excellent and the induced homomorphism $A_{\beta\alpha} \to B \otimes_A A_{\beta\alpha}$ is regular. Then there exists $\beta' \in \mathrm{Spec}_r(B)$ such that $\beta' \to \alpha'$, $\mathrm{ht}(\mathrm{supp}(\beta')) = \mathrm{ht}(\mathrm{supp}(\beta))$ and $\varphi^*(\beta') = \beta$.*

Proof.   Set $C = B \otimes_A A_{\beta\alpha}$. By the universal property of the tensor product there exists a point $\gamma \in \mathrm{Spec}_r(C)$ lying over both α and α'. In fact, $\varphi^*(\alpha') = \alpha$ means that if α' is defined by the homomorphism $\alpha' : B \to R_{\alpha'}$ into a real closed field $R_{\alpha'}$, then $\alpha = \alpha'\varphi$. Thus we have a commutative diagram giving γ:

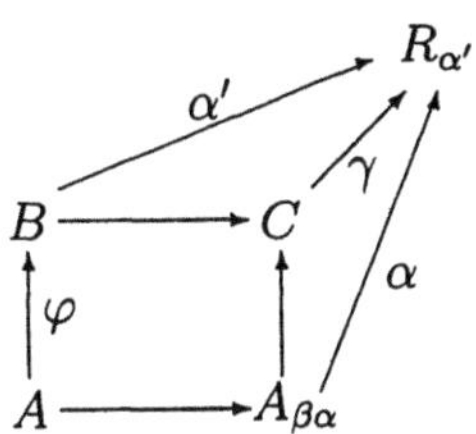

Furthermore, suppose that we find $\delta \in \mathrm{Spec}_r(C)$ solving the problem for the homomorphism $A_{\beta\alpha} \to C$ and the prime cone γ. Then, its restriction to B solves the problem for φ and α', since C is a localization of $B/\mathrm{supp}(\beta)B$. So, replacing A by $A_{\beta\alpha}$ and B by C, we may assume $A = A_{\beta\alpha}$. In particular, A is a local domain with maximal ideal $\mathfrak{m} = \mathrm{supp}(\alpha)$, and $\mathrm{supp}(\beta) = (0)$.

We argue by induction on $d = \dim(\beta \to \alpha)$. Suppose first that $d = 1$. Thus $A = A_{\beta\alpha}$ is a local domain of dimension 1. Let A_1 be the integral closure of A in its quotient field K. The point β, seen as a total ordering of K, defines also a point in $\mathrm{Spec}_r(A_1)$, and by the real going-up (Proposition II.4.3) we get a specialization $\beta \to \alpha_1$ in $\mathrm{Spec}_r(A_1)$ with $\alpha_1 \cap A = \alpha$. Since A_1 is integral over A, $\kappa(\mathrm{supp}(\alpha_1))$ is a finite extension of $\kappa(\mathrm{supp}(\alpha))$, and therefore α_1 can also be defined as a homomorphism $\alpha_1 : A_1 \to R_{\alpha'}$. We put $B_1 = B \otimes_A A_1$, and consider the commutative diagram

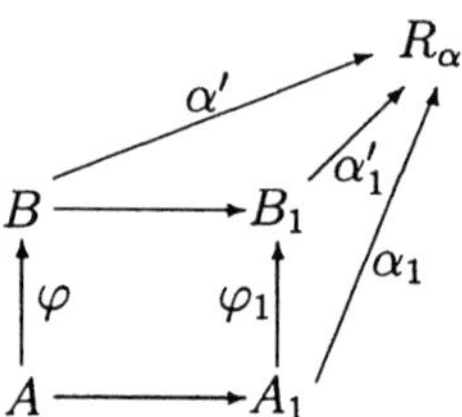

where the point $\alpha_1' \in \mathrm{Spec}_r(B_1)$ is given by the universal property of tensor products, and φ_1 is the homomorphism obtained by base change. Since A is excellent, by Proposition 2.2 $g)$, A_1 is a finite A-module, and therefore by Proposition 1.5 $b)$, φ_1 is regular. Set $\mathfrak{m}_1 = \mathrm{supp}(\alpha_1)$, $\mathfrak{n}_1 = \mathrm{supp}(\alpha_1')$ and $\mathfrak{n} = \mathfrak{n}_1 \cap B$. Then $A_{1,\mathfrak{m}_1}$ is a local noetherian normal domain of dimension 1, whence regular. Let t be a regular parameter such that $t(\beta) > 0$ (note that β is completely determined by this condition). By Corollary 1.11, $B_{1,\mathfrak{n}_1}$ is also regular and t belongs to a system of parameters of $B_{1,\mathfrak{n}_1}$. Let $\beta_1' \in \mathrm{Spec}_r(B_1)$ be any generization of α_1' which is a total ordering with $t(\beta_1') > 0$. Then $\varphi_1^*(\beta_1') = \beta_1$. Finally, let $\beta' = \beta_1' \cap B$. We shall check that β' is the point we sought.

Indeed, by construction $\beta' \to \alpha'$ and $\varphi^*(\beta') = \beta$, so that the only thing to be checked is that $\mathrm{ht}(\mathrm{supp}(\beta')) = 0$. Let $\mathfrak{q} = \mathrm{supp}(\beta')$ and $\mathfrak{q}_1 = \mathrm{supp}(\beta_1')$. Since β_1' lies over β, we get that $\mathfrak{q}_1$ is a zero divisor of B_1, and $\mathrm{ht}(\mathfrak{q}_1) = 0$. On the other hand $\mathrm{ht}(\mathfrak{n}_1/\mathfrak{q}_1) \leq \mathrm{ht}(\mathfrak{n}/\mathfrak{q})$, since B_1 is integral over B. Therefore, using property $i)$ of the definition of excellent rings for B and taking into account once more that B_1 is finite over B we get

$$\mathrm{ht}(\mathfrak{q}) = \mathrm{ht}(\mathfrak{n}) - \mathrm{ht}(\mathfrak{n}/\mathfrak{q}) \leq \mathrm{ht}(\mathfrak{n}) - \mathrm{ht}(\mathfrak{n}_1/\mathfrak{q}_1) = \mathrm{ht}(\mathfrak{n}) - \mathrm{ht}(\mathfrak{n}_1).$$

Now the inclusion

$$B/\mathfrak{m}B \to B_1/\mathfrak{m}_1 B_1 = (B/\mathfrak{m}B) \otimes_{\kappa(\mathfrak{m})} \kappa(\mathfrak{m}_1),$$

is finite and flat, which implies $\mathrm{ht}(\mathfrak{n}/\mathfrak{m}B) = \mathrm{ht}(\mathfrak{n}_1/\mathfrak{m}_1 B_1)$. Finally, also φ and φ_1 are flat, and therefore we have

$$\mathrm{ht}(\mathfrak{n}) = \mathrm{ht}(\mathfrak{m}) + \mathrm{ht}(\mathfrak{n}/\mathfrak{m}B),$$

$$\mathrm{ht}(\mathfrak{n}_1) = \mathrm{ht}(\mathfrak{m}_1) + \mathrm{ht}(\mathfrak{n}_1/\mathfrak{m}_1 B).$$

Since $\mathrm{ht}(\mathfrak{m}) = \mathrm{ht}(\mathfrak{m}_1) = 1$, it follows that $\mathrm{ht}(\mathfrak{n}) = \mathrm{ht}(\mathfrak{n}_1)$ and so $\mathrm{ht}(\mathfrak{q}) = 0$ as claimed. This completes the proof of the case $d = 1$.

Next suppose $d > 1$. Let $\mathfrak{q}_1, \ldots, \mathfrak{q}_s$ be the associated primes of $\mathrm{supp}(\beta)B$. Consider

$$T = \left\{ \beta' \in \mathrm{Spec}_r(B) \,\middle|\, \beta' \to \alpha' \text{ and } \mathrm{supp}(\beta) = \mathfrak{q}_i \text{ for some } i \right\}$$

$$C = T \cap \left(\bigcap_{f(\beta)>0} \{\varphi(f) > 0\} \right).$$

Both T and C are proconstructible. Moreover, any $\beta' \in C$ will solve our problem, and therefore we have to show that $C \neq \emptyset$. Thus, suppose $C = \emptyset$. Since T is compact, there are $f_1, \ldots, f_r \in A$ such that $f_1(\beta) > 0, \ldots, f_r(\beta) > 0$ and $T \cap \{\varphi(f_1) > 0, \ldots, \varphi(f_r) > 0\} = \emptyset$. By Proposition 5.1, there is a chain $\delta \to \gamma \to \alpha$ in $\mathrm{Spec}_r(A)$ such that

   i) $f_i(\delta) > 0$ for all i,
   ii) $\mathrm{supp}(\delta) = \mathrm{supp}(\beta)$, and
   iii) $\dim(\gamma \to \alpha) = 1$.

Furthermore, since $A \to B$ is regular, also $A_{\delta\gamma} \to B \otimes_A A_{\delta\gamma}$ and $A_{\gamma\alpha} \to B \otimes_A A_{\gamma\alpha}$ are regular. By induction hypothesis we find $\gamma' \in \mathrm{Spec}_r(B)$ lying over γ with $\gamma' \to \alpha'$ and such that $\mathrm{ht}(\mathrm{supp}(\gamma')) = \mathrm{ht}(\mathrm{supp}(\gamma))$. Again applying induction to $\delta \to \gamma$, we find δ' with $\delta' \to \gamma'$ and $\mathrm{ht}(\mathrm{supp}(\delta')) = \mathrm{ht}(\mathrm{supp}(\delta))$. We claim that

$$\delta' \in T \cap \{\varphi(f_1) > 0, \ldots, \varphi(f_r) > 0\},$$

which is a contradiction. Indeed, the only non evident fact is that $\mathrm{supp}(\delta') = \mathfrak{q}_i$ for some i. But from

$$\mathrm{ht}(\mathrm{supp}(\delta')) = \mathrm{ht}(\mathrm{supp}(\delta))$$

we get that $\mathrm{supp}(\delta')$ is an associated prime of $\mathrm{supp}(\delta)B = \mathrm{supp}(\beta)B$ and we are done. $\qquad\square$

Corollary 7.2 *Let A be a local excellent ring and let $\widehat{A}$ be its completion. Then over any chain $\alpha_d \to \cdots \to \alpha_0$ in $\mathrm{Spec}_r(A)$ such that $\mathrm{supp}(\alpha_0)$ is the maximal ideal of A, there exits a chain $\widehat{\alpha}_d \to \cdots \to \widehat{\alpha}_0$ in $\mathrm{Spec}_r(\widehat{A})$, with $\mathrm{ht}(\mathrm{supp}(\widehat{\alpha}_i)) = \mathrm{ht}(\mathrm{supp}(\alpha_i))$ for all i.*

Proof. Identifying α_0 with an ordering in the residue field k of A, we say that it has a unique extension $\widehat{\alpha}_0$ to $\widehat{A}$. Now the result follows from the theorem, applied recursively to the specializations $\alpha_i \to \alpha_{i-1}$. $\qquad\square$

8. Connected Components of Constructible Sets

Let A be a commutative ring with unity, and $X = \mathrm{Spec}_r(A)$ its real spectrum In this section we study when the connected components of constructible subsets of X are again constructible, or, in other words, when property I.3.5 **CC** holds for X. Since X is its own Stone space, **CC** holds for X if and only if any constructible subset has finitely many connected components (I.3.5 *d)*). For instance, if A is a field, X is totally disconnected, and **CC** holds if and only if X is finite. On the other hand, **CC** holds for X if and only if it holds for the real spectrum of every finitely generated A-algebra. This follows easily by cylindrical decomposition (Proposition II.6.3), and is left to the reader (a similar argument appears also in the proof of Lemma 8.2 below). Here we shall prove some positive results, for excellent rings. To start with, let us consider the case of complete local rings with real close residue fields, and prove some preliminary lemmas.

Lemma 8.1   *Let k be a real closed field and let $x_1, \ldots, x_n$ be indeterminates. We set $x' = (x_1, \ldots, x_{n-1})$ and $x = (x_1, \ldots, x_n)$. Let $j : A = k[[x']][x_n] \to B = k[[x]]$ be the canonical inclusion, and $j^* : \mathrm{Spec}_r(B) \to \mathrm{Spec}_r(A)$ the corresponding map between real spectra. Let $\alpha \in \mathrm{Spec}_r(A)$ be the image under j^* of the closed point of B. Then:*

a) $j^*(\mathrm{Spec}_r(B)) = U_\alpha$, the set of generizations of α.
b) The family $\{W_\varepsilon\}_{\varepsilon \in k, \varepsilon > 0}$ where

$$W_\varepsilon = \{x_n - \varepsilon > 0,\ x_n + \varepsilon > 0\} \subset \mathrm{Spec}_r(A)$$

is a neighbourhood basis of α.

Proof.   a) We have $\mathrm{supp}(\alpha) = (x', x_n)$. Also j is the composition $A \to A_{\mathrm{supp}(\alpha)} \to B$, and therefore j^* factorizes through $\mathrm{Spec}_r(A_{\mathrm{supp}(\alpha)})$. Since B is the completion of $A_{\mathrm{supp}(\alpha)}$, it follows from the real going-down (Theorem 7.1) that the image of j^* is U_α.

b) Let $f \in A$ such that $f(\alpha) > 0$. We write

$$f = a_r(x')x_n^r + \cdots + a_1(x')x_n + a_0(x') + c,$$

with $a_0(0) = 0$ and $c \in k$. Thus $f(\alpha) = c > 0$. Now, we claim that there exists a constant $M \in k$ such that for any $l = 1, \ldots, r$ and any $\beta \in \mathrm{Spec}_r(A)$ it is

$$|a_l(\beta)| = |a_l(\beta')| < M$$

(where $\beta' \in \mathrm{Spec}_r(k[[x']])$ and β lies over β'). Indeed, we have $a_l = b_l(x') + a_l(0)$ with $b_l(0) = 0$ and since β' makes convex the maximal ideal of $k[[x']]$,

$b_l(\beta)$ is infinitesimal with respect to k and therefore it is enough to take $M = \max\{|a_l(0)|\} + 1$. Then, for $\beta \in W_\epsilon$ we have

$$|f(\beta) - c| \leq \sum_{i=1}^{d} |a_i(\beta)\mathbf{x}_n^i(\beta)| < dM\epsilon.$$

Since $c > 0$, for sufficiently small ϵ we get $f(\beta) > 0$. This shows that $W_\epsilon \subset \{f > 0\}$ and hence $b)$. $\qquad\square$

For the next lemma recall that given a constructible subset C, we denote by $cc(C)$ the number of connected components of C, and by $cc_\alpha(C)$ the number of connected components of C which are adherent to the prime cone α.

Lemma 8.2   *In the situation of the preceding lemma, let $C \subset \mathrm{Spec}_r(A)$ be a constructible set and assume that C is defined by Weierstrass polynomials. Moreover, assume that* **CC** *holds in $k[[\mathbf{x}']]$. Then, for each $\alpha \in \mathrm{Spec}_r(A)$, there exists an integer N such that $cc_\alpha(C \cap W) < N$ for any neighbourhood W of α.*

Proof.   It is enough to show the result for a neighbourhood basis of α. We shall do it for the basis $\{W_\epsilon\}$ of Lemma 8.1 $b)$. Let $P_1, \ldots, P_s$ be a family of Weierstrass polynomials defining C. Let $\{D_i, \xi_{ij}\}_{i=1,\ldots,m;\ j=1,\ldots,l_i}$ be a cylindrical decomposition for $P_1, \ldots, P_s$. Then each D_q is a constructible subset of $\mathrm{Spec}_r(k[[\mathbf{x}']])$ and therefore has a finite number of connected components, say $D_{q1}, \ldots, D_{qm_q}$. Moreover, for each i, $(j^*)^{-1}(D_{ir})$ is a union $\Gamma_1 \cup \cdots \cup \Gamma_u$, where each Γ_d is either a graph of some ξ_{ij} or a slice between two consecutive ξ_{ij} and $\xi_{i(j+1)}$; these sets Γ_d are connected, and obviously there are only finitely many of them.

Now C is also a union of certain Γ_i's and therefore it has only a finite number of connected components, bounded by the total number of Γ's. We claim now that for each $\epsilon > 0$, $C \cap W_\epsilon$ is also the union of the same Γ_i's, where the possible slices of the form $(-\infty, \xi_{q1})$ and $(\xi_{ql_q}, +\infty)$, have been replaced by $(-\epsilon, \xi_{q1})$ and $(\xi_{ql_q}, +\epsilon)$ respectively. Indeed, all we have to show is that the graph Γ_{ij} of each ξ_{ij} is contained in W_ϵ. But let $\beta' \in \mathrm{Spec}_r(k[[\mathbf{x}']])$, and set $\beta = \xi_{ij}(\beta')$. Now ξ_{ij} is a root of some of our Weierstrass polynomials, say of $P_l = \mathbf{x}_n^r + a_1(\mathbf{x}')\mathbf{x}_n^{r-1} + \cdots + a_r(\mathbf{x}')$, with $a_i(0) = 0$ for all i. Thus we have

$$\mathbf{x}_n(\beta)^r + a_1(\beta')\mathbf{x}_n(\beta)^{r-1} + \cdots + a_r(\beta') = 0,$$

and since the $a_i(\beta')$ are infinitesimal with respect to k, $\mathbf{x}_n(\beta)$ is infinitesimal too. In conclusion, $|\mathbf{x}_n(\beta)| < \epsilon$, or equivalently $\beta \in W_\epsilon$. The proof is thus complete. $\qquad\square$

So let us turn to the proof of the main result of the section:

Theorem 8.3   *Let k be a real closed field and $\mathbf{x} = (\mathbf{x}_1, \ldots, \mathbf{x}_n)$ indeterminates. Then* **CC** *holds for $\mathrm{Spec}_r(k[[\mathbf{x}]])$.*

Proof.   By technical reasons which will be understood after reading the proof, we reduce first the problem to the case when k is uncountable. In fact, if k

is countable, we always have $k \subset R$, where R is an uncountable real closed field. Then by Proposition 1.12, the inclusion $k[[\mathbf{x}]] \subset R[[\mathbf{x}]]$ is a regular homomorphism, and therefore the real going-down (Theorem 7.1) holds for the corresponding map $i^* : \operatorname{Spec}_r(R[[\mathbf{x}]]) \to \operatorname{Spec}_r(k[[\mathbf{x}]])$. Thus, since i^* maps the unique closed point of $\operatorname{Spec}_r(R[[\mathbf{x}]])$ to the unique closed point of $\operatorname{Spec}_r(k[[\mathbf{x}]])$, we get that i^* is surjective. But then, given a constructible subset $C \subset \operatorname{Spec}_r(k[[\mathbf{x}]])$, we have $C = i^*(i^*)^{-1}(C)$, and, therefore, if $(i^*)^{-1}(C)$ has finitely many connected components, so does C. Thus, if **CC** holds for $R[[\mathbf{x}]]$ it holds also for $k[[\mathbf{x}]]$.

After this reduction, we assume from now on that k is uncountable. We work by induction on the number n of variables. If $n = 1$ there is nothing to prove. So let $n > 1$ and assume that $C \subset \operatorname{Spec}_r(k[[\mathbf{x}]])$ is a constructible subset with infinitely many connected components. In particular C is not connected and there exist two disjoint open and closed subsets $C_1^{(1)}, C_2^{(1)}$ of C with $C = C_1^{(1)} \cup C_2^{(1)}$. Thus, by Proposition II.1.12, we have $C_i^{(1)} = C \cap U_i^{(1)}$ for some open constructible subset $U_i^{(1)} \subset \operatorname{Spec}_r(k[[\mathbf{x}]])$. Now, in turn, $C_1^{(1)}, C_2^{(1)}$ cannot be both connected. Suppose that $C_2^{(1)}$ is not. Then as above we have $C_2^{(1)} = C_1^{(2)} \cup C_2^{(2)}$ with $C_1^{(2)} \cap C_2^{(2)} = \emptyset$ and $C_i^{(2)} = C_2 \cap U_i^{(2)}$ for some open constructible subset $U_i^{(2)} \subset \operatorname{Spec}_r(k[[\mathbf{x}]])$. Repeating the process we get, for every integer $p \geq 1$, a family $U_1^{(p)}, \ldots, U_p^{(p)}$ of open constructible subsets of $\operatorname{Spec}_r(k[[\mathbf{x}]])$ such that

$i)$ $C \subset U_1^{(p)} \cup \cdots \cup U_p^{(p)}$

$ii)$ $C \cap U_i^{(p)} \neq \emptyset$ for $i = 1, \ldots, p$

$iii)$ $C \cap U_i^{(p)} \cap U_j^{(p)} = \emptyset$ for $i \neq j$.

Now let $\{f_q\}_{q \in \mathbb{N}}$ be a countable collection of elements of $k[[\mathbf{x}]]$ defining C and all the $U_i^{(p)}$'s. Since k is uncountable (here is where this fact is needed!) there exists a linear change of coordinates

$$\begin{aligned}
\mathbf{x}_n &= \tilde{\mathbf{x}}_n, \\
\mathbf{x}_i &= \tilde{\mathbf{x}}_i + a_i\tilde{\mathbf{x}}_n, \quad 1 \leq i \leq n - 1,
\end{aligned}$$

which makes all the f_q's regular in the variable $\tilde{\mathbf{x}}_n$. Indeed, it is enough to take a point $a = (a_1, \ldots, a_{n-1}, 1) \in k^n$ with $f_q^0(a) \neq 0$ for all the initial forms f_q^0 of the series f_q ([Rz12 I.3.1]), which surely exists by the cardinality assumption.

We may therefore assume that all the f_q's are regular in the variable $\mathbf{x}_n$, and by the Weierstrass preparation theorem we may even assume that for all q, $f_q \in k[[\mathbf{x}']][\mathbf{x}_n]$ is a Weierstrass polynomial. Let $j^* : \operatorname{Spec}_r(k[[\mathbf{x}]]) \to \operatorname{Spec}_r(k[[\mathbf{x}']][\mathbf{x}_n])$ be the mapping corresponding to the inclusion $j : k[[\mathbf{x}']][\mathbf{x}_n] \to k[[\mathbf{x}]]$. We denote by $V_i^{(p)}$ (resp. C') the constructible subset of $\operatorname{Spec}_r(k[[\mathbf{x}']][\mathbf{x}_n])$ defined by the same sign conditions on the f_q's that define $U_i^{(p)}$ (resp. C). Then $V_i^{(p)}$ is open for all p, i, and we have $C = (j^*)^{-1}(C')$, and $U_i^{(p)} = (j^*)^{-1}(V_i^{(p)})$.

Let α be the image by j^* of the closed point of $\operatorname{Spec}_r(k[[\mathbf{x}]])$. By Lemma 7.4 $a)$, we have $j^*(\operatorname{Spec}_r(k[[\mathbf{x}]]) = U_\alpha$. Then, for all p, we get

$i')$ $C' \cap U_\alpha \subset V_1^{(p)} \cup \cdots \cup V_p^{(p)}$,

$ii')$ $C' \cap U_\alpha \cap V_i^{(p)} \neq \emptyset$ for $i = 1, \ldots, p$,

$iii')$ $C' \cap U_\alpha \cap U_i^{(p)} \cap U_j^{(p)} = \emptyset$ for $i \neq j$,

and these conditions assert that $C' \cap U_\alpha$ has infinitely many connected components. Therefore, by Proposition II.7.13, for every N there is a neighbourhood W_N of α such that $cc_\alpha(C' \cap W_N) > N$. Since by induction hypothesis, **CC** holds in $k[[\mathbf{x}']]$, Lemma 8.2 can be applied getting the desired contradiction. $\square$

It follows immediately from the preceding theorem that **CC** holds for real spectra of formal algebras over real closed residue fields. In addition, it is clear that the arguments above can be repeated for algebraic and for convergent power series. In fact, we have the following more general result:

Theorem 8.4   *Let A be an excellent ring such that $X = \mathrm{Spec}_r(A)$ has only finitely many closed points. Then **CC** holds for X.*

Proof.   Let $\alpha_1, \ldots, \alpha_n$ be the closed points of X. Then,

$$X = U_{\alpha_1} \cup \cdots \cup U_{\alpha_n}.$$

Now, for each i, let A_{α_i} be the real strict localization of A at α_i, and let B_i be its completion. Since A is excellent, these rings are excellent too (Theorem 2.5). Also, by Cohen's structure theorem, we have $B_i = \kappa(\alpha_i)[[\mathbf{x}]]/I_i$ where $\mathbf{x} = (\mathbf{x}_1, \ldots, \mathbf{x}_n)$ are indeterminates and I_i a certain ideal. Thus, $\mathrm{Spec}_r(B_i)$ can be seen as a constructible subset of $\mathrm{Spec}_r(\kappa(\alpha_i)[[\mathbf{x}]])$ and, by Theorem 7.3, **CC** holds for $\mathrm{Spec}_r(B_i)$. Let us denote by $\psi_i : A \to A_{\alpha_i}$ and $\varphi_i : A_{\alpha_i} \to B_i$ the canonical inclusions. Then we have the corresponding mappings between real spectra

$$\mathrm{Spec}_r(B_i) \xrightarrow{\varphi_i^*} \mathrm{Spec}_r(A_{\alpha_i}) \xrightarrow{\psi_i^*} U_{\alpha_i} \subset \mathrm{Spec}_r(A),$$

where φ_i^* is onto by Corollary 3.3 and ψ_i^* is onto by Theorem II.7.11. Consider any constructible set $C \subset \mathrm{Spec}_r(A)$. Then the constructible set $(\psi_i^* \circ \varphi_i^*)^{-1}(C)$ has a finite number of connected components, say $T_{i1}, \ldots, T_{is_i}$. Thus, since $\psi_i^* \circ \varphi_i^*$ is surjective, we have

$$C \cap U_{\alpha_i} = (\psi_i^* \circ \varphi_i^*)(T_{i1}) \cup \cdots \cup (\psi_i^* \circ \varphi_i^*)(T_{is_i})$$

and these sets are connected. It follows that $C \cap U_{\alpha_i}$ has at most s_i connected components, and therefore

$$C = (C \cap U_{\alpha_1}) \cup \cdots \cup (C \cap U_{\alpha_n})$$

has also a finite number of connected components. This shows that **CC** holds for X. $\square$

Since in the real spectrum of a henselian ring all prime cones specialize to orderings of the residue field, we can generalize our first remark of the section to:

Corollary 8.5   *Let A be en excellent henselian local ring with residue field k. Then, **CC** holds for $\mathrm{Spec}_r(A)$ if and only if $\mathrm{Spec}_r(k)$ is finite.*

We finish with a more detailed study of the behaviour of connected components under completion, which will play an essential role in the geometric applications of Chapter VIII.

Proposition 8.6   *Let A be a ring and let $\alpha \in \mathrm{Spec}_r(A)$ be such that the localization $A_{\mathrm{supp}(\alpha)}$ is excellent. Let $\pi : \mathrm{Spec}_r(\widehat{A}_\alpha) \to \mathrm{Spec}_r(A)$ be the canonical mapping corresponding to the inclusion of A into the completion of the real strict localization A_α. Then, for every constructible set $C \subset \mathrm{Spec}_r(A)$, π induces a one to one correspondence between the connected components of $\pi^{-1}(C)$ and the connected components of $C \cap U_\alpha$. In particular $cc_\alpha(C) \leq cc\left(\pi^{-1}(C)\right)$.*

Proof.   By Theorem II.7.11 and Proposition II.7.13, we may assume that $A = A_\alpha$, and then A is excellent. In particular, α is the only closed point of $\mathrm{Spec}_r(A)$ and $U_\alpha = \mathrm{Spec}_r(A)$. Moreover, by Theorem 8.7, the set C has finitely many connected components, say $C_1, \ldots, C_r$, and these are constructible. Thus

$$\pi^{-1}(C) = \pi^{-1}(C_1) \cup \cdots \cup \pi^{-1}(C_r)$$

is a partition of $\pi^{-1}(C)$ into open and closed subsets and we only have to check that they are connected. In other words, we may assume that C is connected and we have to show that then $\pi^{-1}(C)$ is connected too. Suppose the contrary. Then, there are open constructible sets

i) $\pi^{-1}(C) \subset W_1 \cup W_2,$

ii) $\pi^{-1}(C) \cap W_1 \cap W_2 = \emptyset,$ and

iii) $\pi^{-1}(C) \cap W_k \neq \emptyset,$ $k = 1, 2.$

We shall rewrite these conditions using equations. First we set

$$C = \bigcup_{i=1}^{p} D_i, \qquad \text{with} \quad D_i = \{f_{i1} > 0, \ldots, f_{is} > 0, g_i = 0\},$$

where $f_{ij}, g_i \in A$. Also, by the finiteness theorem we have

$$W_k = \bigcup_{l=1}^{q} W_{kl}, \qquad \text{with} \quad W_{kl} = \{\widehat{h}_{kl1} > 0, \ldots, \widehat{h}_{kls} > 0\},$$

where $\widehat{h}_{klj} \in \widehat{A}$.

Now, condition *i)* above is equivalent to the fact that for any index i and any family of indices $t(k, l)$, we have, in $\mathrm{Spec}_r(\widehat{A})$

$$\{f_{i1} > 0, \ldots, f_{is} > 0, g_i = 0\} \cap \left(\bigcap_{k,l} \{\widehat{h}_{klt(k,l)} \leq 0\} \right) = \emptyset.$$

By the Positivstellensatz (Theorem II.1.14), this equality is, in turn, equivalent to the existence of an equation of the type

$i')$ $\quad \prod_j f_{ij}^{\nu_j} + \sum_{\lambda,\mu} \widehat{a}_{\lambda\mu}^2 (-g_i^2)^{\lambda_0} \prod_j f_{ij}^{\lambda_j} \prod_{k,l} (-\widehat{h}_{klt(k,l)})^{\lambda_{kl}} = 0$

for some $\widehat{a}_{\lambda\mu} \in \widehat{A}$.

In a similar way, condition $ii)$ is equivalent to the fact that for any indices i, m, n we have, in $\mathrm{Spec}_r(\widehat{A})$,

$$\{f_{i1} > 0, \ldots, f_{is} > 0, g_i = 0\} \cap$$
$$\{\widehat{h}_{1m1} > 0, \ldots, \widehat{h}_{1ms} > 0, \widehat{h}_{2n1} > 0, \ldots, \widehat{h}_{2ns} > 0\} = \emptyset,$$

which, by the Positivstellensatz again, is equivalent to

$ii')$ $\quad \prod_{j,u,v} f_{ij}^{\theta_j} \widehat{h}_{1mu}^{\eta_{1u}} \widehat{h}_{2nv}^{\eta_{2v}} + \sum_{\lambda,\mu} \widehat{b}_{\lambda\mu}^2 (-g_i^2)^{\lambda_0} \prod_{j,u,v} f_{i,j}^{\lambda_j} \widehat{h}_{1mu}^{\lambda_{1p}} \widehat{h}_{2nv}^{\lambda_{2v}} = 0$

for some $\widehat{b}_{\lambda\mu} \in \widehat{A}$.

Finally, for $k = 1, 2$, condition $iii)$ implies that for some i_k, l_k, $(k = 1, 2)$, we have, in $\mathrm{Spec}_r(\widehat{A})$,

$$\{f_{i_k1} > 0, \ldots, f_{i_ks} > 0, g_{i_k} = 0, \widehat{h}_{kl_k1} > 0, \ldots, \widehat{h}_{kl_ks} > 0\} \neq \emptyset.$$

Thus, by Proposition 5.1, there is a prime cone $\beta_k \in \mathrm{Spec}_r(\widehat{A})$ such that $\dim(\beta_k) = 1$, and

$iii')$ $\quad f_{i_kj}(\beta_k) > 0, \; g_{i_k}(\beta_k) = 0, \; \widehat{h}_{kl_kp}(\beta_k) > 0$

for all j, p.

In conclusion, conditions $i)$ and $ii)$ are rephrased as a system of polynomial equations with coefficients in A which has a solution $\widehat{a}_{\lambda\mu}, \widehat{b}_{\lambda\mu}, \widehat{h}_{klu}$ in $\widehat{A}$. Then by Rotthaus's Theorem 2.6, there is a solution $a_{\lambda\mu}, b_{\lambda\mu}, h_{klu}$ in A, arbitrarily close to the former in the adic topology. In particular, since β_k is defined by a local homomorphism $\widehat{A} \to \kappa(\alpha)[[t]]$, we may assume from $iii')$ that $h_{kl_ku}(\beta_k) > 0$.

Then, going backwards in the Positivstellensatz, that is, from the existence of a solution for $i')$ and $ii')$ to the set theoretical statements, we get that the open constructible subsets of $\mathrm{Spec}_r(A)$ given by

$$U_k = \bigcup_{l=1}^q U_{kl}, \quad \text{with} \quad U_{kl} = \{\widehat{h}_{kl1} > 0, \ldots, \widehat{h}_{kls} > 0\},$$

verify

$i'')$ $C \subset U_1 \cup U_2$,
$ii'')$ $C \cap U_1 \cap U_2 = \emptyset$, and
$iii'')$ $C \cap U_k \neq \emptyset$, $k = 1, 2$,

which means that C is not connected, contradiction. $\qquad\square$

Notes

The notion of excellent ring was introduced by Grothendieck in his famous [EGA], condensing the fundamental work of Nagata (see [Ng]), and here we follow the more basic reference [Mt]. Our proof of Theorem 2.5 is new as far as we know, although the theorem itself goes back to Seydi ([Sy]). Rotthaus's important theorem on the approximation problem for excellent henselian rings containing $\mathbb{Q}$ appeared in [Rt], solving a conjecture raised by M. Artin in [Ar2]. M. Artin himself proved the approximation property for henselizations of finite algebras and for analytic algebras [Ar1,2]. The final culmination of this approximation theory was achieved by Popescu in [Po1-3], and then clarified by Spivakovsky in [Sk]. We refer the reader to [Te] for a complete survey of the theory and its applications. Among these, there are several concerning complexity and other problems in real geometry ([Co-Rz-Sh], [Qz]).

Extension of orderings under completion was shown in [Rz3], as the essential step towards the solution of Hilbert's 17th problem for compact global analytic sets; the same idea and motivation can be found in [Jw2]. The proof given here is simpler since it uses approximation, not fully available at that time. Curve selection lemmas are classical in algebraic and analytic geometry and appear in a variety of different forms. The formulation of Theorem 4.2 is a quite technical improvement of [Rz8] and implies (in the real case) most algebraic and geometric versions which were scattered in the literature (see [Ls], [Me], [Rn], [Ro]). Proposition 5.1 on real dimension is taken from [Rz6,8], and generalizes the previous results for algebras finitely generated over fields. The existence of real valuations (Proposition 5.2) has been the matter of many papers, among which we quote [Lg1] as a classic and the more recent [Rb], [An] and [An-Rz3]. Constructibility of closures was proved first for power series rings in [Al-An] and then for excellent rings in [An-Br-Rz]; the proof given in this book is slightly shorter since it uses Proposition 5.2. Example 6.2 is due to Gamboa ([Gm]). The real going-down theorem was proved in [Rz9], and it has been a many-sided tool since ([Sch3 §3]). In fact, the result itself was the answer to a question of Alonso-Andradas who needed it to prove Proposition 8.3 on constructibilty of connected components for formal power series rings ([Al-An]). Apart from this use of the real going-down, their proof is very different from ours, and Theorem 8.4 and Corollary 8.5 are new. Finally, Proposition 8.6 is taken from [Rz7]; for a far reaching generalization see [Sch3].

Chapter VIII. Real Analytic Geometry

Summary. In this chapter we apply all the previous results to the study of semianalytic sets in real analytic manifolds. In Section 1 we settle the terminology concerning *global* analytic functions and sets. Sections 2 and 3 are devoted to the local theory, that is, to *germs at points*. We review there several classical results in the framework of real spaces, with some technical suplements that will be needed later. In Section 4 we obtain the algebraic properties of the various rings of *global* analytic functions that will be used in the sequel. Sections 5 to 7 are devoted to the Artin-Lang property, the complexity and the constructibility of topological operations. This is the concrete reward for all preceding abstract work. In Section 8 we put it all together for the nicest case, that of *germs at compact sets*.

1. Semianalytic Sets

Let Ω be a real analytic manifold, which we always suppose paracompact and Hausdorff. Let $\mathcal{O}(\Omega)$ be the ring of global analytic functions on Ω.

Definition 1.1

a) *A* global analytic set *is a subset $X \subset \Omega$ of the form*

$$X = \{x \in \Omega \mid f_1(x) = \cdots = f_r(x) = 0\}$$

for some analytic functions $f_1, \ldots, f_r \in \mathcal{O}(\Omega)$.

b) *A* global analytic function on a closed set $X \subset \Omega$ *is a function $f : X \to \mathbb{R}$ which is the restriction of a global analytic function on Ω.*

These definitions differ from the classical ones, which had local nature. In fact, there are sets (even compact sets) $X \subset \mathbb{R}^n$ that, near every point, are the zero set of an analytic function (depending on the point), but that are not global analytic sets.

(1.2) Real Space Associated to a Global Analytic Set. Let $X \subset \Omega$ be a global analytic set, and let $\mathcal{O}(X)$ denote the ring of global analytic functions on X. For $f \in \mathcal{O}(X)$ define

$$\mathrm{sign}[f] : X \to \mathbb{F}_3 \,;\; x \mapsto \mathrm{sign}[f(x)] = \begin{cases} +1 & \text{if } f(x) > 0 \\ 0 & \text{if } f(x) = 0 \\ -1 & \text{if } f(x) < 0 \end{cases}$$

and let

$$G_X = \{\mathrm{sign}[f] \mid f \in \mathcal{O}(X)\}.$$

Then (X, G_X) is a real space (I.3.2), and therefore we have basic, principal, strictly open and closed sets, constructible sets, as well as the corresponding Zariski notions, and also the invariants $s, \bar{s}, t, \bar{t}$.

More explicitely, a subset $Z \subset X$ is called *global basic open* (resp. *closed*) *semianalytic* if there exist $f_1, \ldots, f_s \in \mathcal{O}(X)$ such that

$$Z = \{x \in X \mid f_1(x) > 0, \ldots, f_s(x) > 0\}$$

$$(\text{resp. } Z = \{x \in X \mid f_1(x) \geq 0, \ldots, f_s(x) \geq 0\} \,).$$

The unions of these are the *global strictly open* (resp. *closed*) *semianalytic sets*. A subset $Z \subset X$ is called *global semianalytic* if it is constructible in (X, G_X), that is, if there exist analytic functions $f_{ij}, g_i \in \mathcal{O}(X)$ such that

$$Z = \bigcup_{i=1}^{r} \{x \in X \mid f_{i1}(x) > 0, \ldots, f_{is_i}(x) > 0, g_i(x) = 0\}.$$

The *(analytic) dimension* $\dim(Z)$ of a global semianalytic set $Z \subset X$ is the maximum d such that Z contains an analytic manifold of dimension d.

A subset $Y \subset X$ is called *global analytic* if it is Zariski closed in (X, G_X), that is, if there exist analytic functions $f_1, \ldots, f_r \in \mathcal{O}(X)$ such that

$$Y = \{x \in X \mid f_1(x) = \cdots = f_r(x) = 0\}.$$

Of course, for the real space (Ω, G_Ω) this coincides with Definition 1.1 *a)*.

We shall denote by $\mathcal{Z}(I)$ the set of all common zeroes of the functions in $I \subset \mathcal{O}(X)$. By a theorem due to Whitney-Bruhat, $\mathcal{Z}(I)$ is a global analytic set ([Wh-Bh nr.8]). We shall denote by $\mathcal{J}(Z)$ the ideal of all functions vanishing on $Z \subset X$; clearly, this is a real ideal.

(1.3) Boundary Bounded Sets. Let $X \subset \Omega$ be a global analytic set and let $\mathcal{C}(X)$ the boolean algebra of its global semianalytic sets. A subset $Z \subset X$ is called *bounded* if its closure is compact. Of course, this means that Z is bounded after any embedding $\Omega \hookrightarrow \mathbb{R}^n$. A set $Z \subset X$ is called *boundary bounded* (*in X*) if its topological boundary (in X)

$$\mathrm{Bd}(Z) = \mathrm{Adh}(Z) \cap \mathrm{Adh}(X \setminus Z) \subset X$$

is bounded. One easily checks that finite unions and finite intersections, complements, closures and interiors of boundary bounded sets are again boundary bounded. In particular, the collection $\mathcal{C}_b(X)$ of all boundary bounded global semianalytic sets of X is a subalgebra of $\mathcal{C}(X)$. In turn, the algebra $\mathcal{C}_b(X)$ is essentially built up from bounded global semianalytic sets. We explain this in the case that Ω is a closed analytic submanifold of some $\mathbb{R}^n$, and $X \subset \Omega$ a global analytic set of $\mathbb{R}^n$ (which is in fact the general case, as we shall see in 4.2). Under these hypotheses, any boundary bounded set $Z \subset X$ can be written as $Z' \cup (B \cap Z)$, where B is an open ball in $\mathbb{R}^n$ that contains the boundary of Z in X and Z' is a union of connected components of $X \setminus B$. This kind of decomposition will be useful later on.

The algebra $\mathcal{C}_b(X)$ does not come from any real space. However, by fully applying the abstract theory developed in the preceding chapters, we shall be able to study complexity and constructibility of topological operations in $\mathcal{C}_b(X)$.

(1.4) Global Analytic Sets and Real Spectra. There is a more sophisticated method to associate a real space to a global analytic set $X \subset \Omega$, which will be the suitable by-pass to elude the difficulties of a direct study of our initial real space (X, G_X). Namely, we consider the real spectrum $\widetilde{X} = \mathrm{Spec}_r(\mathcal{O}(X))$ of the ring $\mathcal{O}(X)$, with its canonical structure of real space (II.1.5). Then, Chapters V to VII work in some way or another for $\widetilde{X}$, and then for X. For the latter transcription, we embed X into $\widetilde{X}$ as follows. Every point $x \in X$ defines a homomorphism $\mathcal{O}(X) \to \mathbb{R}$; $f \mapsto f(x)$, whose kernel is the maximal ideal $\mathfrak{m}_x$ of all functions vanishing at x. Since the residue field of $\mathfrak{m}_x$ is $\mathbb{R}$, there is a unique prime cone α_x with support $\mathfrak{m}_x$ and the map $x \mapsto \alpha_x$ is an embedding from X into $\widetilde{X}$. We can thus compare both real spaces, and the crucial matter will be to what extent X is an Artin-Lang subset of $\widetilde{X}$ (V.5).

2. Semianalytic Set Germs

(2.1) Set Germs. Let Ω be a topological space and fix a point $x \in \Omega$. Two sets $Z_1, Z_2 \subset \Omega$ have the same *germ at x* if $Z_1 \cap U = Z_2 \cap U$ for some neighbourhood U of x. This is an equivalence relation and the corresponding classes are called *set germs at x*. The germ at x of a set $Z \subset \Omega$ is denoted by Z_x. It is clear that inclusions, finite unions, finite intersections and complements of set germs are well defined via representatives. Also via representatives, we define when a set germ Z_x is open (resp. closed) in another Z'_x, and the interior (resp. closure) of Z_x in Z'_x is a well defined set germ. These topological operations behave for germs as they do for sets. On the contrary, connectedness is a quite delicate notion for germs, as we shall see in Section 3.

We are interested in germs in analytic manifolds, and to discuss them we can always reduce to the case that Ω is the affine space $\mathbb{R}^n$ and x is the origin

$O \in \mathbb{R}^n$, which we do henceforth. This reduction can be done locally by means of systems of coordinates, or globally by means of a closed embedding. The first approach is much more elementary, but the second will also be needed, as we shall see in detail later on.

(2.2) Analytic Function Germs. Two analytic functions f and g defined on two neighbourhoods of $O \in \mathbb{R}^n$ have the same *germ at* O if they coincide on a third smaller neigbourhood. The corresponding equivalence class is denoted by f_o or merely f. As usual, we define via representatives all operations (even derivatives) with germs of analytic functions, to get a ring, denoted by $\mathcal{O}_n$. This ring is isomorphic, via Taylor expansions, to the ring of convergent power series $\mathbb{R}\{x_1, \ldots, x_n\}$. Hence, it is a henselian local regular ring of dimension n and its residue field is $\mathbb{R}$; in fact, it is an excellent ring (Examples VII.2.3 *c*)).

(2.3) Analytic Set Germs. Given $f_1, \ldots, f_r \in \mathcal{O}_n$, the set germ $\{f_1 = \cdots = f_r = 0\}$ is well defined, and it is called the *germ of zeroes of* $f_1, \ldots, f_r$. More generally, the *germ of zeroes of* $I \subset \mathcal{O}_n$ is the set germ $\mathcal{Z}(I) = \bigcap_{f \in I}\{f = 0\}$. This is consistent, since, $\mathcal{O}_n$ being noetherian, $\mathcal{Z}(I) = \{f_1 = \cdots = f_r = 0\}$ for any chosen generators $f_1, \ldots, f_r$ of the ideal $I\mathcal{O}_n$. Germs of zeroes are called *analytic set germs*.

Let Z_o be a set germ. A function germ $f \in \mathcal{O}_n$ is > 0 (resp. $\geq 0, = 0$) on Z_o if $Z_o \subset \{f > 0\}$ (resp. $\{f \geq 0\}, \{f = 0\}$). The *ideal* of Z_o, denoted by $\mathcal{J}(Z_o)$ is the ideal of all $f \in \mathcal{O}_n$ which are $= 0$ on Z_o; this is a real ideal. We shall denote by $\mathcal{O}(Z_o)$ the reduced ring $\mathcal{O}_n/\mathcal{J}(Z_o)$.

An analytic set germ X_o is called *irreducible* if it is not the union of two strictly smaller analytic set germs. This holds if and only if the ideal $\mathcal{J}(X_o)$ is prime. Every analytic set germ X_o has a unique (up to permutations) irredundant decomposition into irreducible ones, $X_o = X_o^{(1)} \cup \cdots \cup X_o^{(r)}$. These $X_o^{(i)}$'s are the *irreducible components* of X_o; they are the germs of zeroes of the associated prime ideals of $\mathcal{J}(X_o) \subset \mathcal{O}_n$.

(2.4) Semianalytic Set Germs. Let X_o be an analytic set germ and consider the reduced ring $\mathcal{O}(X_o) = \mathcal{O}_n/\mathcal{J}(X_o)$; this ring is the ring of *analytic function germs of* X_o. Clearly, $\mathcal{Z}$ and $\mathcal{J}$ induce two correspondences between ideals of $\mathcal{O}(X_o)$ and analytic set germs contained in X_o. Via representatives, we can define set germs $\{f > 0\}$, $\{f \geq 0\}$, $\{f = 0\}$, and boolean combinations of these. Then, a *semianalytic set germ* of X_o is a set germ of the form

$$Z_o = \bigcup_{i=1}^{r} \{f_{i1} > 0, \ldots, f_{is_i} > 0, g_i = 0\} \subset X_o,$$

where $f_{ij}, g_i \in \mathcal{O}(X_o)$. We also have *basic, principal, strictly open–* and *basic, principal, strictly closed* semianalytic set germs. We as well have the corresponding Zariski notions; for instance, the Zariski closure of a set germ Z_o is the analytic set germ $\mathrm{Adh}_Z(Z_o) = \mathcal{Z}\mathcal{J}(Z_o)$.

Let now $\mathcal{C}(X_o)$ be the collection of all semianalytic set germs of X_o. This collection $\mathcal{C}(X_o)$ is a boolean algebra, and we are interested in complexity and constructibility matters as described in Section I.3 for real spaces. First we prove:

Proposition 2.5 *In the above situation, the Stone space of the boolean algebra $\mathcal{C}(X_o)$ is the real spectrum $\widetilde{X_o}$ of the ring $\mathcal{O}(X_o)$. In fact, the correspondence*

$$Z_o = \bigcup_{i=1}^{r} \{f_{i1} > 0, \ldots, f_{is_i} > 0, g_i = 0\} \longmapsto$$

$$\widetilde{Z_o} = \bigcup_{i=1}^{r} \{\alpha \in \widetilde{X_o} \mid f_{i1}(\alpha) > 0, \ldots, f_{is_i}(\alpha) > 0, g_i(\alpha) = 0\}$$

is an isomorphism from the boolean algebra of semianalytic set germs of X_o onto that of constructible sets of $\widetilde{X_o}$.

Proof.   We have to see that this is a well defined bijection, which reduces to the fact that $Z_o = \{f_1 > 0, \ldots, f_s > 0, g = 0\} = \emptyset$ if and only if $\widetilde{Z_o} = \emptyset$. Now, if $\widetilde{Z_o} = \emptyset$, the Positivstellensatz (II.1.12) gives an equation

$$(f_1 \cdots f_r)^{2k} + \sum_{\nu} a_\nu f_1^{\nu_1} \cdots f_r^{\nu_r} = hg,$$

where the a_ν's are sums of squares. This equation shows that Z_o is empty. Conversely, let $\widetilde{Z_o} \neq \emptyset$. If Z_o contains the origin, the conclusion is trivial. Otherwise, by the analytic curve selection lemma (Remark VII.4.3 b)), there is an analytic curve $(-\varepsilon, \varepsilon) \to \mathbb{R}^n$; $t \mapsto x(t)$, and a representative Z of Z_o such that $x(0) = O$ and $x(t) \in Z$ for $t > 0$. Hence, O is adherent to Z and $Z_o \neq \emptyset$. $\qquad\square$

Corollary 2.6   (Classical curve selection lemma) *Any non-empty semianalytic set germ contains a half-branch of an analytic curve.*

Proof.   Assume $Z_o \neq \emptyset$. Then, by the proposition, we get $\widetilde{Z_o} \neq \emptyset$ and we obtain the half-branch from the abstract version of Remark VII.4.3 b). $\qquad\square$

Remarks 2.7 $a)$ The above corollary shows that the set Y consisting of the origin and the half-branches of analytic curves, determines completely the semianalytic set germs. Notice that Y is precisely the set of constructible points of $\widetilde{X_o} = \mathrm{Spec}_r(\mathcal{O}(X_o))$, which by Theorem VII.4.2 is an Artin-Lang set. Thus, if we see a semianalytic set germ Z_o as the set of points of Y which it contains, then $(X_o, \mathcal{O}(X_o))$ is a real space and $\mathcal{C}(X_o)$ is the boolean algebra of constructible subsets of Y. In turn, since Y is an Artin-Lang set, $\mathcal{C}(X_o)$ is isomorphic to the boolean algebra of constructible subsets of $\widetilde{X_o}$, recovering the proposition above. This way of looking at semianalytic set germs as sets of half-branches of analytic curves is sometimes useful.

b) Since $\widetilde{X_o}$ is a real space, we can consider the properties **FT**, **AC** and **CC**, as well as the invariants $s, \bar{s}, t, \bar{t}$ (I.3.5, I.3.8). Consequently, all of this applies to X_o.

Once we have identified the Stone space of $\mathcal{C}(X_o)$, we obtain:

Proposition 2.8   *Let X_o be an analytic set germ.*

a) (Positivstellensatz) *A semianalytic set germ of X_o*

$$\{f_1 = 0, \ldots, f_r = 0, g_1 \neq 0, \ldots, g_s \neq 0, h_1 \geq 0, \ldots, h_t \geq 0\}$$

is empty if and only if there exists an equation of the form

$$g_1^{2n_1} \cdots g_s^{2n_r} + \sum_{m_1 + \cdots + m_t \leq d} a_m h_1^{m_1} \cdots h_t^{m_t} = b_1 f_1 + \cdots + b_r f_r,$$

for suitable $n_i, m_j, d \in \mathbb{N}$, $b_i \in \mathcal{O}(X_o)$ and $a_m \in \sum \mathcal{O}(X_o)^2$.
b) (Real Nullstellensatz) *Let I be an ideal of $\mathcal{O}(X_o)$. Then $\mathcal{J}\mathcal{Z}(I) = \sqrt[r]{I}$.*

Proof.   *a)* is the translation of Theorem II.1.14 and *b)* that of Theorem II.2.8.
$\square$

Now we turn to positive semidefinite analytic function germs.

Proposition 2.9   *Let X_o be an analytic set germ. Let $f \in \mathcal{O}(X_o)$ be ≥ 0 on X_o. Then:*

a) (Hilbert's 17th Problem) *There are $h, g_1, \ldots, g_r \in \mathcal{O}(X_o)$, $h \neq 0$, such that*
$$h^2 f = g_1^2 + \cdots + g_r^2.$$
In fact, we can take $h = f^{2n} + h_1^2 + \cdots + h_s^2$ for some $h_1, \ldots, h_s \in \mathcal{O}(X_o)$, and the germ of zeroes of the denominator h is contained in that of f.
b) (Sums of Even Powers of Meromorphic Function Germs) *Let $m \geq 0$. If for every analytic curve germ $\gamma : (-\varepsilon, \varepsilon) \to X_o$ the order of the power series $f \circ \gamma$ is divisible by $2m$, then there are $h, g_1, \ldots, g_r \in \mathcal{O}(X_o)$, $h \neq 0$, such that*
$$h^{2m} f = g_1^{2m} + \cdots + g_r^{2m}.$$

Proof.   *a)* comes from Corollary II.1.15. Let us prove *b)*. Let $X_o^{(1)}, \ldots, X_o^{(s)}$ be the irreducible components of X_o. The total ring of fractions of the ring $\mathcal{O}(X_o)$ is the direct sum of the fields of fractions of the rings $\mathcal{O}(X_o^{(i)})$, and consequently, it is enough to prove the result for each $X_o^{(i)}$ separatedly. In other words, we can assume that X_o is irreducible. Then $\mathcal{O}(X_o)$ is an integral domain, whose field of fractions will be denoted by K. We must show that if f is not a sum of $2m$-th powers in K, then there is an analytic curve germ $\gamma : (-\varepsilon, \varepsilon) \to X_o$ such that the order of the power series $f \circ \gamma$ is not divisible by $2m$. But by *a)*, f is a sum of squares in K, hence, by a theorem due to Becker ([Bel 1.9, p.146]), there is a real valuation v of K, such that the value $v(f)$ is not divisible by $2m$. Our job now is to build up γ out of v.

First of all, we pick an ordering β of K compatible with v, and consider the convex hull V_β of $\mathbb{Q}$ in K with respect to β. Since $V_\beta \subset V$, the value group of V is a homomorphic image of that of V_β, and it follows that $2m$ cannot divide $v_\beta(f)$. Consequently, we can suppose that $V = V_\beta$, and that the residue field of V is $\mathbb{R}$. Now, since $\mathcal{O}(X_o)$ is henselian, β is a generization of the closed point of $\widetilde{X_o}$ (Proposition II.2.4). Thus, V contains $\mathcal{O}(X_o)$, and the maximal ideal $\mathfrak{n}$ of V lies over the maximal ideal $\mathfrak{m}$ of the origin.

Next, a standard application of Hironaka's resolution of singularities to the ring $A = \mathcal{O}(X_o)$ and the ideal fA provides us with a local regular ring $B \subset K$, such that:

i) B is a localization of a finitely generated A-algebra,

ii) The maximal ideal $\mathfrak{n}$ of V lies over the maximal ideal of B, and

iii) There are a regular system of parameters $z_1, \ldots, z_e$ and a unit u of B, such that $f = u z_1^{p_1} \cdots z_e^{p_e}$ for suitable integers $p_i \geq 1$.

(See [Hk1] and also [Hk2 pp.5.8-5.9]; in fact, we only need *local uniformization of valuations.*)

From *iii)*, we get $v(f) = v(u) + p_1 v(z_1) + \cdots + p_e v(z_e)$. Since u is a unit of B, it is a unit of V by *ii)*, and consequently $v(u) = 0$. Thus, as $2m$ does not to divide $v(f)$, there is some p_i, say $i = 1$, such that $2m$ does not divide p_i. On the other hand, by *ii)*, the residue field of B is in between those of A and V, which are both $\mathbb{R}$. We deduce that the residue field of B is also $\mathbb{R}$, and that the completion $\widehat{B}$ is canonically isomorphic to $\mathbb{R}[[\mathbf{x}_1, \ldots, \mathbf{x}_e]]$, via $z_k \mapsto \mathbf{x}_k$. Then, we consider the local homomorphism $A \to B \to \widehat{B} \to \mathbb{R}[[\mathbf{t}]]$ induced by $\mathbf{x}_1 \mapsto \mathbf{t}$ and $\mathbf{x}_i \mapsto \mathbf{t}^{2m}$ for $i > 1$. We have

$$\varphi(f) = a t^p + \cdots, \quad a \neq 0, \quad p = p_1 + 2m \sum_{i=2}^{e} p_i$$

and $2m$ does not divide the order $p = p_1 + 2m \sum_{i=2}^{e} p_i$. If this homomorphism defined an analytic curve germ we would have finished. Thus we still need to approximate φ by another homomorphism $\psi : A \to \mathbb{R}\{\mathbf{t}\}$, so that $\psi(f)$ has the same order as $\varphi(f)$. But this is always possible by M. Artin's approximation theorem ([Tg III.5.1, p.64]). $\qquad\square$

Remark 2.10   In the latter proof, the homomorphism φ could be defined via $\mathbf{x}_i \mapsto a_i \mathbf{t}^{2m}$ for $i > 1$. Then, a suitable choice of the coefficients $a_i \neq 0$ would guarantee $\varphi(h) \neq 0$ for any germ $h \in \mathcal{O}(X_o)$ given in advance. In this form, the criterion gives rise to a characterization.

Next, we obtain complexity bounds for semianalytic germs. To that end, we recall some facts concerning dimension.

(2.11) Dimension. The *(analytic) dimension* of a set germ Z_o of $\mathbb{R}^n$, denoted $\dim(Z_o)$, is the Krull dimension of the ring $\mathcal{O}(Z_o)\mathcal{O}_n/\mathcal{J}(Z_o)$; hence, $\dim(Z_o) = n - \operatorname{ht}(\mathcal{J}(Z_o))$. It is also clear that

$$\dim(Z_o \cup Z_o') = \max\{\dim(Z_o), \dim(Z_o')\}.$$

This algebraic definition of dimension has a natural geometric meaning when Z_o is semianalytic. Namely, $\dim(Z_o)$ is in that case *the biggest d such that any representative of Z_o contains an analytic manifold of dimension d.*

To give a quick proof of this, we can suppose that $Z_o = \{f_1 > 0, \ldots, f_s > 0, g = 0\}$ and that $\mathcal{J}(Z_o)$ is a prime ideal $\mathfrak{p}$. We have $\operatorname{ht}(\mathfrak{p}) = n - d$, where $d = \dim(Z_o)$. Then, after a linear change of coordinates, there are $h_1, \ldots, h_{n-d} \in \mathfrak{p}$ such that $D = \frac{D(h_1, \ldots, h_{n-d})}{D(x_{d+1}, \ldots, x_n)} \notin \mathfrak{p}$ and, for suitable $q \geq 1$, $D^q \mathfrak{p}$ is contained in the ideal generated by the h_i's ([Rz12 II.2, II.3]). Since $Z_o \setminus \{D = 0\} \neq \emptyset$, by the curve selection lemma we get an analytic half-branch $t \mapsto x(t)$ such that $x(0) = O$ and $x(t) \in Z \setminus \{D = 0\}$ for any representative Z of Z_o and $t > 0$ small enough. But then, near $x(t)$, $Z \setminus \{D = 0\}$ coincides with $\{h_1 = \cdots = h_{n-d} = 0\}$, which is an analytic manifold of dimension d by the implicit functions theorem. $\square$

Immediately from this, we see that, for any global semianalytic set Z,

$$\dim(Z) = \max\{\dim(Z_x) : x \in \operatorname{Adh}(Z)\}.$$

Here there is another familiar property of dimension:

$$\dim(\operatorname{Adh}(Z_o) \setminus Z_o) < \dim(Z_o),$$

if $Z_o \neq \emptyset$.

Assume again that $\mathfrak{p} = \mathcal{J}(Z_o)$ is prime and $Z_o = \{f_1 > 0, \ldots, f_s > 0\} \cap \mathcal{Z}(\mathfrak{p})$. Then, $f = f_1 \cdots f_s \notin \mathfrak{p}$, and f vanishes on $\operatorname{Adh}(Z_o) \setminus Z_o$, so that $\mathfrak{p}$ is smaller than $\mathcal{J}(\operatorname{Adh}(Z_o) \setminus Z_o)$ and the inequality of dimensions follows. $\square$

On the other hand, dimensions are preserved by the tilde operator:

$$\dim(Z_o) = \dim(\widetilde{Z_o})$$

(the second dimension as defined in arbitrary real spectra, II.2.5).

Indeed, let $Z_o = \{f_1 > 0, \ldots, f_s > 0, g = 0\}$ and let $\mathfrak{p} = \mathcal{J}(Z_o)$ be prime. Then the support $\mathfrak{q}$ of any $\alpha \in \widetilde{Z_o}$ contains $\mathfrak{p}$, and so

$$\dim(Z_o) = \dim(\mathcal{O}_n/\mathfrak{p}) \geq \dim(\mathcal{O}_n/\mathfrak{q}) = \dim(\alpha),$$

which gives the inequality $\dim(Z_o) \geq \dim(\widetilde{Z_o})$. Conversely, from Proposition II.2.9 we get a prime cone $\alpha \in \widetilde{Z_o}$ such that $\mathfrak{p} = \operatorname{supp}(\alpha)$, so that $\dim(Z_o) = \dim(\alpha) \leq \dim(\widetilde{Z_o})$. $\square$

Finally, here there are the announced complexity bounds. We shall use the combinatorial function τ of Corollary IV.7.9 and Section V.2: $\tau(i) = i$ for $0 \leq i \leq 2$ and $\tau(i) = \binom{4^{i-1} - 2^{i-1} + 1}{2 \cdot 4^{i-2} - 2^{i-2} + 1}$ for $i \geq 3$. In addition, we define $t(j) = \sum_{i=0}^{j} \tau(i)$ and $t'(j) = \sum_{i=0}^{j} i\tau(i)$ for $j \geq 0$.

Theorem 2.12   *Let X_o be an analytic set germ of dimension d.*

    a) All basic open semianalytic set germs of X_o can be described with d strict inequalities, and some cannot with less than d.
    In symbols, $s(X_o) = d$.

    b) All basic closed semianalytic set germs of X_o can be described with $\frac{1}{2}d(d+1)$ relaxed inequalities, and some cannot with less than $\frac{1}{2}d(d+1) - 1$.
    In symbols, $\frac{1}{2}d(d+1) - 1 \le \bar{s}(X_o) \le \frac{1}{2}d(d+1)$.

    c) Every strictly open semianalytic set germ of X_o is a union of $t(d)$ basic open semianalytic set germs. By complementation, every strictly closed semianalytic germ of X_o is a union of $d^{t(d)}$ basic closed semianalytic set germs.
    In symbols, $t(X_o) \le t(d)$, and $\bar{t}(X_o) \le d^{t(d)}$.

    d) Every semianalytic set germ Z_o of X_o has a description

$$Z_o = \bigcup_{j=1}^{r} \{f_{j1} > 0, \ldots, f_{jk_j} > 0, g_j = 0\},$$

where the number of g_j's is $\le t(d)$ and the number of f_{jk}'s is $\le t'(d)$.

Proof.   *a)* and *b)* are translations of Remark VII.5.6 *a)*. Then, *c)* and *d)* follow from Theorems V.2.14 and V.2.17 respectively. $\square$

The exact value of $\bar{s}$ is known to be $2 = \frac{1}{2}d(d+1) - 1$ in the regular case $X_0 = \mathbb{R}_o^2$. This is based on the explicit analysis of planar semianalytic set germs ([D-C]), something difficult to carry on in singular analytic surface germs. Let us stress that from the abstract theory one can deduce $\bar{s}(X_0) = \frac{1}{2}d(d+1) - 1$ for all X_0's if $\bar{s}(X_0) = 2$ for all X_0's of dimension 2.

3. Cylindrical Decomposition of Germs

In this section we discuss the topological behaviour of semianalytic set germs, in relation with the extra conditions **FT**, **AC** and **CC** (I.3.5). Again, we consider germs at the origin $O \in \mathbb{R}^n$ and identify $\mathcal{O}_n$ with the ring $\mathbb{R}\{x_1, \ldots, x_n\}$ of convergent power series. The *open polycylinder of radius* $\rho = (\rho_1, \ldots, \rho_n)$, $\rho_i > 0$, *centered at a point* $a \in \mathbb{R}^n$ is the open set

$$\Delta = \{(x_1, \ldots, x_n) \in \mathbb{R}^n \mid \rho_i > |x_i - a_i|, \quad i = 1, \ldots, n\}.$$

Notice that $\Delta \cap \mathbb{R}^{n-1} = \Delta'$, where Δ' is the open polycylinder of radius $\rho' = (\rho_1, \ldots, \rho_{n-1})$ centered at $a' = (a_1, \ldots, a_{n-1})$. The image of an open polycylinder by a linear isomorphism of $\mathbb{R}^n$ will be called an *open box*. Note that both open polycylinders and boxes are global strictly open semianalytic subsets of $\mathbb{R}^n$. The main result here is the following:

Proposition 3.1	*Let $Z_1, \ldots, Z_r$ be global semianalytic subsets of an open neighbourhood U of the origin $O \in \mathbb{R}^n$. Then, after shrinking U, there is a partition $\Gamma_1, \ldots, \Gamma_p$ of U such that:*

a) Every Γ_k is a global semianalytic subset of U adherent to the origin.
b) The closure of every Γ_k is a basic closed global semianalytic subset of U.
c) The closure of every Γ_k is a union of Γ_ℓ's.
d) Every Z_i is a union of Γ_k's.

Furthermore, at every point $a \in U$ there are centered arbitrarily small open boxes $\Delta \subset U$ such that

e) Every intersection $\Gamma_k \cap \Delta$ is connected.

Proof.	We argue by induction on n. The case $n = 1$ is trivial, since there are very few semianalytic germs at $O \in \mathbb{R}$, namely, $\{x > 0\}$, $\{x = 0\}$, $\{x < 0\}$, $\{x \geq 0\}$ and $\{x \leq 0\}$. So, let $n > 1$, and let $f_1, \ldots, f_s \in \mathcal{O}(U)$ be analytic functions defining the Z_i's. Up to a linear change of coordinates, we can assume that their germs at the origin are regular with respect to the last variable x_n, and then, by Weierstrass's Preparation Theorem, after shrinking U to $U' \times (-\varepsilon, \varepsilon)$, we find analytic functions $u_1, \ldots, u_s \in \mathcal{O}(U)$, strictly positive on U, and monic polynomials $P_1, \ldots, P_s \in \mathcal{O}(U')[x_n]$ of degrees $d_1, \ldots, d_s$ such that

$$P_i(0, x_n) = x_n^{d_i} \quad \text{and} \quad f_i = \pm u_i P_i$$

for all $i = 1, \ldots, s$. Clearly, we can simply suppose that $f_i = \pm P_i$. Moreover, by adding the missing derivatives, we may assume that the family of polynomials $f_1, \ldots, f_s$ is closed under derivation. Now we apply Proposition II.6.1 to $A = \mathcal{O}(U')$, obtaining a partition of $\mathrm{Spec}_r(A)$ into constructible sets $C_1, \ldots, C_m$ (here there is no C_0 because all polynomials are monic) and abstract semialgebraic functions $\zeta_{i1}, \ldots, \zeta_{iq_i}$ on each C_i, $i = 1, \ldots, m$. After embedding U' into $\mathrm{Spec}_r(\mathcal{O}(U'))$ (as was explained in 1.4), we consider the global semianalytic sets $Z_i' = C_i \cap U'$ and the restrictions $\xi_{i\ell}$ of the $\zeta_{i\ell}$'s to Z_i'. Then, directly from Proposition II.6.1 and Proposition and Definition II.6.2, we get:

i) For every $x' \in Z_i'$, $\xi_{i1}(x') < \cdots < \xi_{iq_i}(x')$ are the different real roots of the polynomials $f_1(x', x_n), \ldots, f_r(x', x_n) \in \mathbb{R}[x_n]$.

ii) For $k = 1, \ldots, s$ and $(x', x_n) \in U' \times \mathbb{R}$, the sign of $f_k(x', x_n)$ depends only on the signs of $x_n - \xi_{i1}(x'), \ldots, x_n - \xi_{iq_i}(x')$.

iii) For each $i = 1, \ldots, m$, the *graphs*

$$G_{i\ell} = \{(x', x_n) \in Z_i' \times \mathbb{R} \,|\, x_n = \xi_{i\ell}(x')\}, \quad l = 1, \ldots, q_i,$$

and the *slices*

$$T_{i\ell} = \{(x', x_n) \in Z_i' \times \mathbb{R} \,|\, \xi_{i\ell}(x') < x_n < \xi_{il+1}(x')\}, \quad l = 0, \ldots, q_i$$

are global semianalytic subsets of $U' \times \mathbb{R}$, where we put $\xi_{i0} = -\infty$ and $\xi_{iq_i+1} = \infty$.

(Here we extend the canonical embedding $U' \hookrightarrow \mathrm{Spec}_r(A)$ to another canonical embedding $U' \times \mathbb{R} \hookrightarrow \mathrm{Spec}_r(A[\mathbf{x}_n])$.) Next, we have:

iv) Every $\xi_{i\ell}$ extends continuously to $\mathrm{Adh}(Z_i') \cap U'$.

Let $a' \in U'$ be adherent to Z_i'. Let (a_p') be any sequence of points of Z_i' converging to a'. Since $\xi_{i\ell}(x')$ is a root of a monic polynomial $\pm f_k = \mathbf{x}_n^{d_k} + c_1 \mathbf{x}_n^{d_k-1} + \cdots + c_{d_k} \in \mathcal{O}(\Delta')[\mathbf{x}_n]$ whose coefficients are continuous at a', it follows that the sequence $(\xi_{i\ell}(a_p'))$ of roots of the polynomials $f_k(a_p', \mathbf{x}_n)$ is bounded, and consequently, contains a converging subsequence, whose limit a_n will be $\xi_{i\ell}(a') = \lim_{x' \to a'} \xi_{i\ell}(x')$. In fact, it suffices to show now that for any sequence (b_p') of points of Z_i' converging to a' such that there exists $b_n = \lim_p \xi_{i\ell}(b_p')$, we have $a_n = b_n$. To show it, consider the signs of the f_i's at $(b_p', \xi_{i\ell}(b_p'))$, which by *ii)* coincide with those at $(a_p', \xi_{i\ell}(a_p'))$. By continuity, the polynomials $f_i(a', \mathbf{x}_n)$ have the same *relaxed* signs at b_n and at a_n. Since at least $f_k(a', b_n) = f_k(a', a_n) = 0$, the conclusion follows immediately from *Thom's lemma* ([B-C-R 2.5.4, p.33]), stated below:

Let $h_1, \ldots, h_s \in \mathbb{R}[t]$ be a family of polynomials which is stable under derivation, and consider the semialgebraic sets

$$S = \left\{ t \in \mathbb{R} \ \middle| \ \begin{array}{l} h_1(t) > 0, \ldots, h_\lambda(t)) > 0 \\ h_{\lambda+1}(t) < 0, \ldots, f_\mu(t)) < 0 \\ h_{\mu+1}(t) = \cdots = f_s(t)) = 0 \end{array} \right\}$$

and

$$S^* = \left\{ t \in \mathbb{R} \ \middle| \ \begin{array}{l} h_1(t) \geq 0, \ldots, h_\lambda(t) \geq 0 \\ h_{\lambda+1}(t) \leq 0, \ldots, f_\mu(t) \leq 0 \\ h_{\mu+1}(t) = \cdots = f_s(t) = 0 \end{array} \right\}.$$

Then, S is either empty, or a point, or an open interval. If $S \neq \emptyset$, then $S^* = \mathrm{Adh}(S)$, and if $S = \emptyset$, then S^* is either empty or a point.

Now, by induction, we find a partition $\Pi_1, \ldots, \Pi_q$ of a maybe smaller U' verifying *a)-d)* for $Z_1', \ldots, Z_m'$. In this situation, every Π_j is contained in some Z_i', and we can construct graphs and slices over it by restricting the functions $\xi_{i\ell}$. These graphs and slices are the Γ_k's we sought. First of all, since the origin $O' \in \mathbb{R}^{n-1}$ is adherent to all the Π_j's, by *iv)* all the limits $\lim_{x' \to o'} \xi_{i\ell}(x')$ exist, and since $\pm f_k(0, \mathbf{x}_n) = \mathbf{x}_n^{d_k}$, all of them are zero. In particular, the origin $O \in \mathbb{R}^n$ is adherent to all the Γ_k's. Moreover, we can assume that all the $G_{i\ell}$'s are contained in U. Thus, since the Π_j's are a partition of U', the Γ_k's are a partition of U. Furthermore, every Γ_k is the intersection of a $G_{i\ell}$ or a $T_{i\ell}$ with $\Pi_j \times (-\varepsilon, \varepsilon)$, and by *iii)* Γ_k is a global semianalytic subset of Δ. Next, let us look at the closure of Γ_k. By construction, up to a permutation of $f_1, \ldots, f_s$ which we omit to simplify notations,

$$\Gamma_k = \left\{ (x', x_n) \in \Pi_j \times (-\varepsilon, \varepsilon) \ \middle| \ \begin{array}{l} f_1(x', x_n) > 0, \ldots, f_\lambda(x', x_n) > 0 \\ f_{\lambda+1}(x', x_n) < 0, \ldots, f_\mu(x', x_n) < 0 \\ f_{\mu+1}(x', x_n) = \cdots = f_s(x', x_n) = 0 \end{array} \right\}.$$

Then, a straightforward computation using Thom's lemma, gives

$$\mathrm{Adh}(\Gamma_k) = \left\{ (x', x_n) \in \mathrm{Adh}(\Pi_j) \times (-\varepsilon, \varepsilon) \;\middle|\; \begin{array}{l} f_1(x', x_n) \geq 0, \ldots, f_\lambda(x', x_n) \geq 0 \\ f_{\lambda+1}(x', x_n) \leq 0, \ldots, f_\mu(x', x_n) \leq 0 \\ f_{\mu+1}(x', x_n) = \cdots = f_s(x', x_n) = 0 \end{array} \right\}.$$

From this and the corresponding properties of Π_j it follows that $\mathrm{Adh}(\Gamma_k)$ is basic closed, and a union of Γ_p's. The property that every Z_i is a union of Γ_k's comes from *ii)*.

Finally, let us see *e)*. Fix $a = (a', a_n) \in U$, and $\eta > 0$ arbitrarily small. By induction, we find arbitrarily small open boxes Δ' centered at a' such that all the $\Pi_j \cap \Delta'$'s are connected. We are going to check that for $\Delta = \Delta' \times (a_n - \eta, a_n + \eta)$, all the $\Gamma_k \cap \Delta$'s are connected too. Of course, this is trivial if a is not adherent to Γ_k, hence we assume $a \in \mathrm{Adh}(\Gamma_k)$. As Γ_k is a graph or a slice over a Π_j we also have $a' \in \mathrm{Adh}(\Pi_j)$. We must distinguish several cases.

Case 1: Γ_k is the graph $G_{i\ell}$ over Π_j. Then, by *iv)*, we can choose Δ' so that $|\xi_{i\ell}(x') - \xi_{i\ell}(a')| < \eta$ for $x' \in \Delta'$. Then, $\Gamma_k \cap \Delta$ is the graph of $\xi_{i\ell}$ over $\Pi_j \cap \Delta'$, and, consequently, it is connected.

Case 2: Γ_k is the slice $T_{i\ell}$ over Π_j, and $\xi_{i\ell}(a') < a_n < \xi_{i\ell+1}(a')$. Then we can suppose $\xi_{i\ell}(a') < a_n - \eta < a_n < a_n + \eta < \xi_{i\ell+1}(a')$, and Δ' small enough to have $\xi_{i\ell}(x') < a_n - \eta$ and $a_n + \eta < \xi_{i\ell+1}(x')$ for all $x' \in \Delta'$. In this situation $\Gamma_k \cap \Delta = (\Pi_j \cap \Delta') \times (a_n - \eta, a_n + \eta)$ is connected.

Case 3: Γ_k is the slice $T_{i\ell}$ over Π_j, and $\xi_{i\ell}(a') = a_n < \xi_{i\ell+1}(a')$ or $\xi_{i\ell}(a') < a_n = \xi_{i\ell+1}(a')$. We solve the first possibility, the second being similar. We can suppose that $a_n + \eta < \xi_{i\ell+1}(a')$ and $a_n - \eta < \xi_{i\ell}(x') < a_n + \eta < \xi_{i\ell+1}(x')$ for $x' \in \Delta'$. Then, $\Gamma_k \cap \Delta = \{x \in \Delta \mid \xi_{i\ell}(x') < x_n\}$ is connected.

Case 4: Γ_k is the slice $T_{i\ell}$ over Π_j, and $\xi_{i\ell}(a') = a_n = \xi_{i\ell+1}(a')$. Here we may assume that $a_n - \eta < \xi_{i\ell}(x') < \xi_{i\ell+1}(x') < a_n + \eta$, so that $\Gamma_k \cap \Delta$ is the slice $G_{i\ell}$ over $\Pi_j \cap \Delta'$, which is connected.

We are done. $\qquad\square$

Corollary 3.2   *Let X_o be an analytic set germ, and let $Z_o \subset X_o$ be a semianalytic set germ. Then:*

a) **FT** *holds for X_o: If Z_o is open (resp. closed) in X_o, then it is a strictly open (resp. closed) semianalytic set germ of X_o.*

b) **AC** *holds for X_o: The closure and the interior of Z_o in X_o are also semianalytic.*

Proof.   We prove the assertion for closures and closed set germs; for interiors and open set germs it follows by complementation. We can assume that $X_o = \mathbb{R}^n_o$. Then, by the preceding proposition, we find a representative Z of Z_o contained in some open neighbourhood U of the origin, which is a union of Γ_k's verifying *a)-c)*. Then *a)* and *b)* imply that $\mathrm{Adh}(Z)$ is a global semianalytic subset of U, and $\mathrm{Adh}(Z_o)$ is semianalytic. If Z_o is closed, then we can choose Z closed, and then it is a union of closures of Γ_k's. By *c)*, Z is a union of global basic closed semianalytic subsets of U, and Z_o is strictly closed. $\qquad\square$

Now we come to the notion of connectedness for semianalytic germs:

Proposition and Definition 3.3   *Let $Z_o \subset \mathbb{R}_o^n$ be a semianalytic set germ. Then there are a neighbourhood basis of O consisting of open boxes*

$$\Delta = \Delta_0 \supset \Delta_1 \supset \Delta_2 \supset \cdots,$$

and a representative $Z \subset \Delta$ of Z_o, such that:

a) Z has finitely many connected components $Z^{(1)}, \ldots, Z^{(m)}$, which are global semianalytic subsets of Δ adherent to O.

b) All the $Z^{(i)} \cap \Delta_p$'s are non-empty and connected.

The germs $Z_o^{(i)}$ depend only on Z_o and are called the connected components of *Z_o. Thus, $\mathbf{CC}$ holds for X_o.*

Proof.   The data in the statement come readily from Proposition 3.1. Now, to show that the germs $Z_o^{(i)}$ depend only on Z_o, suppose we are given two lists $Z, Z^{(i)}, \Delta_j$ and $Z^*, Z^{*(k)}, \Delta_\ell^*$. Since the $Z^{(i)}$'s are finitely many, they are open and closed in Z, which is also true for $Z^{(i)} \cap \Delta_\ell^*$ in $Z \cap \Delta_\ell^* = Z^* \cap \Delta_\ell^*$ (for ℓ big enough). Hence $Z^{(i)} \cap \Delta_\ell^*$ is a union of connected components of $Z^* \cap \Delta_\ell^*$, that is, a union of some $Z^{*(k)} \cap \Delta_\ell^*$'s. Similarly, for $\Delta_j \subset \Delta_\ell^*$, we find that each $Z^{*(k)} \cap \Delta_j$ is a union of some $Z^{(i')} \cap \Delta_j$'s. Hence, $Z^{(i)} \cap \Delta_j$ is the union of those $Z^{(i')} \cap \Delta_j$'s, which implies that there is only one k, one i' and $i' = i$. We are done. $\square$

Remarks 3.4   *a)* If a semianalytic set germ Z_o has m connected components, then any representative Z of Z_o has at most m connected components adherent to the origin, and exactly m if Z is small enough.

b) Let Z be a global semianalytic set and let $cc(Z_x)$ denote the number of connected components of the germ Z_x. Then the function $x \mapsto cc(Z_x)$ is locally bounded.

Indeed, by Proposition 3.2 *e)* and with its notations, $cc(Z_x)$ is bounded in U by the number of Γ_k's. $\square$

c) If Z is a global semianalytic set, the family of its connected components is locally finite. Also, we deduce that Z is locally connected.

Finally, we can read the previous results in terms of the tilde operator for germs:

Proposition 3.5   *Let X_o be an analytic set germ, $\widetilde{X_o} = \mathrm{Spec}_r(\mathcal{O}(X_o))$. Then the tilde operator $Z_o \mapsto \widetilde{Z_o}$ preserves closures, interiors and connected components.*

Proof.   Although X_o is not a true real space, we can mimic the proofs of I.3.5 using Corollary 3.2 and Proposition and Definition 3.3. $\square$

4. Rings of Global Analytic Functions

Let Ω be a real analytic manifold and $\mathcal{O}_\Omega$ its sheaf of germs of analytic functions. Then the ring of global sections of $\mathcal{O}_\Omega$ is the ring $\mathcal{O}(\Omega)$ of global analytic functions on Ω. Let $X \subset \Omega$ be a global analytic set and let $\mathcal{O}(X)$ denote the ring of global analytic functions on X. Straight from the definition we get $\mathcal{O}(X) = \mathcal{O}(\Omega)/\mathcal{J}(X)$ (recall that $\mathcal{J}(X) \subset \mathcal{O}(\Omega)$ is the ideal of all functions which vanish on X). Even more, any analytic function defined in a neighbourhod of X represents an analytic function on X:

Proposition 4.1   *We have*
$$\mathcal{O}(X) = \Gamma(U, \mathcal{O}_\Omega/\mathcal{J}(X)\mathcal{O}_\Omega)$$
for every open neighbourhood U of X.

Proof.   It is well-known that the canonical homomorphism
$$\mathcal{O}(\Omega)/\mathcal{J}(X) \to \Gamma(\Omega, \mathcal{O}_\Omega/\mathcal{J}(X)\mathcal{O}_\Omega),$$
is an isomorphism. Indeed, the sheaf $\mathcal{J}(X)\mathcal{O}_\Omega$ is a coherent sheaf of $\mathcal{O}_\Omega$-ideals, since it is generated by a subset of $\mathcal{O}(\Omega)$ ([Fr I.8]). Then, by Cartan's Theorem B ([Ca nr.6, Th.3 B]), the exact sequence of sheaves
$$0 \to \mathcal{J}(X)\mathcal{O}_\Omega \to \mathcal{O}_\Omega \to \mathcal{O}_\Omega/\mathcal{J}(X)\mathcal{O}_\Omega \to 0$$
induces an exact sequence on global sections:
$$0 \to \mathcal{J}(X) \to \mathcal{O}(\Omega) \to \Gamma(\Omega, \mathcal{O}_\Omega/\mathcal{J}(X)\mathcal{O}_\Omega) \to 0.$$
Thus, the assertion follows. Finally, the canonical homomorphism
$$\Gamma(\Omega, \mathcal{O}_\Omega/\mathcal{J}(X)\mathcal{O}_\Omega) \to \Gamma(U, \mathcal{O}_\Omega/\mathcal{J}(X)\mathcal{O}_\Omega)$$
is an isomorphism, since the stalks of $\mathcal{O}_\Omega/\mathcal{J}(X)\mathcal{O}_\Omega$ vanish off the closed set X.
$\square$

(4.2) Embeddings into Affine Spaces. As was said before, Ω is always assumed to be paracompact and Haussdorff. Then, by a deep theorem due to Grauert ([Gr nr.3, Th.3]), *there is a closed analytic embedding $\Omega \hookrightarrow \mathbb{R}^n$ (for n big enough).* Consequently, we can simply suppose that $\Omega \subset \mathbb{R}^n$ is a closed analytic submanifold of the affine space $\mathbb{R}^n$. On the other hand, another important theorem due to Cartan ([Ca nr.10, Cor. to Prop.15]), says that *every closed analytic submanifold $\Omega \subset \mathbb{R}^n$ is the zero set of finitely many global analytic functions,* that is, Ω is a global analytic subset of $\mathbb{R}^n$. Then, seemingly, Ω is endowed with two different rings of global analytic functions: the one coming from its abstract structure of real analytic manifold, and the one coming from its embedded structure of global analytic set. However, these two rings are in fact the same by Proposition 4.1.

Thus, we have:

$$\mathcal{O}(\Omega) = \mathcal{O}(\mathbb{R}^n)/J, \quad \mathcal{O}(X) = \mathcal{O}(\mathbb{R}^n)/I,$$

where $J \subset \mathcal{O}(\mathbb{R}^n)$ is the ideal of all functions vanishing on Ω, and $I \subset \mathcal{O}(\mathbb{R}^n)$ that of those vanishing on X. With the notation at the beginning of the section, we have $\mathcal{J}(X) = I/J$. Moreover, after Cartan's Theorem A ([Ca nr.6, Th.3 A]), we know that *the ideal J generates every ideal $\mathcal{J}(\Omega_x)$, $x \in \mathbb{R}^n$,* but this is not necessarily the case for I and the $\mathcal{J}(X_x)$'s.

A useful remark is that we can easily construct proper analytic functions $h : \Omega \to [0,1)$. Take, after a closed embedding into $\mathbb{R}^n$, $h(x) = \|x\|^2/(1 + \|x\|^2)$. Note that by this construction and a translation in $\mathbb{R}^n$, we can prescribe $h^{-1}(0)$ to be any chosen point $x_0 \in X$ and the sets $W_\rho = \{x \in X \mid h(x) < \rho\}$, $0 < \rho < 1$, to form a neighbourhood basis of x_0 in X.

(4.3) Local Rings of Global Analytic Functions and Regularity. Let $d = \dim(X)$. We fix a point $x \in X \subset \Omega$. We then have the maximal ideal $\mathfrak{m}$ of all functions vanishing at x, and the local ring $\mathcal{O}_x(X) = \mathcal{O}(X)_\mathfrak{m}$. To study this ring we need some additional constructions. Let $\mathcal{O}_x$ denote the ring of germs at x of analytic functions of Ω. There is a canonical inclusion $\mathcal{O}(\Omega) \subset \mathcal{O}_x$, and the ideal $\mathcal{J}(X) \subset \mathcal{O}(\Omega)$ extends to an ideal $\mathcal{J}_x \subset \mathcal{O}_x$; we shall denote by $\mathcal{A}_x(X)$ the local henselian excellent ring $\mathcal{O}_x/\mathcal{J}_x$. We obtain a homomorphism $\mathcal{O}(X) \to \mathcal{A}_x(X)$ that induces a local homomorphism $\mathcal{O}_x(X) \to \mathcal{A}_x(X)$. Finally, we consider the ideal $\mathcal{J}(X_x) \subset \mathcal{O}_x$ of the analytic set germ X_x, and the ring $\mathcal{O}(X_x) = \mathcal{O}_x/\mathcal{J}(X_x)$ of analytic function germs of X_x. Clearly, $\mathcal{J}(X_x) \supset \mathcal{J}_x$ and we recall that $\mathcal{J}_x$ need not coincide with $\mathcal{J}(X_x)$; in general, by the real Nullstellensatz for germs (Proposition 2.7 *b)*), we have $\mathcal{J}(X_x) = \sqrt[r]{\mathcal{J}_x}$. Thus, we get a second local homomorphism $\mathcal{A}_x(X) \to \mathcal{O}(X_x)$ which is surjective but need not be an isomorphism. Now we say that x is a *regular point of X* if the local ring $\mathcal{A}_x(X)$ is a regular ring of dimension d; otherwise, we say that x is a *singular point of X*. An essential result by Whitney-Bruhat ([Wh-Bh nr.10, Prop.16]) says that *the set of singular points of X is a global analytic subset of dimension $< d$.* In particular, if $\dim(X_x) = d$, then x is a limit of regular points of X. Clearly, if x is regular, then X is a manifold of dimension d near x, but not conversely. However, note that if the whole X is a submanifold of dimension d of Ω, then the canonical epimorphism $\mathcal{A}_x(X) \to \mathcal{O}(X_x)$ is an isomorphism, and since the latter ring is regular of dimension d, every point of X is regular.

Proposition 4.4   *Let $x \in X \subset \Omega$ be as above. Then*

a) *The local ring $\mathcal{O}_x(X)$ is excellent.*

b) *The canonical homomorphism $\mathcal{O}_x(X) \to \mathcal{A}_x(X)$ is a regular homomorphism which extends to an isomorphism of the respective completions.*

c) $\dim(X_x) \leq \dim(\mathcal{O}_x(X)) \leq d$.

d) *If x is a regular point of X, then the local ring $\mathcal{O}_x(X)$ is regular of dimension d and $\mathcal{J}_x = \mathcal{J}(X_x)$.*

Proof. We first consider the case $\Omega = \mathbb{R}^n$. Then by the base change $- \otimes_{\mathcal{O}(\mathbb{R}^n)} \mathcal{O}(\mathbb{R}^n)/\mathcal{J}(X)$, it is enough to prove *a)* and *b)* for $X = \mathbb{R}^n$. To do this, we start by showing that the homomorphism $\mathcal{O}(\mathbb{R}^n) \to \mathcal{O}_x$ is flat. By Proposition VII.1.3, we have to show that given $f_i \in \mathcal{O}(\mathbb{R}^n)$ and $g_{i,x} \in \mathcal{O}_x$ with $\sum_{i=1}^r f_i g_{i,x} = 0$, there are $h_{ji} \in \mathcal{O}(\mathbb{R}^n)$, $a_{j,x} \in \mathcal{O}_x$ such that

$$g_{i,x} = \sum_j h_{j,i} a_{j,x} \qquad \text{for all } i,$$

and

$$\sum_i h_{j,i} f_i = 0 \qquad \text{for all } j.$$

(notice that the first equation must hold in the ring $\mathcal{O}_x$, while the second must hold in the ring $\mathcal{O}(\mathbb{R}^n)$). By Cartan-Oka's Coherence Theorem ([Gu-Ro IV.C.1, IV.B.7], [Tg II.6.4, II.6.5]), the sheaf $\mathcal{R}$ of relations among the $f_i's$ is coherent, and therefore by Cartan's Theorem A, its stalk at x is generated by global sections. Since $(g_{1,x}, \dots, g_{r,x}) \in \mathcal{R}_x$, we may write

$$(g_{1,x}, \dots, g_{r,x}) = \sum_j a_{j,x}(h_{j1}, \dots, h_{jr}),$$

where $a_{j,x} \in \mathcal{O}_x$ and $h_{ji} \in \Gamma(\mathbb{R}^n, \mathcal{R})$, showing flatness.

Now, localizing at the maximal ideal $\mathfrak{m}$ of x, we deduce that the local homomorphism $\mathcal{O}_x(\mathbb{R}^n) \to \mathcal{O}_x$ is faithfully flat. In particular, the first ring is noetherian and the homomorphism extends uniquely to the adic completions $\widehat{\mathcal{O}_x(\mathbb{R}^n)} \to \widehat{\mathcal{O}_x}$. Then we notice that the n projections $\mathbf{x}_i - x_i : \mathbb{R}^n \to \mathbb{R}$ generate the maximal ideal of $\mathcal{O}_x$, and by faithful flatness they generate that of $\mathcal{O}_x(\mathbb{R}^n)$. Hence, the above homomorphism between the completions is onto, and in fact an isomorphism. Thus, the ring $\mathcal{O}_x(\mathbb{R}^n)$ is regular of dimension n. Next, we prove that $\mathcal{O}_x(\mathbb{R}^n)$ is excellent by means of Theorem VII.2.4. Here the residue field $\mathbb{R}$ of the unique maximal ideal is if fact contained in the ring, and the needed derivations are the partial derivatives $D_i = \partial/\partial \mathbf{x}_i$. Finally, since the two rings $\mathcal{O}_x(\mathbb{R}^n)$ and $\mathcal{O}_x$ are excellent the two homomorphisms into their common completion are regular, and, by descent (Proposition VII.1.10 *b)*), so is $\mathcal{O}_x(\mathbb{R}^n) \to \mathcal{O}_x$.

Now, we prove the assertion concerning dimensions. For one inequality, note that $\dim(X_x) = \dim(\mathcal{O}(X_x)) \leq \dim(\mathcal{A}_x(X))$, since we have an epimorphism $\mathcal{A}_x(X) \to \mathcal{O}(X_x)$, and $\dim(\mathcal{O}_x(X)) = \dim(\mathcal{A}_x(X))$ since the two rings have the same completion. For the other inequality, an argument similar to that of 2.10 shows that near x the set X contains an analytic manifold of dimension $e = \dim(\mathcal{O}_x(X))$, hence $d = \dim(X) \geq e$. If x is a regular point of X, then the local ring $\mathcal{A}(X_x)$ is regular of dimension d, and consequently so is $\mathcal{O}_x(X)$. Furthermore, $\mathcal{J}_x$ is a prime ideal of height $n - d$ contained in the kernel $\mathcal{J}(X_x)$ of the canonical epimorphism $\mathcal{A}_x(X) \to \mathcal{O}(X_x)$. But this kernel is an ideal of height $n - d$, since $\dim(X_x) = d$. Hence, $\mathcal{J}_x = \mathcal{J}(X_x)$.

For an arbitrary manifold, the proposition follows via a closed embedding into $\mathbb{R}^n$. $\qquad\qquad\square$

Now, we describe a useful construction involving bounded analytic functions.

(4.5) Filter Basis Associated to a Prime Cone of Bounded Analytic Functions. Let $\mathcal{B}(X) \subset \mathcal{O}(X)$ denote the ring of bounded analytic functions on X; note that a global analytic function which is bounded on X is also bounded on a neighbourhood of X. By the well-known trick that any function f becomes bounded after division by $1 + f^2$, many questions concerning the real space associated to X can be reduced to statements concerning the ring $\mathcal{B}(X)$. To be precise, let $\mathfrak{p} \subset \mathcal{O}(X)$ be a prime ideal, and set $\mathfrak{q} = \mathfrak{p} \cap \mathcal{B}(X)$. Then we have:

a) *The residue fields of $\mathfrak{p}$ and $\mathfrak{q}$ coincide. In fact, the canonical homomorphism $\kappa(\mathfrak{q}) \to \kappa(\mathfrak{p})$ is an isomorphism.*

Any element $\frac{f+\mathfrak{p}}{g+\mathfrak{p}}$ with $g \notin \mathfrak{p}$ can always be written in the form

$$\left(\frac{f}{1 + f^2 + g^2} + \mathfrak{q} \right) \Big/ \left(\frac{g}{1 + f^2 + g^2} + \mathfrak{q} \right).$$

$\square$

In particular, the inclusion $\mathcal{B}(X) \subset \mathcal{O}(X)$ induces another between real spectra $\mathrm{Spec}_r(\mathcal{O}(X)) \subset \mathrm{Spec}_r(\mathcal{B}(X))$. Hence, although what follows next concerns prime cones of bounded analytic functions, it also applies to prime cones of arbitrary analytic functions.

Let $\mathfrak{q} \subset \mathcal{B}(X)$ be the support of a prime cone $\beta \in \mathrm{Spec}_r(\mathcal{B}(X))$, and let $V \subset \kappa(\mathfrak{q})$ be the smallest real valuation ring of $\kappa(\mathfrak{q})$ compatible with β. We recall that V is the convex hull of $\mathbb{Q}$ in $\kappa(\mathfrak{q})$ with respect to β. Then:

b) *The ring $B = \mathcal{B}(X)/\mathfrak{q}$ is contained in V, and the maximal ideal of V lies over a maximal ideal $\mathfrak{n}/\mathfrak{q}$ of B. The residue fields of these maximal ideals are both $\mathbb{R}$.*

Let $f \in \mathcal{B}(X)$ be bounded by $N > 0$. Then $\sqrt{N - f}$ and $\sqrt{N + f}$ are well-defined bounded analytic functions in some neighbourhood of X, and by Proposition 4.1 they belong to $\mathcal{B}(X)$. Consequently, $(N - f)(\beta) \geq 0$ and $(N + f)(\beta) \geq 0$. Thus $-N \leq f(\beta) \leq N$, and $f + \mathfrak{q} \in V$. On the other hand, we have the sequence $\mathbb{R} \subset \mathcal{B}(X) \to \mathcal{B}(X)/\mathfrak{n} \to k$, where the residue field k of V is archimedean over $\mathbb{R}$. Hence, k is in fact $\mathbb{R}$. $\square$

We denote by λ_β the canonical homomorphism $\mathcal{B}(X) \to \mathcal{B}(X)/\mathfrak{n} \to \mathbb{R}$. Notice that λ_β is compatible with β: $f(\beta) \geq 0$ implies $\lambda_\beta(f) \geq 0$ for every $f \in \mathcal{B}(X)$. Finally, let ϕ_β be the collection of all sets $\Lambda_{f,\varepsilon} = \{x \in X \mid -\varepsilon \leq f(x) \leq \varepsilon\}$, for $\varepsilon > 0$ and $f \in \mathfrak{n}$. Then:

c) *ϕ_β is a filter basis.*

For, suppose that some $\Lambda_{f,\varepsilon} = \emptyset$. Then $f^2(x) > \varepsilon^2 > 1/N$ for all $x \in X$ and $N > 0$ big enough. It follows that $1/f^2$ is defined and bounded by $N > 0$ in a neighbourhood of X. Hence, $1/f^2 \in \mathcal{B}(X)$ and $f \notin \mathfrak{n}$. Now consider two sets $\Lambda = \Lambda_{f,\varepsilon}$ and $\Lambda' = \Lambda_{g,\eta}$. An easy computation shows that $\Lambda \cap \Lambda' \supset \Lambda_{h,\delta}$ with $h = f^2 + g^2$ and $0 < \delta < \sqrt{\varepsilon}, \sqrt{\eta}$. $\qquad\square$

These filter bases will be needed to deal with prime cones β "at infinity". Note however that different β's can have the same associated ϕ_β's (when the λ_β's coincide).

Proposition and Definition 4.6   *Let β be a prime cone of $\mathcal{B}(X)$ and ϕ_β the associated filter basis. The following assertions are equivalent:*

a) $\bigcap_{\Lambda \in \phi_\beta} \Lambda \neq \emptyset$.
b) $\beta \to a$ *for some point $a \in X$.*
c) *For every (some) proper analytic mapping $h : \Omega \to [0,1)$, there is $\rho > 0$ such that $h(\beta) < \rho < 1$.*

If that is the case, $\bigcap_{\Lambda \in \phi_\beta} \Lambda$ consists exactly of the specialization $a \in X$ of β and we say that β is bounded. *Otherwise, we say that β is* free.

Proof.   *a)* $\Rightarrow$ *b)* Pick any point $a \in \bigcap_{\Lambda \in \phi_\beta} \Lambda$ and let us see that $\beta \to a$. By uniqueness of specializations, this already implies that the intersection of the $\Lambda_{f,\varepsilon}$'s is the singleton $\{a\}$. We shall also use the homomorphism $\lambda_\beta : \mathcal{B}(X) \to \mathbb{R}$ associated to β. We must see that $g(\beta) \geq 0$ implies $g(a) \geq 0$ for $g \in \mathcal{B}(X)$. But $g(\beta) \geq 0$ implies $t = \lambda_\beta(g) \geq 0$. Then $f = g - t \in \mathfrak{n}$, and $a \in \bigcap_{\varepsilon > 0} \Lambda_{f,\varepsilon}$. Hence, $t - \varepsilon \leq g(a) \leq t + \varepsilon$ for all $\varepsilon > 0$, and $g(a) = t \geq 0$. We are done.

b) $\Rightarrow$ *c)* Let $\beta \to a \in X$. For any (proper) analytic mapping $h : \Omega \to [0,1)$ we can take $\rho > 0$ such that $h(a) < \rho < 1$. Then $h(\beta) < \rho$ by continuity of specializations.

c) $\Rightarrow$ *a)* Let $h : \Omega \to [0,1)$ be a proper analytic mapping and $\rho > 0$ such that $h(\beta) < \rho < 1$. We have $t = \lambda_\beta(h) \leq \rho$. Then $f = h - t \in \mathfrak{n}$ and $\Lambda_{f,\varepsilon} \in \phi_\beta$ for all $\varepsilon > 0$. Now, for ε small, $t + \varepsilon \leq \rho + \varepsilon < 1$, and $\Lambda_{f,\varepsilon} = X \cap h^{-1}([0, t + \varepsilon])$ is a compact set. Thus ϕ_β is a filter basis that contains at least one compact set, which implies *a)*. $\qquad\square$

We end with a geometric evaluation criterion:

Proposition 4.7   *Let β be a prime cone of $\mathcal{B}(X)$, and ϕ_β its associated filter basis. Let $f \in \mathcal{B}(X)$ be > 0 on some closed set T which meets every $\Lambda \in \phi_\beta$. Then, for every integer $p \geq 1$ there are $u, g \in \mathcal{B}(X)$ such that $\lambda_\beta(u) > 0$, $g(\beta) > 0$, and $uf = g^p$.*

Proof.   Let us denote also by $f : \Omega \to \mathbb{R}$ an analytic extension of f. The function $f^{\frac{p-1}{p}}$ is well-defined, analytic and > 0 in the open set $U = \{x \in \Omega \mid f(x) > 0\}$. By hypothesis, $U \supset T$, and we can pick another open set $W \supset T$ such that $\mathrm{Adh}(W) \subset U$. Now, $Z = \Omega \setminus W$ and T are disjoint closed subsets of Ω, and there is a continuous function $\chi : \Omega \to [0,1]$ which is $\equiv 1$ on Z and $\equiv 0$ on T. Then, we consider the continuous function $\psi = \chi +$

$(1 - \chi)f^{\frac{p-1}{p}}$. By construction, ψ is $\equiv 1$ on Z, coincides with $f^{\frac{p-1}{p}}$ on T, and is > 0 everywhere. Now, by Whitney's approximation theorem ([Nh 1.6.5, p.34]), there is an analytic function $v : \Omega \to \mathbb{R}$ such that $|v(x) - \psi(x)| \leq \frac{1}{2}\psi(x)$ for all $x \in \Omega$. It follows that

$$0 < \tfrac{1}{2}\psi(x) \leq v(x) \leq \tfrac{3}{2}\psi(x)$$

for $x \in \Omega$. In particular, v is bounded and has a bounded square root in $\mathcal{O}(\Omega)$; moreover, v is a unit and consequently $v(\beta) > 0$. Let $p \geq 1$. Since the function $f^{2p-2} + v^{2p}$ is bounded and > 0 everywhere, it has a p-th root w which is > 0 everywhere; in particular, $w(\beta) > 0$. Now, $u = f^{p-1}v^p/w^p$ is a bounded analytic function and, setting $\lambda_\beta(u) = t$, we have $h = u - t \in \mathfrak{n}$. In this situation, $\Lambda_{h,\varepsilon} \in \phi_\beta$ for all $\varepsilon > 0$. By hypothesis, T meets every $\Lambda_{h,\varepsilon}$, and we pick $x_\varepsilon \in \Lambda_{h,\varepsilon} \cap T$. Since $x_\varepsilon \in T$, $\psi(x_\varepsilon) = f^{\frac{p-1}{p}}(x_\varepsilon)$, and the bounds above give

$$\frac{1}{2^p}f(x_\varepsilon)^{p-1} \leq v(x_\varepsilon)^p \leq \frac{3^p}{2^p}f(x_\varepsilon)^{p-1}.$$

From these bounds we deduce:

$$u(x_\varepsilon) = \frac{f(x_\varepsilon)^{p-1}v(x_\varepsilon)^p}{f(x_\varepsilon)^{2p-2} + v(x_\varepsilon)^{2p}} \geq \frac{(1/2^p)f(x_\varepsilon)^{2p-2}}{(1 + 3^{2p}/2^{2p})f(x_\varepsilon)^{2p-2}} = \frac{2^p}{4^p + 9^p} = c > 0.$$

On the other hand, $x_\varepsilon \in \Lambda_{h,\varepsilon}$ implies $u(x_\varepsilon) \leq t + \varepsilon$, and it follows $t \geq c - \varepsilon$. This being valid for all $\varepsilon > 0$, we get $t \geq c$. Whence, $\lambda_\beta(u) > 0$, and $u(\beta) > 0$. Finally, we take $g = fv/w$, and since w, v and $u = f^{p-1}v^p/w^p$ are positive in β, $g(\beta) > 0$. $\qquad\square$

Corollary 4.8 *Let β be a prime cone of $\mathcal{B}(X)$, and ϕ_β its associated filter basis. Let $f_1, \ldots, f_r \in \mathcal{B}(X)$ be such that $f_1(\beta) \geq 0, \ldots, f_r(\beta) \geq 0$. Then the closed set $\{x \in X \mid f_1(x) \geq 0, \ldots, f_r(x) \geq 0\}$ meets every $\Lambda \in \phi_\beta$.*

Proof. Suppose that $\{x \in X \mid f_1(x) \geq 0\}$ does not meet a given $\Lambda \in \phi_\beta$. Then $f = -f_1$ is > 0 on $T = \Lambda$, and by the proposition (for $p = 2$), we deduce $f_1(\beta) < 0$, contradiction. By induction, suppose that $\{x \in X \mid f_1(x) \geq 0, \ldots, f_{r-1}(x) \geq 0\}$ meets every $\Lambda \in \phi_\beta$. If the set $\{x \in X \mid f_1(x) \geq 0, \ldots, f_r(x) \geq 0\}$ does not meet a given $\Lambda \in \phi_\beta$, then $f = -f_r$ is > 0 on $T = \{x \in X \mid f_1(x) \geq 0, \ldots, f_{r-1}(x) \geq 0\} \cap \Lambda$ and, by the proposition again, $f_r(\beta) < 0$, contradiction. $\qquad\square$

5. Hilbert's 17th Problem and Real Nullstellensatz

Let X be a global analytic subset of the real analytic manifold Ω. Let $\mathcal{O}(X)$ be the ring of global analytic functions and let $\widetilde{X}$ be the real spectrum of $\mathcal{O}(X)$; we already saw that X embeds in $\widetilde{X}$ (1.4). Notations could suggest that $\widetilde{X}$ is the Stone space of X, but this is not the case even for the real line $X = \mathbb{R}$: there are non-empty constructible sets $C \subset \tilde{\mathbb{R}}$ such that $C \cap \mathbb{R} = \emptyset$. Yet, we have the following result:

Proposition 5.1   *Let $C \subset \tilde{X}$ be a non-empty closed constructible set. Then $C \cap X \neq \emptyset$*

Proof.   Clearly, we can assume $C = \{\beta \in \tilde{X} \mid f_1(\beta) \geq 0, \dots, f_r(\beta) \geq 0\}$. As $C \neq \emptyset$, we find $\beta \in C$, and by Corollary 4.8, $C \cap X$ meets every $\Lambda \in \phi_\beta$. In particular, $C \cap X \neq \emptyset$. $\square$

From this weak Artin-Lang property we deduce a weak Positivstellensatz:

Proposition 5.2   *Let $f_1, \dots, f_r, h_1, \dots, h_t \in \mathcal{O}(X)$. Then the following assertions are equivalent:*

a) The global semianalytic set

$$\{x \in X \mid f_1(x) = 0, \dots, f_r(x) = 0, h_1(x) \geq 0, \dots, h_t(x) \geq 0\}$$

is empty.

b) There exists an equation of the form

$$1 + \sum_{m_1 + \cdots + m_t \leq d} a_m h_1^{m_1} \cdots h_t^{m_t} = b_1 f_1 + \cdots + b_r f_r,$$

for suitable $m_j, d \in \mathbb{N}$, $b_i \in \mathcal{O}(X)$ and $a_m \in \sum \mathcal{O}(X)^2$.

Proof.   By Proposition 5.1, *a)* is equivalent to

$$\{\beta \in \tilde{X} \mid f_1(\beta) = 0, \dots, f_r(\beta) = 0, h_1(\beta) \geq 0, \dots, h_t(\beta) \geq 0\} = \emptyset.$$

Then, this is equivalent to *b)* by the abstract Positivstellensatz without $\neq$'s (Theorem II.1.14 with $g_1 = \cdots = g_s = 1$). $\square$

Anyhow, despite the lack of a full Positivstellensatz, the relationship between the boolean algebras $\mathcal{C}(X)$ of global semianalytic sets and $\mathcal{C}(\tilde{X})$ of constructible sets is rich enough to draw interesting geometric consequences.

Let U_X denote the collection of all bounded prime cones $\beta \in \tilde{X}$: U_X is the set of all generizations of points $x \in X$. This set U_X is an open neighbourhood of X in $\tilde{X}$, and its complement $\tilde{X} \setminus U_X$ is the set of all free prime cones (Proposition and Definition 4.6). The set U_X can be seen as an infinitesimal neighbourhood of X, where we have in fact the Artin-Lang property:

Proposition 5.3   *Let $C \subset \tilde{X}$ be a constructible set. If $C \cap U_X \neq \emptyset$, then $C \cap X \neq \emptyset$.*

Proof.   We can assume that $C = \{\beta \in \tilde{X} \mid f_1(\beta) > 0, \dots, f_s(\beta) > 0, g(\beta) = 0\}$ with $f_i, g \in \mathcal{O}(X)$. Then suppose that there is some $\beta \in C$ such that

$\beta \to a \in X$. Since the homomorphism $\mathcal{O}_a(X) \to \mathcal{A}_a(X)$ is regular (Proposition 4.4), there is a generization $\beta_a \in \mathrm{Spec}_r(\mathcal{A}_a(X))$ of a lying over β. Then $f_1(\beta_a) > 0, \ldots, f_s(\beta_a) > 0, g(\beta_a) = 0$, and by the analytic curve selection lemma (Remarks VII.4.3), we find, near a, points $x \in X$ such that $f_1(x) > 0, \ldots, f_s(x) > 0, g(x) = 0$. We are done. $\qquad\square$

To progress further, we need the typical sufficient condition for convergency:

Proposition 5.4   *Let $C \subset \widetilde{X}$ be a closed constructible set such that $C \cap X$ is compact. Then $C \subset U_X$.*

Proof.   We can assume that $C = \{\beta \in \widetilde{X} \mid f_1(\beta) \geq 0, \ldots, f_r(\beta) \geq 0\}$. Let $\beta \in C$. By Corollary 4.8, the set $C \cap X$ meets every $\Lambda \in \phi_\beta$, and this set $C \cap X$ being compact, the intersection $\bigcap_{\Lambda \in \phi_\beta} C \cap \Lambda$ is not empty. Thus β is bounded. $\qquad\square$

Now we deduce some applications under compactness asumptions:

Corollary 5.5   *Let $f_1, \ldots, f_r \in \mathcal{O}(X)$ be such that the set*
$$S = \{x \in X \mid f_1(x) \geq 0, \ldots, f_r(x) \geq 0\}$$
is compact; let $d = \dim(S)$. Then
$$\widetilde{S} = \{\alpha \in \widetilde{X} \mid f_1(\alpha) \geq 0, \ldots, f_r(\alpha) \geq 0\},$$
is a noetherian subspace of dimension d of $\widetilde{X}$ and S is an Artin-Lang subset of $\widetilde{S}$. In particular, $\widetilde{S}$ depends only on S and not on the particular choice of the f_i's.

Proof.   The set $\widetilde{S}$ is a subspace by the very definition. That it is noetherian of dimension d follows from two facts proved by Whitney-Bruhat [Wh-Bh nr.8, Prop 11, Cor.]. Namely, that any global analytic subset of X has a locally finite decomposition into irreducible components, and that an irreducible global analytic set contains no proper global analytic subset of its same dimension. Finally, to show that S is an Artin-Lang subset of $\widetilde{S}$ let $C \subset \widetilde{S}$ be constructible. Since S is compact, $C \subset U_X$ by Proposition 5.4. Then, by Proposition 5.3 we have $C \cap S \neq \emptyset$ as wanted. $\qquad\square$

Theorem 5.6 (Positivstellensatz)   *Let $f_1, \ldots, f_r, g_1, \ldots, g_s, h_1, \ldots, h_t \in \mathcal{O}(X)$ be such that the set*
$$\{x \in X \mid f_1(x) = 0, \ldots, f_r(x) = 0, h_1(x) \geq 0, \ldots, h_t(x) \geq 0\}$$
is compact. Then the following assertions are equivalent:

a) The global semianalytic set
$$\left\{ x \in X \ \middle| \ \begin{array}{l} f_1(x) = 0, \ldots, f_r(x) = 0 \\ g_1(x) \neq 0, \ldots, g_s(x) \neq 0 \\ h_1(x) \geq 0, \ldots, h_t(x) \geq 0 \end{array} \right\}$$
is empty.

b) There exists an equation of the form

$$g_1^{2n_1} \cdots g_s^{2n_r} + \sum_{m_1+\cdots+m_t \le d} a_m h_1^{m_1} \cdots h_t^{m_t} = b_1 f_1 + \cdots + b_r f_r,$$

for suitable $n_i, m_j, d \in \mathbb{N}$, $b_i \in \mathcal{O}(X)$ and $a_m \in \sum \mathcal{O}(X)^2$.

Proof.   Again by the abstract Positivstellensatz, it is enough to show that *a)* fails if there is some $\beta \in \widetilde{X}$ such that

$$f_1(\beta) = \cdots = f_r(\beta) = 0, g_1(\beta) \neq 0, \ldots, g_s(\beta) \neq 0, h_1(\beta) \ge 0, \ldots, h_t(\beta) \ge 0.$$

But such a β must be bounded by the compactness assumption and Proposition 5.4, and then *a)* fails by Proposition 5.3. $\qquad\square$

In particular, we have:

Theorem 5.7 (Real Nullstellensatz)   *Let $I \subset \mathcal{O}(X)$ be a finitely generated ideal whose zero set is compact. Then $\mathcal{J}\mathcal{Z}(I) = \sqrt[r]{I}$.*

Proof.   Apply the preceding result with $h_1 = \cdots = h_s \equiv 1$. $\qquad\square$

We also get:

Theorem 5.8 (Hilbert's 17th Problem)   *Let $f \in \mathcal{O}(X)$ have compact zero set. If $f \ge 0$ on X, then there are $h, g_1, \ldots, g_r \in \mathcal{O}(X)$, $h \neq 0$, such that*

$$h^2 f = g_1^2 + \cdots + g_r^2.$$

In fact, we can take $h = f^{2n} + h_1^2 + \cdots + h_s^2$ for some $h_1, \ldots, h_s \in \mathcal{O}(X)$, and the zero set of the denominator h is contained in that of f.

Proof.   By hypothesis $\{x \in X \mid -f(x) \ge 0\} = \{x \in X \mid f(x) = 0\}$ and the latter set is compact. Thus we can apply Theorem 5.6 to the conditions $f \neq 0, -f \ge 0$, to obtain an equation showing that $f(\beta) \ge 0$ for all prime cones $\beta \in \widetilde{X}$. Then the result follows from Corollary II.1.15. $\qquad\square$

As for germs, the next question is for which integers m a positive semidefinite global analytic function is a sum of $2m$-th powers. Quite remarkably, this is actually a local question:

Theorem 5.9 (Sums of Even Powers of Meromorphic Functions)   *Let $f \in \mathcal{O}(X)$ have compact zero set and be ≥ 0 on X. Let $m \ge 0$. If for every point $x \in X$ there are analytic function germs $h_x, g_{1,x}, \ldots, g_{r,x} \in \mathcal{O}(X_x)$, $h_x \neq 0$, (the function germs and their number r depending on the point) such that*

$$h_x^{2m} f_x = g_{1,x}^{2m} + \cdots + g_{r,x}^{2m},$$

then there are global analytic functions $h, g_1, \ldots, g_r \in \mathcal{O}(X)$, $h \neq 0$, such that

$$h^{2m} f = g_1^{2m} + \cdots + g_r^{2m}.$$

Proof. We first reduce the problem to the case when X is irreducible. Since the decomposition of X into irreducible components is locally finite and the zero set of f is compact, there are finitely many components $X^{(1)}, \ldots, X^{(s)}$ that meet that zero set, and the union $X^{(0)}$ of all the others is a global analytic set on which f is > 0. By Proposition 4.1, f has a $2m$-th root g_0 in $\mathcal{O}(X^{(0)})$. Now suppose that for every $X^{(i)}$, $1 \leq i \leq s$, we have expressions

$$h_i^{2m} f = g_{i1}^{2m} + \cdots + g_{ir}^{2m}$$

with the same number of summands after adding enough 0^{2m}'s. We choose for each $i = 1, \ldots, s$, a function $d_i \in \mathcal{O}(X)$, ≥ 0 on X and whose zero set is $\bigcup_{j \neq i} X^{(j)}$. In this situation:

$$(d_0 + d_1 h_1 + \cdots + d_s h_s)^{2m} f = (d_0 g_0)^{2m} + \sum_{i=1}^{s} d_i^{2m}(g_{i1}^{2m} + \cdots + g_{ir}^{2m}).$$

Thus, we assume henceforth that X is irreducible, and denote by K the field of fractions of the integral domain $\mathcal{O}(X)$. If f is not a sum of $2m$-th powers in K, then there is a real valuation v of K, such that the value $v(f)$ is not divisible by $2m$ ([Bel 1.9, p.146]). Let β be an ordering of K compatible with v. As for germs (Proposition 2.8 b)), we can replace V by the convex hull V_β of $\mathbb{Q}$ in K with respect to β. Now, suppose that β is free. Then, since the zero set of f is compact, there is $\Lambda \in \phi_\beta$ that does not meet that zero set, and consequently f is > 0 on Λ. But then, by Proposition 4.7, we can write $uf = g^{2m}$ with $\lambda_\beta(u) > 0$. This latter condition means that u is a unit in $V = V_\beta$, hence its value is zero, and we have $v(f) = 2mv(g)$, contradiction. Consequently, β is not free. Thus, $\beta \to x \in X$, and this implies that V contains $\mathcal{O}(X)$, and the maximal ideal $\mathfrak{n}$ of V lies over the maximal ideal $\mathfrak{m}$ of the point x. We are going to show that the assumption of the statement fails for this x.

Let V^h be the henselization of V, which is a valuation ring with the same residue field and value group as V (Proposition II.7.9). Furthermore the inclusion $\mathcal{O}_x(X) \subset V \subset V^h$ is a local homomorphism, which, by the universal property of henselizations (Proposition and Definition II.7.7 c)), extends to $A = \mathcal{O}_x(X)^h$. Let $\mathfrak{p}$ denote the kernel of $A \to V^h$, which is a real prime ideal. By Proposition II.7.6 c), $\mathfrak{p}$ is a minimal prime of A, and $A/\mathfrak{p} \to V^h$ is an inclusion. Thus we have constructed a real valuation of the quotient field of $A/\mathfrak{p}$ such that $2m$ does not divide the value of $\bar{f} = f$ mod $\mathfrak{p}$. Consequently, $\bar{f}$ is not a sum of $2m$-th powers in that quotient field. On the other hand, the canonical homomorphism $\mathcal{O}_x(X) \to \mathcal{A}_x(X)$ also extends to A, and we get $\mathcal{O}_x(X) \to A \to \mathcal{A}_x(X)$. Since all rings involved are excellent and have the same completion, we deduce from Proposition VII.3.1 that $\mathfrak{p}$ generates a minimal prime ideal $\mathfrak{q}$ of $\mathcal{A}_x(X)$. Thus, we have an inclusion $A/\mathfrak{p} \to \mathcal{A}_x(X)/\mathfrak{q}$ which extends to an isomorphism of the respective completions. As the smaller ring is henselian, it follows easily from Rotthaus's Theorem II.2.8 and from the properties of $\mathfrak{p}$, that $\mathfrak{q}$ is a real ideal and $\bar{f}$ is not a sum of $2m$-th powers in $\kappa(\mathfrak{q})$. But $\mathcal{O}(X_x) = \mathcal{A}_x(X)/\sqrt[r]{(0)}$, so that $\mathfrak{q}$ is in fact a minimal prime of $\mathcal{O}(X_x)$. We are done. $\qquad\square$

Remarks 5.10   *a)* Let f and m as above. The last theorem leads to the global version of Proposition 2.9 *b)*: *If for every analytic curve* $\gamma : (-\varepsilon, \varepsilon) \to X$ *the order of the Taylor expansion of* $f \circ \gamma$ *at 0 is divisible by* $2m$, *then there are* $h, g_1, \ldots, g_r \in \mathcal{O}(X)$, $h \neq 0$, *such that*

$$h^{2m} f = g_1^{2m} + \cdots + g_r^{2m}.$$

b) The modification described in Remark 2.10 to get a characterization of sums of even powers can be made here too. However, this depends on the function we are studying. One gets a more intrinsic characterization, changing Theorem 5.4 into *"for every analytic curve* $\gamma : (-\varepsilon, \varepsilon) \to X$ *not contained in the singular locus of* X *the order..."* (see [Rz11]).

Finally, note that in case X is compact, all compactness assumptions are redundant, and we get the best possible statements. In fact, in this case $U_X = \widetilde{X}$, X is an Artin-Lang subset of $\widetilde{X}$, and $\widetilde{X}$ is the Stone space of $\mathcal{C}(X) = \mathcal{C}_b(X)$.

6. Minimal Generation of Global Semianalytic Sets

Let Ω be a real analytic manifold and $X \subset \Omega$ a global analytic set of dimension d. In this section we want to study the complexity of global semianalytic sets. We start with the compact case, where the machinery developed in the preceding chapters apply directly. First we prove:

Lemma 6.1   *Let* F *be a fan of* $\widetilde{X}$ *and assume that some* $\beta \in F$ *specializes to a point* $a \in X$. *Then* $\#(F) \leq 2^d$

Proof.   We apply Theorem VI.4.2 with $A = \mathcal{O}(X)$, $\gamma = \beta$, $L = K = \kappa(\mathrm{supp}(\beta))$, $\alpha = a$, and $\kappa(\alpha) = \mathbb{R}$. Then $A_{\mathrm{supp}(\alpha)} = \mathcal{O}_a(X)$, which is noetherian of dimension $\leq d$, so that $\dim(\beta \to \alpha) \leq d$. In conclusion, $s(F) \leq d$, or, in other words, $\#(F) = 2^s$ with $s \leq d$. $\qquad\square$

Proposition 6.2   *Let* $f_1, \ldots, f_r \in \mathcal{O}(X)$ *be such that the set*

$$S = \{x \in X \mid f_1(x) \geq 0, \ldots, f_r(x) \geq 0\}$$

is compact and assume that $d = \dim(S)$. *Then, the space of signs*

$$\widetilde{S} = \{\beta \in \widetilde{X} \mid f_1(\beta) \geq 0, \ldots, f_r(\beta) \geq 0\}.$$

has the invariants $s(\widetilde{S}) = d$ *and* $\bar{s}(\widetilde{S}) = \frac{1}{2}d(d+1)$.

Proof.   Consider any fan $F \subset \widetilde{S}$. Since S is compact, any $\beta \in F$ specializes to some $x \in S$ (Proposition 5.4) and from the lemma we get $\#(F) \leq 2^d$. Thus, by the global stability formula $s(\widetilde{S}) \leq d$. For the converse inequality it is enough to take a regular point x of X such that $f_1(x) > 0, \ldots, f_r(x) > 0$, which exists by the assumption on the dimension of S. Then, $\mathcal{O}_x(X)$ is a regular local ring of dimension d (Proposition 4.4 d)) and we can construct a fan F with $\#(F) = 2^d$ which specializes to x. Hence $F \subset \widetilde{S}$ and we are done.

To prove the statement for $\bar{s}$, since $\widetilde{S}$ is a noetherian space of dimension d (Corollary 5.5), Theorem V2.9 gives the upper bound $\bar{s}(\widetilde{S}) \leq \frac{1}{2}d(d+1)$. For the lower bound, we take again a regular point x of X such that $f_1(x) > 0, \ldots, f_r(x) > 0$, and let $u_1, \ldots, u_d \in \mathcal{O}(X)$ be a regular system of parameters of the local ring $A = \mathcal{O}_x(X)$. We define a sequence $Y_0 \subset Y_1 \subset \cdots \subset Y_d$ of subvarieties of X by the following conditions: each Y_i is the irreducible component of $\{u_{i+1} = \cdots = u_d = 0\}$ (the whole space for $i = d$) that contains the point x. Notice that Y_i contains Y_{i-1}, and Y_0 is the point x. Since $A_{(u_i)}/(u_{i+1}, \ldots, u_d)$ is a rank 1 discrete valuation ring with local parameter u_i, we see that $\widetilde{Y_{i-1}}$ is a divisor of $\widetilde{Y_i}$ in the sense of Proposition and Definition V.4.1, and also in that sense, u_i is a uniformizer of $\widetilde{Y_{i-1}}$. After this preparation, we are ready to repeat the arguments used in Propositions VI.5.3, VI.5.4 and VII.5.5 to bound $\bar{s}$ from below. We shall construct by induction a basic closed set $C_i \subset \widetilde{Y_i}$ with $\dim_x(C_i) = i$ which cannot be written with less than $m_i = \frac{1}{2}i(i+1)$. First we show:

> *Let Z be a global semianalytic subset of Y_i and C the corresponding constructible subset of $\widetilde{Y_i}$. Suppose that $\dim(Z_x) = i$. Then C contains a fan F_i with 2^i elements.*

This is trivial for $i = 0$. Then, for $i = 1$, we are only saying that C contains at least two elements. But in this case, Z contains a half-branch at x, hence C contains at least two closed points of $\widetilde{Y_1}$. We stress that this was the missing starting point in the proof of Proposition VII.5.5 (Remark VII.5.6 b)). For $i \geq 2$, the result is, via the tilde operator for germs, a particular case of the existence of fans in excellent rings (Proposition VII.5.4).

Now, let $C_0 = Y_0$, and suppose C_{i-1} is already constructed. We pick the fan $F_{i-1} \subset C_{i-1}$ of the preceding statement. Using Proposition V.4.3, we find a basic closed set $C_i \subset Y_i$ with $\dim_x(C) = i$ which cannot be written with less than $(i-1) + m_{i-1} + 1 = m_i$. $\square$

To continue we recall that a set $Z \subset X$ is called *boundary bounded* if its topological boundary in X is compact (1.3).

Theorem 6.3   *Let $Z \subset X$ be a boundary bounded global basic open semianalytic set. Then, there are global analytic functions $f_1, \ldots, f_d \in \mathcal{O}(X)$ such that*

$$Z = \{x \in X \mid f_1(x) > 0, \ldots, f_d(x) > 0\}.$$

Proof. Let $g_1, \ldots, g_s \in \mathcal{O}(\Omega)$ be such that $Z = \{x \in X \mid g_1(x) > 0, \ldots, g_s(x) > 0\}$. Replacing each g_i by $g_i/(1 + g_i^2)$, we assume that g_i is bounded. Now, let $h : \Omega \to [0, 1)$ be a proper analytic mapping. Since Z is boundary bounded, $\mathrm{Bd}(Z) \subset \{x \in X \mid h(x) < \rho\}$, with $0 < \rho < 1$. Let $T = \{x \in X \mid h(x) \geq \rho\}$. This is a global semianalytic set, whose connected components will be denoted by T_k. They are closed in Ω, and form a locally finite family (Remark 3.4 *b)*). On the other hand, since no T_k contains boundary points of Z, either $T_k \subset Z$ or $Z \cap T_k = \emptyset$. Thus, $T \cap Z$ is a union of T_k's, and consequently, $T \cap Z$ is closed in Ω. Since the sets $T \cap Z$ and $X \setminus Z$ are closed in Ω and disjoint, we can find a bounded continuous function g which is ≥ 1 on $T \cap Z$ and ≤ -1 on $X \setminus Z$ (in case $T \cap Z = \emptyset$, we take $g \equiv -1$). Then, after replacing g by a close analytic approximation, we get $g \in \mathcal{B}(X)$ separating these two sets. By this construction, we have:

$$Z = \{x \in X \mid g_1(x) > 0, \ldots, g_s(x) > 0, h(x) < \rho\} \cup \{x \in X \mid g(x) > 0\}.$$

Next, we consider the following constructible sets of $\widetilde{X} = \mathrm{Spec}_r(\mathcal{O}(X))$:

$$C = \{\beta \in \widetilde{X} \mid g_1(\beta) > 0, \ldots, g_s(\beta) > 0\},$$

$$C_1 = \{\beta \in \widetilde{X} \mid g_1(\beta) > 0, \ldots, g_s(\beta) > 0, h(\beta) < \rho\}$$

and

$$C_2 = \{\beta \in \widetilde{X} \mid g(\beta) > 0\}.$$

After embedding X into $\widetilde{X}$, we have

$$Z = C \cap X = (C_1 \cup C_2) \cap X,$$

Thus we get a new description of Z which, as we will see, has the advantage that we can control the behaviour at free precones, but has the disadvantage that that we do not know whether $C_1 \cup C_2$ is basic (warning: it is not known whether $C = C_1 \cup C_2$!). We will show that $C_1 \cup C_2$ is indeed a basic open subset of $\widetilde{X}$ which can be described with d strict inequalities, from which the theorem will follow.

Now let $\beta \in \mathrm{Spec}_r(\mathcal{B}(X))$; we have:

i) If β is free, and $g(\beta) \geq 0$, then $g_1(\beta) > 0, \ldots, g_s(\beta) > 0$.
ii) If $g(\beta) = 0$, then β is not free.

Indeed, let ϕ_β be the filter basis associated to β. To show *i)*, let $t = \lambda_\beta(g)$. If $g(\beta) \geq 0$, then $t \geq 0$, and for all small $\varepsilon > 0$ we get $\Lambda_{g-t,\varepsilon} \subset \{x \in X \mid g(x) > -1\}$. Since each g_i is > 0 on the latter set, we conclude from Proposition 4.7 that $g_i(\beta) > 0$.

To show *ii)*, note that if $g(\beta) = 0$, then g belongs to the maximal ideal of the valuation V_β, and consequently $\Lambda = \Lambda_{g,\frac{1}{2}} \in \phi_\beta$. But, by construction,

$$\Lambda \subset \{x \in X \mid h(x) \leq \rho\}.$$

Since h is proper, the latter set is compact, and since Λ is closed, we conclude that Λ is compact too. This implies that β is not free.

Note that by $i)$ and Proposition 5.1, $C \supset C_1 \cup C_2$, and by Proposition 5.2, $C \cap U_X = (C_1 \cap C_2) \cap U_X$.

Henceforth, we proceed in several steps.

Step 1: $(C_1 \cup C_2) \cap \mathrm{Adh}_Z(\mathrm{Bd}(C_1 \cup C_2)) = \emptyset.$

Since C is basic, $C \cap \mathrm{Adh}_Z(\mathrm{Bd}(C)) = \emptyset$, and since $C \supset C_1 \cup C_2$, it is enough to show that $\mathrm{Adh}_Z(\mathrm{Bd}(C)) \supset \mathrm{Adh}_Z(\mathrm{Bd}(C_1 \cup C_2))$, even that

$$\mathrm{Bd}(C) \supset \mathrm{Bd}(C_1 \cup C_2).$$

To start with, by Proposition and Definition 4.6 $c)$, the constructible set $H_\rho = \{\beta \in \widetilde{X} \mid h(\beta) < \rho\}$ contains no free prime cone. Thus, since $C \cap U_X = (C_1 \cup C_2) \cap U_X$ we get

$$C \cap H_\rho = (C_1 \cup C_2) \cap H_\rho.$$

From this, since H_ρ is open, we deduce that

$$\mathrm{Adh}(C) \cap H_\rho = \mathrm{Adh}(C_1 \cup C_2) \cap H_\rho,$$

and consequently

$$\mathrm{Bd}(C) \supset \mathrm{Bd}(C) \cap H_\rho = \mathrm{Bd}(C_1 \cup C_2) \cap H_\rho.$$

Finally,
$$\mathrm{Bd}(C_1 \cup C_2) \subset H_\rho.$$

Indeed, if $\beta \in \mathrm{Bd}(C_1 \cup C_2)$, then either $h(\beta) \leq \rho$ or $g(\beta) = 0$. By $ii)$, this latter condition implies that β is not free, and then, by Proposition 5.2 again and the fact that $g(x) = 0$ implies $h(x) \leq \rho$ for all $x \in X$, we get what we wanted.

Step 2: There is no 4-element fan obstruction to basicness.

Let $F = \{\beta_1, \beta_2, \beta_3, \beta_4\} \subset \widetilde{X}$ be a fan. Then $\#(F \cap (C_1 \cup C_2)) \neq 3$. Indeed, there are three cases to distinguish:

Case 1: All the β_i's are free. Then $F \cap (C_1 \cup C_2) = F \cap C_2$, and, C_2 being basic, $\#(F \cap C_2) \neq 3$.

Case 2: No β_i is free. Then, $F \cap (C_1 \cup C_2) = F \cap C$, and, C being basic, $\#(F \cap C) \neq 3$.

Case 3: $\beta_1 \to a \in X$ and β_2 is free. Let $\mathfrak{p} \subset \mathcal{O}(X)$ be the common support of all the β_i's, and let $\mathfrak{q} = \mathfrak{p} \cap \mathcal{B}(X)$; we know that $K = \kappa(\mathfrak{p}) = \kappa(\mathfrak{q})$. By the fan trivialization theorem (Theorem VI.1.6), there is a valuation ring V of K, compatible with all the β_i's, and such that F specializes to a trivial fan $\{\alpha_1, \alpha_2\}$ of the residue field k of V. By 4.5 $b)$, $\mathcal{B}(X)/\mathfrak{q} \subset V$, and the specializations $\beta_i \to \alpha_j$ restrict to specializations of $\mathcal{B}(X)$, which we denote by the same letters. Let, say, $\beta_1 \to \alpha_1$. Then, $\alpha_1 \to a$, and since β_2 is free, it cannot specialize to α_1. Thus we have the situation:

$$\beta_1, \beta_3 \to \alpha_1 \to a, \quad \beta_2, \beta_4 \to \alpha_2.$$

Furthermore, β_4 must be free, since otherwise, by the argument above, β_2 would not be free. Now, we assume that three of the β_i's belong to $C_1 \cup C_2$, and shall prove that the fourth also belongs. To that end, we first notice that if any of β_2, β_4 belongs to C_2, then both belong. For, let $\beta_2 \in C_2$ and $\beta_4 \notin C_2$, that is, $g(\beta_2) > 0$ and $g(\beta_4) \leq 0$. Then, by specialization, $g(\alpha_2) = 0$ and by $ii)$, α_2 would not be free, a contradiction. By this remark, $\beta_2, \beta_4 \in C_2$. Then, if β_1 or β_3 are in C_2, since C_2 is principal, the other also belongs, and $\#(F \cap (C_1 \cup C_2)) = 4$. Thus we can suppose $\beta_1, \beta_3 \notin C_2$, and consequently $\beta_1 \in C_1$ (or $\beta_3 \in C_1$, which would be analogous). In this situation, by $i)$, $\beta_1, \beta_2, \beta_4 \in C$, and, C being basic, $\beta_3 \in C$. As $\beta_3 \in U_X$, we conclude $\beta_3 \in C_1 \cup C_2$, and again $\#(F \cap (C_1 \cup C_2)) = 4$. We are done.

Step 3: Conclusion.

Let $F \subset \widetilde{X}$ be a finite fan. Then $F \cap (C_1 \cup C_2)$ is a basic subset of F, by Theorem IV.7.2 and Step 2. Hence, $\#(F) \equiv 0 \mod \#(F \cap (C_1 \cup C_2))$ (Proposition III.3.8 $b)$). According to the global generation formula (Theorem V.1.4), it remains to show that

$$2^d \#(F \cap (C_1 \cup C_2)) \equiv 0 \mod \#(F).$$

We distinguish two cases:
   Case 1: All the $\beta \in F$ are free. Then, $F \cap (C_1 \cup C_2) = F \cap C_2$, and since C_2 is principal, the congruence holds for $d = 1$ (Proposition III.3.8 $a)$).
   Case 2: Some $\beta \in F$ specializes to a point $a \in X$. Then, by Lemma 6.1, we have $\#(F) = 2^s$ with $s \leq d$, and the above congruence is immediate.

Thus, Step 3 and the proof of the theorem are complete.   □

Remark 6.4   Although the stability index of the boolean algebra $C_b(X)$ is not properly defined, we can denote it by $s_b(X)$, and then the preceding theorem says that $s_b(X)$ is bounded by the dimension. In fact, we have $s_b(X) = d$.

   Indeed, choose any regular point $x \in X$. Then, there are global analytic functions $f_1, \ldots, f_d \in \mathcal{O}(X)$ which are a regular system of parameters of the local ring $A = \mathcal{O}(X_x)$. We know that the formula $f_1 > 0, \ldots, f_d > 0$ defines a basic open set C in $\mathrm{Spec}_r(A)$ which cannot be described with less than d inequalities (Remark VI.1.5 $b)$). Now, let $h : \Omega \to [0, 1)$ be an analytic proper mapping with $h(x) = 0$, and consider the boundary bounded global basic open semianalytic set Z defined by the formula $f_1 > 0, \ldots, f_d > 0, \frac{1}{2} - h > 0$. If Z could be described with less that d inequalities, so could the germ Z_x, and by the tilde operator for germs, also $\widetilde{Z_x}$. But $\frac{1}{2} - h$ being a positive unit in $\mathcal{O}(X_x)$, $\widetilde{Z_x} = C$, contradiction.   □

Next, we look at the other complexity invariants. We use the combinatorial functions $t(j)$ and $t'(j)$ introduced for Theorem 2.12.

Corollary 6.5   *Let Z be a boundary bounded global semianalytic subset of X. Then:*

a) There are global analytic functions on Ω, f_{jk} and g_j, such that

$$Z = \bigcup_{j=1}^{r} \{x \in X \mid f_{j1}(x) > 0, \ldots, f_{jk_j}(x) > 0, g_j(x) = 0\},$$

where the number of g_j's is $\leq t(d)$ and the number of f_{jk}'s is $\leq 1 + t(d) + t'(d)$.

b) If Z is open, then

$$Z = \bigcup_{j=1}^{t} \{x \in X \mid f_{j1}(x) > 0, \ldots, f_{js}(x) > 0\},$$

where $t \leq 1 + t(d)$ and $s \leq d$.

c) If Z is closed, then

$$Z = \bigcup_{j=1}^{\bar{t}} \{x \in X \mid f_{j1}(x) \geq 0, \ldots, f_{j\bar{s}}(x) \geq 0\},$$

where $\bar{t} \leq 1 + d^{t(d)}$ and $\bar{s} \leq 1 + \frac{1}{2}d(d+1)$.

In particular, we can write $t_b(X) \leq 1 + t(d)$ and $\bar{t}_b(X) \leq 1 + d^{t(d)}$.

Proof.   Let $h : \Omega \to [0, 1)$ be a proper analytic map. Since Z is boundary bounded,

$$\mathrm{Bd}(Z) \subset W_\rho = \{x \in X \mid h(x) < \rho\} \subset W_{\bar{\rho}} = \{x \in X \mid h(x) \leq \rho\}$$

with $0 < \rho < 1$. As we have explained in the proof of Theorem 6.3, in this situation $Z \setminus W_\rho$ is a union of connected components of $X \setminus W_\rho$, and a closed subset of Ω. On the other hand, the closed sets $\mathrm{Adh}(X \setminus Z)$ and $Z \setminus W_\rho$ are disjoint (their intersection would consists of boundary points of Z outside W_ρ). Thus, there is an analytic function $g : \Omega \to \mathbb{R}$ which is ≥ 1 on $Z \setminus W_\rho$ and ≤ -1 on $\mathrm{Adh}(X \setminus Z)$. Hence:

$$Z = (Z \cap W_\rho) \cup \{x \in X \mid g(x) > 0\} = (Z \cap W_{\bar{\rho}}) \cup \{x \in X \mid g(x) \geq 0\}.$$

a) We consider the first of these expressions. By Proposition 6.2, $\widetilde{W_\rho}$ is a subspace of $\widetilde{X}$ with $s(\widetilde{W_\rho}) = d$, which by Theorem V.2.17 implies that $Z \cap W_\rho$ can be written in W_ρ as in the statement, where the number of g_i's is $\leq t(d)$ and the number of f_{jk}'s is $\leq t'(d)$. Now, the final bounds come from the appearances of the functions h and g.

b) We use again the first expression above. Arguing as above, $Z \cap W_\rho$ is the union of no more than $t(d)$ global basic open semianalytic sets

$$Z_j = \{x \in X \mid h(x) < \rho, f_{j1}(x) > 0, \ldots, f_{jr}(x) > 0\}.$$

Now these Z_j's are basic open and boundary bounded, hence, by Theorem 6.3, they can be written with $s \leq d$ strict inequalities. Finally, Z is the union of the Z_j's and the principal set $\{x \in X \mid g(x) > 0\}$.

c) Here we need the second expression for Z. By Proposition 6.2 and Theorem V.2.14, $Z \cap W_{\bar\rho}$ is the union of no more than $d^{t(d)}$ global basic closed semianalytic sets

$$Z_j = \{x \in X \mid h(x) \leq \rho, f_{j1}(x) \geq 0, \ldots, f_{jr}(x) \geq 0\}.$$

Now, again by Proposition 6.2, we can take $r \leq \frac{1}{2}d(d+1)$, and the conclusion follows. $\square$

Remark 6.6   We see in the last proof that the main difficulty concerning boundary bounded global semianalytic sets is the computation of $\bar s$. What we get from the preceding arguments is that *any bounded global basic closed semianalytic set can always be described by* $1 + \frac{1}{2}d(d+1)$ *non strict inequalities.*

7. Topology of Global Semianalytic Sets

As in the preceding section, let Ω be a real analytic manifold and $X \subset \Omega$ a global analytic set of dimension d. We shall discuss here the topology of boundary bounded global semianalytic sets. First we prove a general local-global result:

Proposition 7.1   *Let $Z \subset X$ be a global semianalytic set, and $x \in X$. Then there is a strictly closed global semianalytic neighbourhood U of x such that* $\mathrm{Adh}(Z) \cap U$ *is a global strictly closed semianalytic set.*

Proof.   After a closed embedding $\Omega \hookrightarrow \mathbb{R}^n$, we can suppose that $X = \Omega = \mathbb{R}^n$. Then it is enough to show that there are a neighbourhood $U \subset \mathbb{R}^n$ and global analytic functions $f_{ij} \in \mathcal{O}(\mathbb{R}^n)$ such that

$$\mathrm{Adh}(Z) \cap U = \bigcup_{i=1}^{r} \{x \in U \mid f_{i1}(x) \geq 0, \ldots, f_{is_i}(x) \geq 0\}.$$

Indeed, we can replace U by a closed ball $\{y \in \mathbb{R}^n \mid \varepsilon - \|y - x\| \geq 0\}$, so that U is a global strictly closed semianalytic set.

Consequently, consider the semianalytic set germ $\mathrm{Adh}(Z_x)$. This is strictly closed by Corollary 3.2, that is,

$$\mathrm{Adh}(Z_x) = \bigcup_{i=1}^{r} \{f_{i1} \geq 0, \ldots, f_{is_i} \geq 0\},$$

where $f_{ij} \in \mathcal{O}(\mathbb{R}_x^n)$, and the problem here is to find the f_{ij}'s in $\mathcal{O}(\mathbb{R}^n)$. To do it, we consider the regular homomorphism $\mathcal{O}_x(\mathbb{R}^n) \to \mathcal{A}_x(\mathbb{R}^n) = \mathcal{O}(\mathbb{R}_x^n)$ (Proposition 4.4 b) and d)), and the associated map between real spectra $\varphi : \mathrm{Spec}_r(\mathcal{O}(\mathbb{R}_x^n)) \to \mathrm{Spec}_r(\mathcal{O}_x(\mathbb{R}^n))$. By the properties of the tilde operator for germs (Proposition 3.4), we have:

$$\widetilde{\mathrm{Adh}(Z_x)} = \mathrm{Adh}(\widetilde{Z_x}) = \mathrm{Adh}(\varphi^{-1}(C)),$$

where $C \subset \mathrm{Spec}_r(\mathcal{O}_x(\mathbb{R}^n))$ is the constructible set defined by the same formula that defines Z (this formula involves global analytic functions). Now the real going-down holds for φ (Theorem VII.7.1), and this implies (Proposition II.4.2):

$$\mathrm{Adh}(\varphi^{-1}(C)) = \varphi^{-1}(\mathrm{Adh}(C)).$$

On the other hand, the image of φ is the set $Y \subset \mathrm{Spec}_r(\mathcal{O}_x(\mathbb{R}^n))$ of all generizations of x (the real going-down again). Thus we get

$$\widetilde{\mathrm{Adh}(Z_x)} = \varphi^{-1}(Y \cap \mathrm{Adh}(C)),$$

and our problem reduces to show that $Y \cap \mathrm{Adh}(C)$ is strictly closed in Y. Since Y is proconstructible in $\mathrm{Spec}_r(\mathcal{O}_x(\mathbb{R}^n))$, it is enough to show that $Y \cap \mathrm{Adh}(C)$ is constructible in Y (Proposition II.1.12). But $Y \cap \mathrm{Adh}(C)$ is the closure of $Y \cap C$ in Y. Indeed, if $\beta \in Y \cap \mathrm{Adh}(C)$, then $\beta \to x$ and there is $\gamma \in C$ with $\gamma \to \beta$ (Proposition II.2.3). Thus, $\gamma \to x$ and $\gamma \in Y$. Finally, the closure of $Y \cap C$ in Y is constructible because Y is homeomorphic to the real spectrum of the henselization of $\mathcal{O}_x(\mathbb{R}^n)$ at x (Theorem II.7.11), which is excellent (Theorem VII.2.5), and closures are constructible in excellent rings (Theorem VII.6.1).

The preceding argument can be tracked through the two diagrams below:

$$\begin{array}{ccc}
\mathcal{O}_x(\mathbb{R}^n) \longrightarrow \mathcal{O}_x(\mathbb{R}^n)^h & \qquad & C \longleftarrow Y \cap C \\
\downarrow \qquad\qquad \downarrow & \qquad & \uparrow \qquad\qquad \uparrow \\
\mathcal{A}_x(\mathbb{R}^n) \;\overset{=}{\Rrightarrow}\; \mathcal{O}(\mathbb{R}_x^n) & \qquad & Z_x \;\overset{=}{\Lleftarrow}\; Z_x
\end{array} \qquad \square$$

From this we readily get:

Theorem 7.2   *Let $Z \subset X$ be a boundary bounded global semianalytic set. Then $\mathrm{Adh}(Z)$ is a global strictly closed semianalytic set of X, that is:*

$$\mathrm{Adh}(Z) = \bigcup_{i=1}^{r} \{x \in X \mid f_{i1}(x) \geq 0, \ldots, f_{is_i}(x) \geq 0\}$$

for some $f_{ij} \in \mathcal{O}(X)$.

Proof.   Again, we can suppose that $X = \Omega = \mathbb{R}^n$. Let $h : \mathbb{R}^n \to [0,1)$ be a proper analytic map. Since $\mathrm{Bd}(Z)$ is bounded, h is $< \rho$ on $\mathrm{Bd}(Z)$ with $0 < \rho < 1$. Let $T = \{x \in \mathbb{R}^n \mid h(x) \geq \rho\}$. Since T does not meet the boundary of Z, the set $Z \cap T$ is a union of connected components of T. Such a union is closed, as well as the union $T \setminus Z$ of the remaining components, since the family of connected components of T is locally finite. Thus, we can find an analytic function $g \in \mathcal{O}(\mathbb{R}^n)$ which is ≥ 0 on $Z \cap T$ and < 0 on $T \setminus Z$. Thus:

$$\mathrm{Adh}(Z) = \mathrm{Adh}(Z \setminus T) \cup \{x \in \mathbb{R}^n \mid h(x) \geq \rho, g(x) \geq 0\}.$$

Now, $\mathrm{Adh}(Z \setminus T) \subset \{x \in X \mid h(x) \leq \rho\}$, and by Proposition 7.1 and compactness, $\mathrm{Adh}(Z \setminus T)$ is a finite union of global strictly closed semianalytic sets.

$\square$

By complementation we deduce:

Theorem 7.3   *Let $Z \subset X$ be a boundary bounded global semianalytic set. Then the interior $\mathrm{Int}(Z)$ of Z in X is a global strictly open semianalytic set of X, that is:*

$$\mathrm{Int}(Z) = \bigcup_{i=1}^{r} \{x \in X \mid f_{i1}(x) > 0, \ldots, f_{is_i}(x) > 0\}$$

for some $f_{ij} \in \mathcal{O}(X)$.

Remarks 7.4   Let $Z \subset X$ be a global semianalytic set.

a) If $\mathrm{Adh}(Z) \setminus Z$ is bounded, then $\mathrm{Adh}(Z)$ is global semianalytic.

For, consider a compact global semianalytic set $W \supset \mathrm{Adh}(Z) \setminus Z$ (say, a finite union of closed balls after some embedding into $\mathbb{R}^n$, or a typical set $\{x \in X \mid h(x) \leq \rho\}$). Then $\mathrm{Adh}(Z \cap W)$ is global semianalytic by the theorem, and $\mathrm{Adh}(Z) = Z \cup \mathrm{Adh}(Z \cap W)$.

$\square$

b) If $Z \setminus \mathrm{Int}(Z)$ is bounded, then $\mathrm{Int}(Z)$ is global semianalytic.
This follows by complementation from *a)*.

$\square$

Notice anyway, that in these two remarks the conclusions are weaker than in the theorems, since we cannot guarantee that $\mathrm{Adh}(Z)$ is global strictly closed or that $\mathrm{Int}(X)$ is global strictly open.

We now turn to connected components. Again, we start with a local-global result:

Proposition 7.5   *Let $Z \subset X$ be a global semianalytic set, and $x \in X$. Let m be the number of connected components of the semianalytic set germ Z_x. Then, there is a global strictly open semianalytic neighbourhood U of x such that $Z \cap U$ is the union of m connected global semianalytic sets.*

Proof. Clearly, we can assume that $X = \Omega$, and after a closed embedding, also that $\Omega = \mathbb{R}^n$. Then we have the following commutative diagram:

$$\mathcal{O}(\mathbb{R}^n) \longrightarrow \mathcal{O}_x(\mathbb{R}^n) \longrightarrow \mathcal{O}_x(\mathbb{R}^n)^h \longrightarrow \widehat{\mathcal{O}_x(\mathbb{R}^n)}$$

$$\mathcal{A}_x(\mathbb{R}^n) \xrightarrow{\ \equiv\ } \mathcal{O}(\mathbb{R}^n_x) \longrightarrow \widehat{\mathcal{O}(\mathbb{R}^n_x)}$$

Here, $\mathcal{O}_x(\mathbb{R}^n)^h$ is the henselization of $\mathcal{O}_x(\mathbb{R}^n)$, and the homomorphism into $\mathcal{O}(\mathbb{R}^n_x)$ comes from the fact that this latter ring is henselian; we stress that the latter homomorphism is not an isomorphism, which makes essential the use of completions. From this diagram we deduce another for real spectra, and the formula that defines Z defines constructible sets in all the real spectra involved. With the obvious notations we get the following diagram of inverse images of constructible sets

$$\widetilde{Z} \longleftarrow (\widetilde{Z})_x \longleftarrow (\widetilde{Z})^h_x \longleftarrow \widehat{Z}_x$$

$$\widetilde{Z}_x \xLeftarrow{\ \equiv\ } \widetilde{Z}_x \longleftarrow \widehat{Z}_x$$

In this situation, from Proposition VII.8.6 we get

$$cc((\widetilde{Z})^h_x) = cc(\widehat{Z}_x) \quad \text{and} \quad cc(\widetilde{Z}_x) = cc(\widehat{Z}_x).$$

Thus,

$$m = cc((\widetilde{Z})^h_x).$$

Now, by Proposition II.7.13 $a)$, we find an open neighbourhood $\widetilde{U}$ of x in $\operatorname{Spec}_r(\mathcal{O}(\mathbb{R}^n))$ and m constructible sets $\widetilde{C}_1, \ldots, \widetilde{C}_m$ wich are disjoint, open and closed in $\widetilde{Z} \cap \widetilde{U}$ and cover this set. Since $\mathbb{R}^n$ is embedded in $\operatorname{Spec}_r(\mathcal{O}(\mathbb{R}^n))$, we obtain an open neighbourhood U of x in $\mathbb{R}^n$ and m global semianalytic subsets of $\mathbb{R}^n$, $C_1, \ldots, C_m$, which are disjoint, open and closed in $Z \cap U$, and cover $Z \cap U$. Now, by Proposition 3.5, $m = cc(\widetilde{Z}_x) = cc(Z_x)$, and, by Proposition and Definition 3.3, we find an open box $\Delta \subset U$ such that $Z \cap \Delta$ has exactly m connected components. Clearly, these components must be $C_1 \cap \Delta, \ldots, C_m \cap \Delta$. Since Δ is a global semianalytic set, so are all the $C_i \cap \Delta$'s, and the proof is complete. $\qquad\square$

We next deduce:

Proposition 7.6 *Let $Z \subset X$ be a global semianalytic set, such that $\operatorname{Adh}(Z) \setminus Z$ is bounded. Then any union of connected components of Z is a global semianalytic set.*

Proof. We choose a proper analytic function $h : \Omega \to [0, 1)$ and $0 < \rho < 1$ such that $\operatorname{Adh}(Z) \setminus Z \subset \{x \in X \mid h(x) < \rho\}$. Since $\{x \in X \mid h(x) \leq \rho\}$ is compact, from the preceding proposition we deduce that $Z \cap \{x \in X \mid h(x) < \rho\}$

is a union of finitely many connected global semianalytic sets $T_1, \ldots, T_r$. On the other hand, the set $Z' = Z \cap \{x \in X \mid h(x) \geq \rho\}$ is closed in Ω, and its connected components are a locally finite family of disjoint closed subsets of Ω. Hence, any union T of them is closed, and disjoint from the union $Z' \setminus T$ of the others, which is also closed. Thus there is an analytic function $g \in \mathcal{O}(\Omega)$ which is > 0 on T and < 0 on $Z' \setminus T$. We conclude that $T = Z' \cap \{x \in X \mid g(x) > 0\}$ is global semianalytic. Finally, any union C of connected components of Z is the union of some T_i's and a T; hence C is global semianalytic. $\quad\square$

From this we immediately get:

Theorem 7.7   *Let $Z \subset X$ be a boundary bounded global semianalytic set. Then the connected components of Z are global semianalytic sets.*

Proof.   Note that $\mathrm{Bd}(Z) \supset \mathrm{Adh}(Z) \setminus Z$. $\quad\square$

Remark 7.8   Again, although not properly defined for $\mathcal{C}_b(X)$, we can use the terminology of real spaces concerning the extra conditions of I.3.5. Then, Theorems 7.2, 7.3 and 7.7 say that **FT**, **AC** and **CC** hold for $\mathcal{C}_b(X)$.

8. Germs at Compact Sets

In this section we consider a different bounded situation, which generalizes at a time semianalytic germs at a point and semianalytic sets in a compact global analytic set. Let Ω be a real analytic manifold, and let $K \subset \Omega$ be a compact global semianalytic set.

(8.1) Germs. Two sets $Z_1, Z_2 \subset \Omega$ have the same *germ at K* if $Z_1 \cap U = Z_2 \cap U$ for some neighbourhood U of K. This is an equivalence relation and the corresponding classes are called *set germs at K*. The germ at K of a set $Z \subset \Omega$ is denoted by Z_K. It is clear from this starting definition that all generalities in Section 2 for germs at a point $x \in \Omega$ go through for germs at $K \subset \Omega$. The ring of analytic function germs at K will be denoted by $\mathcal{O}(\Omega_K)$. Also, we define when an analytic function germ $f \in \mathcal{O}(\Omega_K)$ is $> 0, \geq 0, = 0$ on a set germ Z_K, and the *ideal $\mathcal{J}(Z_K)$ of Z_K*. We set $\mathcal{O}(Z_K) = \mathcal{O}(\Omega_K)/\mathcal{J}(Z_K)$. The (*analytic*) *dimension* $\dim(Z_K)$ of Z_K is the Krull dimension of the ring $\mathcal{O}(Z_K)$.

It is an important theorem by Frisch ([Fr I.9]) that $\mathcal{O}(\Omega_K)$ is a noetherian ring. Hence, we can consistently define the *germ of zeroes $\mathcal{Z}(I)$* of any $I \subset \mathcal{O}(\Omega_K)$. These germs of zeroes are the *analytic set germs at K*. Moreover, we have the notions of *irreducibility* and *irreducible component* of an analytic set germ X_K, which correspond to primeness and associated prime of the ideal $\mathcal{J}(X_K)$.

Let X_K be an analytic set germ. The ring $\mathcal{O}(X_K)$ is *the ring of analytic function germs of X_K*. The operators $\mathcal{Z}$ and $\mathcal{J}$ define two correspondences between ideals of $\mathcal{O}(X_K)$ and analytic set germs contained in X_K. We consider the boolean algebra $\mathcal{C}(X_K)$ of all *semianalytic set germs* of X_K, and the notions of *basic, principal, strictly open* and *basic, principal, strictly closed* semianalytic set germ, as well as the Zariski counterparts. We have:

Proposition 8.2   *In the above situation, the Stone space of the boolean algebra $\mathcal{C}(X_K)$ is the real spectrum $\widetilde{X_K}$ of the ring $\mathcal{O}(X_K)$. In fact, the correspondence*

$$Z_K = \bigcup_{i=1}^{r} \{f_{i1} > 0, \ldots, f_{is_i} > 0, g_i = 0\} \longmapsto$$

$$\widetilde{Z_K} = \bigcup_{i=1}^{r} \{\alpha \in \widetilde{X_K} \mid f_{i1}(\alpha) > 0, \ldots, f_{is_i}(\alpha) > 0, g_i(\alpha) = 0\}$$

is an isomorphism from the boolean algebra of semianalytic set germs of X_K onto that of constructible sets of $\widetilde{X_K}$.

Proof.   We must see that $Z_K = \{f_1 > 0, \ldots, f_s > 0, g = 0\} = \emptyset$ if and only if $\widetilde{Z_K} = \emptyset$. Now, if $\widetilde{Z_K} = \emptyset$ we can apply the Positivstellensatz as in the proof of Proposition 2.5. Conversely, let $\gamma \in \widetilde{Z_K}$. Let $\Omega' \subset \Omega$ be an open set containing K on which the f_i's and g are defined, and write $X' = \Omega' \cap X$. Also, we pick a proper analytic function $h' : \Omega' \to [0, 1)$. Since K is compact, we have

$$K \subset \{x \in X' \mid h'(x) < \rho\} \subset W' = \{x \in X' \mid h'(x) \leq \rho\} \subset \Omega'.$$

This W' inherits from Ω' a structure of real space, whose Stone space $\widetilde{W'}$ is a subspace of $\mathrm{Spec}_r(\mathcal{O}(X'))$ (Corollary 5.5). Then, via the homomorphism $\mathcal{O}(X') \subset \mathcal{O}(X_K)$, γ lies over a prime cone of

$$\{\beta \in \widetilde{W'} \mid f_1(\beta) > 0, \ldots, f_s(\beta) > 0, g(\beta) = 0\}.$$

Indeed, since $\rho - h > 0$ on K, the square root $\sqrt{\rho - h}$ is a well defined unit of the ring $\mathcal{O}(X_K)$, and $(\rho - h)(\gamma) > 0$.

Since $\widetilde{W'}$ is the Stone space of W', we conclude that $\{x \in W' \mid f_1(x) > 0, \ldots, f_s(x) > 0, g = 0\} \neq \emptyset$. This works for arbitrarily small Ω', and we conclude that the germ Z_K is not empty. $\square$

Straight from this proposition, the reader can state and prove for germs at K the *real Nullstellensatz*, the *Positivstellensatz*, the positive solution to *Hilbert's 17th problem*, and the *valuative criterion for sums of even powers*, as we did for germs at a point in Section 2. Moreover, we now have the real space $\widetilde{X_K}$, which is a noetherian space of signs. Thus we shall study the invariants $s, \bar{s}, t, \bar{t}$ both for $\widetilde{X_K}$ and X_K. But before doing that we take a closer look at the ring $\mathcal{O}(X_K)$.

(8.3) Rings of Analytic Function Germs at a Compact Set. Let X_K be an analytic set germ and $\mathcal{O}(X_K)$ its ring of analytic function germs. To every point $x \in K$ corresponds the maximal ideal $\mathfrak{m}$ of all function germs $f \in \mathcal{O}(X_K)$

vanishing at x. In fact, these are all the maximal ideals of $\mathcal{O}(X_K)$. Indeed, if I is any ideal of that ring, and $f_1, \ldots, f_s \in I$, then, in some neigbourhood U of K, the set $Z = \{x \in U \mid f_1(x) = \cdots f_s(x) = 0\}$ is well-defined. This set meets K, since otherwise, $U' = U \setminus Z$, on which the function $f = f_1^2 + \cdots + f_s^2$ is > 0, would be a neighbourhood of K, hence f would be a unit in I. Once we know that $Z \cap K \neq \emptyset$ for any $f_1, \ldots, f_s \in I$, by compactness we find a point $x \in K$ at which every $f \in I$ vanishes. Thus, I is contained in the maximal ideal $\mathfrak{m}$ of that point. (Note that in this argument we do not use that $\mathcal{O}(X_K)$ is noetherian.)

Now we repeat the construction of 4.3. The localization of the ring $\mathcal{O}(X_K)$ at the maximal ideal $\mathfrak{m}$ of the point $x \in K$ will be denoted by $\mathcal{O}_x(X_K)$. Let $\mathcal{O}_x$ stand for the ring of germs at x of analytic functions of Ω. There is a canonical inclusion $\mathcal{O}(\Omega_K) \subset \mathcal{O}_x$, and the ideal $\mathcal{J}(X_K) \subset \mathcal{O}(\Omega_K)$ extends to an ideal $\mathcal{J}_x \subset \mathcal{O}_x$; we shall denote $\mathcal{A}_x(X_K)$ the local henselian excellent ring $\mathcal{O}_x/\mathcal{J}_x$. We obtain a homomorphism $\mathcal{O}(X_K) \to \mathcal{A}_x(X_K)$ that induces a local homomorphism $\mathcal{O}_x(X_K) \to \mathcal{A}_x(X_K)$. Finally, let X be any representative of X_K; clearly the set germ X_x depends on X_K rather than on X. We consider the ideal $\mathcal{J}(X_x) \subset \mathcal{O}_x$, and the ring $\mathcal{O}(X_x) = \mathcal{O}_x/\mathcal{J}(X_x)$. Since, $\mathcal{J}_x \subset \mathcal{J}(X_x)$, we get a local epimorphism $\mathcal{A}_x(X_K) \to \mathcal{O}(X_x)$ (which need not be an isomorphism).

Proposition 8.4   *In the situation above, we have:*

  a) *The ring $\mathcal{O}(X_K)$ is excellent.*
  b) *The canonical homomorphism $\mathcal{O}_x(X_K) \to \mathcal{A}_x(X_K)$ is regular and extends to an isomorphism of the completions.*
  c) $\dim(X_K) = \max\{\dim(X_x) : x \in K\}$

Proof.   After a closed embedding $\Omega \hookrightarrow \mathbb{R}^n$, we may assume $\Omega = \mathbb{R}^n$. We claim that the homomorphism $\mathcal{O}_x(X_K) \to \mathcal{A}_x(X_K)$ is faithfully flat and extends to an isomorphism of the completions. Let $f_1, \ldots, f_s$ be analytic functions on an open neighbourhood U of K, which generate the ideal $\mathcal{J}(X_K) \subset \mathcal{O}(\mathbb{R}^n_K)$, and set $X = \{x \in U \mid f_1(x) = \cdots = f_s(x) = 0\}$. Then for all open neighbourhoods $W \subset U$ of K we have $\mathcal{A}_x(X \cap W) = \mathcal{A}_x(X_K)$, the ring $\mathcal{O}_x(X_K)$ is the inductive limit of the rings $\mathcal{O}_x(X \cap W)$, and the homomorphism $\mathcal{O}_x(X_K) \to \mathcal{A}_x(X_K)$ is the inductive limit of the homomorphisms $\mathcal{O}_x(X \cap W) \to \mathcal{A}_x(X \cap W)$. Since the latter are faithfully flat and induce isomorphisms between the respective completions (Proposition 4.4 $b)$), the former is faithfully flat and induces an isomorphism between the respective completions. Now by descent (Proposition VII.1.7 $a)$), we have the faithfully flat local homomorphisms of noetherian rings

$$\mathcal{O}_x(X) \to \mathcal{O}_x(X_K) \to \mathcal{A}_x(X_K) \to \widehat{\mathcal{O}_x(X)}.$$

From this our claim follows.

Now we show $a)$ for $X_K = \mathbb{R}^n_K$. In this case, every localization $\mathcal{O}_x(\mathbb{R}^n_K)$ is regular of dimension n with residue field $\mathbb{R}$ (by the preceding claim). Thus we can apply Theorem VII.2.4 with the projections $\mathbf{x}_i : \mathbb{R}^n \to \mathbb{R}$ and the derivations $D_j = \partial/\partial x_j$, and conclude that $\mathcal{O}(\mathbb{R}^n_K)$ is excellent. In general, $\mathcal{O}(X_K)$ is a homomorphic image of $\mathcal{O}(\mathbb{R}^n_K)$, hence also excellent. Moreover, now we know that the homomorphism $\mathcal{O}_x(X_K) \to \widehat{\mathcal{O}_x(X_K)}$ is regular, and by descent, so is $\mathcal{O}_x(X_K) \to \mathcal{A}_x(X_K)$, which settles $b)$.

Finally, we have

$$\dim(X_K) = \dim(\mathcal{O}(X_K)) = \max\{\dim(\mathcal{O}_x(X_K)) : x \in K\}.$$

Then, by $b)$ and Proposition 4.4 $c)$, we have

$$\dim(X_x) \leq \dim(\mathcal{O}_x(X_K)) \leq \dim(X \cap W)$$

for any W as above. Thus:

$$\max\{\dim(X_x) : x \in K\} \leq \dim(X_K) \leq \max\{\dim(X_x) : x \in X \cap W\},$$

and taking the W's very small, we get $c)$. $\qquad\qquad\square$

Remarks 8.5 Let X_K be an analytic germ. After Propositions 8.2 and 8.4, we can deduce several interesting facts concerning the space of signs $\widetilde{X_K}$.

$a)$ Every prime cone of $\widetilde{X_K}$ has a specialization $x \in K$. Hence $K = (\widetilde{X_K})_{\mathrm{max}}$.

Let $\beta \in \widetilde{X_K}$. By Proposition 8.2, if $f_1(\beta) \geq 0, \ldots, f_r(\beta) \geq 0$ the semianalytic germ $Z_K = \{f_1 \geq 0, \ldots, f_r \geq 0\}$ is non-empty. Since that germ is closed, we deduce that the intersection with K of any representative is a non-empty closed subset of K. By compactness, we conclude that the intersection of all the representatives of all Z_K's is not empty. Clearly, any point in that intersection is a specialization of β (which in particular implies that the intersection is a singleton). $\qquad\qquad\square$

$b)$ Every point of K is constructible, but there are others. Namely, pick any analytic curve germ $(-\varepsilon, \varepsilon) \to X; t \mapsto x(t)$, $x(0) \in K$, such that the half-branch $\gamma = \{x(t) \mid t > 0\}$ does not meet K. This corresponds to a prime cone $\gamma \notin \widetilde{K}$ wich specializes to $x \in K$.

Firstly, by $a)$ and Proposition VII.6.3, any constructible point $\beta \in \widetilde{X_K}$ which is not closed is a generization of a point $x \in K$ with $\dim(\beta \to x) = 1$. Then, by Theorem VII.7.1, we find a prime cone β_x of the ring $\mathcal{A}_x(X_K)$ lying over β, and this β_x can be realized by a half-branch γ as above (Remark VII.4.3 $b)$). Thus, it remains to see that γ does not meet K. But, otherwise, since the germ K_x is semianalytic, γ would be completely contained in K for small ε, and, as a prime cone, γ would be a constructible point of $\widetilde{K}$. This is impossible by Proposition 8.2. It remains to show that a γ as above is constructible. To that end, we pick generators $g_1, \ldots, g_r$ of the support of γ and any h which is > 0 on γ and < 0 on $\gamma' = \{x(t) \mid t < 0\}$. Then, γ is described by the formula $h > 0, g_1 = \cdots = g_r = 0$. $\qquad\qquad\square$

$c)$ The set $Y = (\widetilde{X_K})_{\mathrm{const}}$ is an Artin-Lang subset of $\widetilde{X_K}$. Geometrically, $Y \setminus K$ consists of all analytic half-branches coming towards K from outside, an aura as depicted below.

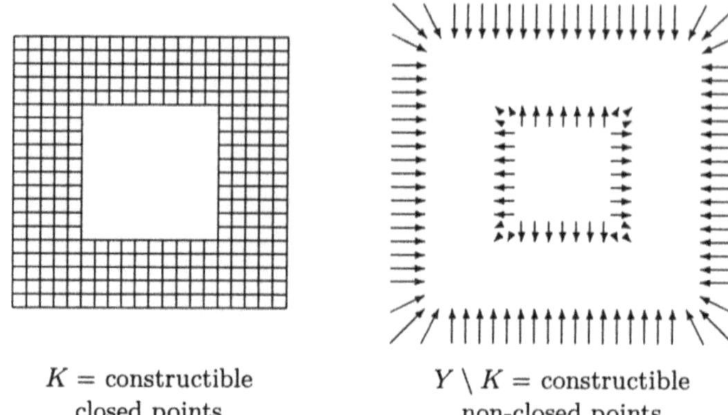

$$K = \text{constructible} \qquad\qquad Y \setminus K = \text{constructible}$$
$$\text{closed points} \qquad\qquad\qquad \text{non-closed points}$$

Let C be a constructible set of $\widetilde{X_K}$, and $\beta \in C$. By a), $\beta \to x \in K$, and by the curve selection lemma (Theorem VII.4.2), we can suppose that β is a half-branch $\{x(t) \mid t > 0\}$. If β is constructible, we are done. If not, $\{x(t) \mid t > 0\} \subset K$, and $x(t) \in C$ for t small enough. $\square$

Now, we come to the announced complexity bounds:

Proposition 8.6   *Let X_K be an analytic set germ of dimension d. Then*

a) All basic open semianalytic set germs of X_K can be described with d strict inequalitites, and some cannot with less than d.
In symbols, $s(X_K) = d$.

b) All basic closed semianalytic set germs of X_K can be described with $\frac{1}{2}d(d+1)$ relaxed inequalities, and some cannot with less than $\frac{1}{2}d(d+1) - 1$.
In symbols, $\frac{1}{2}d(d+1) - 1 \leq \bar{s}(X_K) \leq \frac{1}{2}d(d+1)$.

c) Every strictly open semianalytic set germ of X_K is a union of $t(d)$ basic open semianalytic set germs, By complementation, every strictly closed semianalytic germ of X_K is the union of $d^{t(d)}$ basic closed semianalytic set germs.
In symbols, $t(X_K) \leq t(d)$, and $\bar{t}(X_K) \leq d^{t(d)}$.

d) Every semianalytic set germ Z_K of X_K has a description

$$Z_K = \bigcup_{j=1}^{r} \{f_{j1} > 0, \ldots, f_{jk_j} > 0, g_j = 0\},$$

where the number of g_j's is $\leq t(d)$ and the number of f_{jk}'s is $\leq t'(d)$.

Proof. By Proposition VI.4.6, we have

$$s(\widetilde{X_K}) \leq s_0(\widetilde{X_K}) + d \quad \text{and} \quad \bar{s}(\widetilde{X_K}) \leq s_0(\widetilde{X_K})(d+1) + \frac{1}{2}d(d+1).$$

But since in $\widetilde{X_K}$ all closed points have residue field $\mathbb{R}$, it is $s_0(\widetilde{X_K}) = 0$, and we get the two upper bounds for s and $\bar{s}$. The lower bounds will follow from

Propositions VII.5.3 and 5.5, by finding a specialization $\beta \to x \in K$ with $\dim(\beta \to x) = d$. But let $x \in K$ be such that $\dim(X_x) = d$. Then, all the rings in the sequence

$$\mathcal{O}_x(X_K) \to \mathcal{A}_x(X_K) \to \mathcal{O}(X_x)$$

have the same dimension d (Propositions 4.4 and 8.4). In particular, there is a prime cone β_x of $\mathcal{A}_x(X_K)$ such that $\dim(\beta_x \to x) = d$. This β_x lies over a prime cone β of the ring $\mathcal{O}_x(X_K)$, which is a generization of x, and we claim that $\dim(\beta \to x) = d$. Indeed, set $\mathfrak{p}_x = \mathrm{supp}(\beta_x)$ and $\mathfrak{p} = \mathrm{supp}(\beta)$. Then $\mathfrak{p}_x$ is a minimal prime of $\mathcal{A}_x(X_K)$. Now, since the homomorphism $\mathcal{O}_x(X_K) \to \mathcal{A}_x(X_K)$ induces an isomorphim between the completions and the two rings are excellent, we conclude that $\mathfrak{p}$ is also a minimal prime, and $\dim(\mathcal{O}_x(X_K)/\mathfrak{p}) = \dim(\mathcal{A}_x(X_K)/\mathfrak{p}_x) = d$. This completes the argument for s and $\bar{s}$. The other assertions follow from Theorems V.2.14 and V.2.17. $\qquad\square$

As was said at the end of Section 3, in general the lower bound for $\bar{s}$ cannot be improved. However, the reader will find no difficulty in adapting the proof of Proposition 8.2 to obtain the following more precise lower bound.

Proposition 8.7   *Let X_K be an analytic set germ of dimension d, and suppose that there is a regular point x of some representative X of X_K of the same dimension d such that $\dim(K_x) \geq 1$. Then $\bar{s}(X_K) \geq \frac{1}{2}d(d+1)$.*

Remarks 8.8   We complete the overview concerning complexity with analytic versions of the multilocal criteria given in Propositions VI.7.7, VI.7.8 and VI.7.9 for algebraic varieties. This is possible because by Remark 8.5 $a)$, the ring $\mathcal{O}(X_K)$ is totally archimedean (Definition VI.6.4). The resulting criteria are as follows. Let X_K be an analytic set germ of dimension d.

$a)$ A semianalytic set germ $Z_K \subset X_K$ is basic open (and can be described with s strict inequalities) if and only if any two points $x, y \in K$ are contained in an open set U such that $Z_K \cap U_K$ is basic open (and can be described with s strict inequalities) in U_K.

$b)$ A semianalytic set germ $Z_K \subset X_K$ is basic closed if and only if any two points $x, y \in K$ are contained in an open set U such that $Z_K \cap U_K$ is basic closed in U_K.

$c)$ Two closed semianalytic germs $Z_K, Z_K' \subset X_K$ can be separated if and only if every finite set $E \subset K$ of 2^{d-1} points is contained in an open set U such that $Z_K \cap U_K$ and $Z_K' \cap U_K$ can be separated.

Next, we turn to topology. It is clear from Proposition 7.2 that **FT** and **AC** hold for X_K and, then, for $\widetilde{X_K}$. Of course, **FT** must hold because $\widetilde{X_K}$ is a space of signs, and, since $\mathcal{O}(X_K)$ is excellent, **AC** holds by Proposition VII.6.1. We also see that the tilde operator $Z_K \mapsto \widetilde{Z}_K$ preserves closures and interiors. Thus, we are left with the study of connected components. This is our very last occasion to play the game with real spectra and tilde operators.

Proposition and Definition 8.9   *Let $Z_K \subset \Omega_K$ be a semianalytic set germ.*

   a) There exist a neighbourhood basis of K consisting of global strictly open semianalytic sets
$$U = U_0 \supset U_1 \supset U_2 \supset \cdots \, ,$$
   and a representative $Z \subset U$ of Z_K with finitely many connected components $Z^{(1)}, \ldots, Z^{(m)}$, which are global semianalytic subsets of U adherent to K and such that all the intersections $Z^{(i)} \cap U_p$ are non-empty and connected. The semianalytic germs $Z_K^{(1)}, \ldots, Z_K^{(m)}$ depend only on Z_K and are called the connected components of Z_K.

   b) The connected components of the constructible set $\widetilde{Z_K}$ are the sets $\widetilde{Z_K^{(1)}}, \ldots, \widetilde{Z_K^{(m)}}$.

*In other words, **CC** holds both for X_K and $\widetilde{X_K}$, and the tilde operator preserves connected components.*

Proof.   Let U be any bounded global semianalytic neigbourhood of K, and pick a global semianalytic set Z of U which is a representative of Z_K. By Theorem 7.7, Z has finitely many connected components which are all global semianalytic sets of U. Let T be the union of those among them which are not adherent to K. Then $\mathrm{Adh}(T)$ is a global semianalytic set of U by Proposition 7.2, and $U' = U \setminus \mathrm{Adh}(T)$ is a bounded global semianalytic neighbourhood of K such that $Z' = Z \cap U'$ has finitely many connected components which are all global semianalytic and adherent to K. After this remark, it is clear that *a)* will follow by showing that the number of connected components of $Z \cap U$ adherent to K is bounded for all U as above. But this and *b)* will follow if we show that $\widetilde{Z_K}$ has finitely many connected components. Indeed, let $C_1, \ldots, C_m$ be those components. If we could choose U such that $Z \cap U$ has $n > m$ connected components adherent to K, say $Z^{(1)}, \ldots, Z^{(n)}$, then each $Z^{(i)}$ would be global semianalytic and open in Z. Hence, the $\widetilde{Z_K^{(i)}}$'s would be a partition of $\widetilde{Z_K}$ by open sets, and $\widetilde{Z_K}$ would have at least n connected components.

Consequently, let us show that $\widetilde{Z_K}$ has finitely many connected components. By compactness and Proposition 3.1, we can cover K with finitely many open sets U, each U covered by finitely many connected sets Γ, each Γ a global semianalytic subset of U either contained in Z or in $U \setminus Z$, and each germ Γ_x at each $x \in U$, connected. Moreover, we can shrink the covering to another consisting of open sets $U' \subset \mathrm{Adh}(U') \subset U$, so that the compact sets $K' = K \cap \mathrm{Adh}(U')$ cover K. Now, we consider the canonical restriction homomorphisms $\varphi : \mathcal{O}(\Omega_K) \to \mathcal{O}(U_{K'})$, and the corresponding continuous maps $\varphi^* : \widetilde{U_{K'}} \to \widetilde{\Omega_K}$. Next, we consider the constructible sets $\widetilde{Z_{K'}} \subset \widetilde{U_{K'}}$ and their images $T = \varphi^*(\widetilde{Z_{K'}}) \subset \widetilde{U_{K'}}$. We claim that the constructible set $\widetilde{Z_K} \subset \widetilde{\Omega_K}$ is the union of the above T's.

Clearly, all T's are contained in $\widetilde{Z_K}$. Conversely, let $\beta \in \widetilde{Z_K}$, and let $x \in K$ be its closed specialization. Then x belongs to some K', and it follows that β belongs to the corresponding T. Indeed, since the homomorphism $\mathcal{O}_x(\Omega_K) \to$

$\mathcal{O}(\Omega_x)$ is regular (Proposition 4.4 $b)$), there is a prime cone β_x of the ring $\mathcal{O}(\Omega_x)$ lying over β (real going-down, Theorem VII.7.1). Then, since $\mathcal{O}(\Omega_x) = \mathcal{O}(U_x)$ the above homomorphism factorizes through $\mathcal{O}_x(U_{K'})$, and β_x lies over a prime cone β' of $\mathcal{O}_x(U_{K'})$. By construction, $\beta' \in T$ and $\varphi(\beta') = \beta$. The claim is proved.

By this claim and the continuity of the φ's, it suffices to show that the $\widetilde{Z_{K'}}$'s have all finitely many connected components. To do it, fix U, U' and K'. By the tilde operator (Proposition 9.2) for germs at K', $\widetilde{Z_{K'}}$ is the union of finitely many $\widetilde{\Gamma_{K'}}$'s, hence, it is enough to see that each $\widetilde{\Gamma_{K'}}$ has finitely many connected components. For the proof of this, we need the closed continuous map

$$\mu : \widetilde{U_{K'}} \to K' ; \ \beta \mapsto x = \text{the closed specialization of } \beta$$

(Proposition and Definition II.2.2 $d)$). The set $\mu(\widetilde{\Gamma_{K'}}) = \text{Adh}(\Gamma) \cap K'$ is a global semianalytic subset of U, and consequently it has finitely many, say m, connected components (Remark 3.4 $c)$). We are going to show that $\widetilde{\Gamma_{K'}}$ has at most m connected components. First, we see that each fiber $\mu^{-1}(x) \cap \widetilde{\Gamma_{K'}}$ is connected.

Fix x, and consider the canonical homomorphism $\psi : \mathcal{O}(U_{K'}) \to \mathcal{O}(U_x)$. By the real going-down again, the image of the associated map $\psi^* : \widetilde{U_x} \to \widetilde{U_{K'}}$ consists of the prime cones $\beta \in \widetilde{U_{K'}}$ such that $\beta \to x$, that is, the image is $\mu^{-1}(x)$. Consequently, $\mu^{-1}(x) \cap \widetilde{\Gamma_{K'}} = \psi^*(\widetilde{\Gamma_x})$. But the germ Γ_x is connected and by Proposition and Definition 3.3, the constructible set $\widetilde{\Gamma_x}$ is connected too. Then, so is its image $\mu^{-1}(x) \cap \widetilde{\Gamma_{K'}}$.

Finally we complete the argument as follows. Suppose that $\widetilde{\Gamma_{K'}}$ has more that m connected components. Then we can write $\widetilde{\Gamma_{K'}} = C_1 \cup \cdots \cup C_n$, where $n > m$, and the C_i's are non-empty, open, closed and disjoint in $\widetilde{\Gamma_{K'}}$. Since each fiber $\mu^{-1}(x) \cap \widetilde{\Gamma_{K'}}$ is connected, it is contained in some C_i, which implies that the sets $\mu(C_1), \ldots, \mu(C_n)$ are disjoint. Furthermore, they cover $\mu(\widetilde{\Gamma_{K'}})$, which has m connected components. Thus, it remains to show that each $\mu(C_i)$ is closed in $\mu(\widetilde{\Gamma_{K'}})$ to have a contradiction. But C_i is open and closed in the constructible set $\widetilde{\Gamma_{K'}}$, hence constructible. In particular, C_i is compact in the Harrison topology, and consequently, so is its image $\mu(C_i)$. As $\mu(\widetilde{\Gamma_{K'}}) \subset K'$ is Haussdorff, we conclude that $\mu(C_i)$ is closed. $\qquad\square$

We quote again the best case in which all these results apply: that of a compact global analytic set X. Then, setting $K = X$, every assertion in this section is in fact global. Thus, the theory of global semianalytic sets in X behaves remarkably well concerning complexity and constructibility of topological operations, as well as it does (Section 5) concerning the Positivstellensatz and Hilbert's 17th Problem. Here we arrive at the end of our journey from combinatorics to geometry. On the last stage, we find very satisfactory results in the compact case. In fact we can go a little bit beyond of that compact case as we have seen for boundary bounded sets and germs at compact sets. However, the study of arbitrary non-compact analytic sets seems to require new ideas and much work for the future.

Notes

The study of real analytic sets was started in the fifties by Bruhat, Cartan and Whitney, [Bh-Ca1,2], [Wh-Bh], who settled the basic properties concerning complexification, dimension, irreducible components, and regularity. Cartan's paper [Ca], where the important theorems A and B are proved for real manifolds, is the threshold of the study of global real analytic sets. This paper contains also a detailed exposition of the coherence problems in the real case. Other important results concerning global analytic sets and functions are Grauert's embedding theorem ([Gr]) and Frisch's finiteness theorems for coherent sheaves ([Fr]) which imply the fact that the rings of real analytic function germs at compact semianalytic sets are noetherian.

Semianalytic sets where first systematically studied by Lojasiewicz in [Lo1, 2], whose motivation was the division problem for distributions. The local basic properties of semianalytic sets were later condensed in Hironaka's rectiliniarization theorem ([Hk2]), a byproduct of his desingularization theorems ([Hk1]). This in fact led Hironaka to the study of subanalytic sets, a category discovered also by Hardt ([Ht1,2]) and the matter of a great interest ever since. Our approach, based on cylindrical decompositions (Proposition 3.1), follows the pattern of [Bt-Mm], [Rz4] and [Fz-Rc-Rz]. The bijection between semianalytic set germs and constructible subsets of the real spectrum of the ring of analytic function germs appeared in the two latter papers. The main improvement here is Proposition 3.3, which describes very carefully the behaviour of connected components of set germs in order to get information useful in the global setting.

The real Nullstellensatz and the solution to Hilbert's 17th problem for germs were obtained by Risler ([Rs]), and then a series of similar or somehow improved versions were obtained by different authors ([Ls], [Me], [Rn], [Rz1]). The global counterparts of these results came much later, after the first attempts by Adkins and Leahy ([Ad], [Ad-Lh]) and the answers in dimension 2 by Bochnak-Risler ([Bo-Rs]) and Jaworski ([Jw1]), or under strong additional assumptions by Bochnak-Kucharz-Shiota ([Bo-Ku-Sh]). Then, Jaworski ([Jw2]) and Ruiz ([Rz3]) solved independently the two problems in the compact case; the presence of real spectra in these solutions was clarified in [Rz5,6]. The characterization of sums of $2n$-th powers comes from [Rz11], although there the global result was proved directly, without describing explicitely the connection local-global. For non-singular compact analytic surfaces the same characterization had been obtained by Kucharz ([Ku]). The study of the topology of global semianalytic sets was first done in the compact case in ([Rz7]). The only known results without any compactness assumption correspond to dimension 2 ([Cs-An]). Our statements are given consistently under weakened compactness assumptions, in the form already proposed in [Rz3] and [Rz10]. We do it by means of filters of semianalytic sets associated to prime cones. This idea, inspired by Gillman-Jerison's famous book [Gi-Je], was first used in dimension 1 in [An-Be], and afterwards in arbitrary dimension in [Cs1,2], [Cs-An] and [Jw3]. We present here a mixed simplified version of Castilla's and Jaworski's formulations.

The first results on minimal generation of global semianalytic sets appeared in [An-Br-Rz] for global basic open semianalytic subsets of compact global analytic sets. The germ case was treated there as a preliminary tool. Our exact computations were not possible then, since the break through for spaces of orderings had not been done yet. Furthermore, the treatment given in Section 6 is much simpler, as well as more general. It yields very precise results for boundary bounded sets. Quite remarkably, the exact value of the invariant $\bar{s}$ for basic closed semianalytic set germs remains unknown up to one unit, except for planar set germs ([D-C]). Recently, Delzell ([Dz2]) found a surprising application of basicness of semianalytic germs to a classical question on Hilbert's 17th problem (see [Dz1]). Finally we can only mention the problem of comparing the different complexities defined respectively by regular, Nash or analytic functions. For this involved topic, closely related to M. Artin's approximation theory, we refer the reader to [An-Rz5,6], [Rz-Sh] and [Co-Rz-Sh].

Bibliography

[Ab]   S. Abhyankar: On the valuations centered in a local domain. *Amer. J. Math.* **78**, 321-348 (1956).

[Ac-An-Bg]   F. Acquistapace, C. Andradas, F. Broglia: Geometric obstructions to separation of semialgebraic sets. *To appear.*

[Ac-Bg-Vz]   F. Acquistapace, F. Broglia, P. Vélez: An algorithmic criterion for basicness in dimension 2. *Manuscripta math.* **85**, 45-66 (1994)

[Ad]   W.A. Adkins: A real analytic Nullstellensatz for two dimensional manifolds. *Boll. U.M.I.* **14B**, 888-903 (1977)

[Ad-Lh]   W.A. Adkins, J.V. Leahy: A global real analytic Nullstellensatz. *Duke Math. J.* **43**, no. 1, 81-86 (1976)

[Al-An]   M.E. Alonso, C. Andradas: Real spectra of complete local rings. *Manuscripta math.* **58**, 155-177 (1987)

[Al-Gm-Rz]   M.E. Alonso, J.M. Gamboa, J. Ruiz: Ordres sur les surfaces réelles. *C.R. Acad. Sci. Paris* **298**, 17-19 (1984)

[Al-Ry]   M.E. Alonso, M.-F. Roy: Real strict localizations. *Math. Z.* **194**, 429-441 (1987)

[An]   C. Andradas: Specialization chains of real valuation rings. *J. Algebra* **124**, 437-446 (1989)

[An-Be]   C. Andradas, E. Becker: A note on the real spectrum of analytic functions on an analytic manifold of dimension 1. *In* Real analytic and algebraic geometry, Proceedings, Trento 1988. *Springer LNM* **1420**, 1-21 (1990)

[An-Br-Rz]   C. Andradas, L. Bröcker, J.M. Ruiz: Minimal generation of basic open semianalytic sets. *Invent. math.* **92**, 409-430 (1988)

[An-Rz1]   C. Andradas, J.M. Ruiz: More on basic semialgebraic sets. *In* Real algebraic geometry, Proceedings, Rennes 1991. *Springer LNM* **1524**, 128-139 (1992)

[An-Rz2]   C. Andradas, J.M. Ruiz: Low dimensional sections of basic semialgebraic sets. *Illinois J. Math.* **38**, no. 2, 303-326 (1994)

[An-Rz3]   C. Andradas, J.M. Ruiz: On local uniformization of orderings. *In* Recent advances in real algebraic geometry and quadratic forms, Proceedings of the RAGSQUAD Year, Berkeley 1990-1991. *AMS Contemporary Math.* **155**, 19-46 (1994)

[An-Rz4]   C. Andradas, J.M. Ruiz: Ubiquity of Lojasiewicz's example of a nonbasic semialgebraic set. *Michigan Math. J.* **41**, 465-472 (1994)

[An-Rz5]   C. Andradas, J.M. Ruiz: Algebraic versus analytic basicness. *In* Real analytic and algebraic geometry, Proceedings, Trento 1992, 1-19. Walter de Gruyter, Berlin New York 1995

[An-Rz6]   C. Andradas, J.M. Ruiz: Algebraic and analytic geometry of fans. *Memoirs AMS* **553** (1995)

[A]   E. Artin: Über die Zerlegung definiter Funktionen in Quadrate. *Abh. Math. Sem. Univ. Hamburg* **5**, 100-115 (1927)

[A-S]   E. Artin, O. Schreier: Algebraische Konstruktion reeller Körper. *Abh. Math. Sem. Univ. Hamburg* **5**, 85-99 (1927)

[Ar1]   M. Artin: On the solutions of analytic equations. *Invent. math.* **5**, 277-291 (1968)

[Ar2]   M. Artin: Algebraic approximations of structures over complete local rings. *Publ. Math. I.H.E.S.* **36**, 23-58 (1969)

[Ar-Mz]   M. Artin, B. Mazur: On periodic points. *Annals of Math.* **81**, 82-99 (1965)

[At-Mc]   M.F. Atiyah, I.G. Macdonald: Introduction to commutative algebra. Addison-Wesley Publ. Co., Reading Massachusetts 1969

[Ba]   R. Baer: Über nicht-archimedisch geordnete Körper (Beiträge zur Algebra 1). *Sitz. Ber. der Heidelberger Academie* **8**, Abh. 1927

[Be1]   E. Becker: The real holomorphy ring and sums of $2n$-th powers. *In* Géométrie algébrique réelle et formes quadratiques, Proceedings, Rennes 1981. *Springer LNM* **959**, 139-181 (1982)

[Be2]   E. Becker: On the real spectrum of a ring and its applications to semi-algebraic geometry. *Bulletin AMS* **15**, 19-60 (1986)

[Be-Br]   E. Becker, L. Bröcker: On the description of the reduced Witt ring. *J. Algebra* **52**, 328-346 (1978)

[Be-Kö]   E. Becker, E. Köpping: Reduzierte quadratische Formen und Semiordnungen reeller Körper. *Abh. Math. Sem. Univ. Hamburg* **46**, 143-177 (1977)

[Bl-Sl]   J.L. Bell, A.R. Slomson: Models and ultraproducts: an introduction (2nd edition). North Holland, Publishing Company, Amsterdam London 1971

[B-O]   M. Ben-Or: Lower bounds for algebraic computation trees. *In* 15th ACM Annual Symp. Theory of Computation, Proceedings, Boston 1983, 80-86.

[Bt-Mm]   E. Bierstone, P.D. Milman: Semianalytic and subanalytic sets. *Publ. Math. I.H.E.S.* **67**, 5-42 (1988)

[B-C-R]   J. Bochnak, M. Coste, M.-F. Roy: Géométrie algébrique réelle. *Ergeb. Math. 12*, Springer-Verlag, Berlin Heidelberg New York 1987

[Bo-Ku-Sh]   J. Bochnak, W. Kucharz, M. Shiota: On equivalence of ideals of real global analytic functions and the 17th Hilbert problem. *Invent. math.* **63**, 403-421 (1981)

[Bo-Rs]   J. Bochnak, J.-J. Risler: Le théorème des zéros pour les variétés analytiques réelles de dimension 2. *Ann. Sc. Éc. Norm. Sup. Paris* **8**, 343-364 (1975)

[Bk TE]   N. Bourbaki: Théorie des ensembles. Hermann, Paris 1970

[Bk CA]   N. Bourbaki: Commutative algebra. Hermann, Paris 1972

[Br1]   L. Bröcker: Zur Theorie der quadratischen Formen über formal reellen Körpern. *Math. Ann.* **210**, 233-256 (1974)

[Br2]   L. Bröcker: Characterization of fans and hereditarily pythagorean fields. *Math. Z.* **151**, 149-163 (1976)

[Br3]   L. Bröcker: Über die Anzahl der Anordnungen eines kommutativen Körpers. *Arch. Math.* **29**, 458-464 (1977)

[Br4]   L. Bröcker: Positivbereiche in kommutativen Ringen. *Abh. Math. Sem. Univ. Hamburg* **52**, 170-178 (1982)

[Br5]   L. Bröcker: Real spectra and distributions of signatures. *In* Géométrie algébrique réelle et formes quadratiques, Proceedings, Rennes 1981. *Springer LNM* **959**, 249-272 (1982)

[Br6]   L. Bröcker: Minimale Erzeugung von Positivbereichen. *Geom. Dedicata* **16**, 335-350 (1984)

[Br7]   L. Bröcker: Spaces of orderings and semialgebraic sets. *In* Quadratic and hermitian forms, Proceedings, Hamilton 1983. *Canad. Math. Conf. Proc.* **4**, 231-248 (1984)

[Br8]   L. Bröcker: On the separation of basic semialgebraic sets by polynomials. *Manuscripta math.* **60**, 497-508 (1988)

[Br9]   L. Bröcker: On the stability index of noetherian rings. *In* Real analytic and algebraic geometry, Proceedings, Trento 1988. *Springer LNM* **1420**, 72-80 (1990)

[Br10]   L. Bröcker: On basic semialgebraic sets. *Expo. Math.* **9**, 289-334 (1991)

[Br-Schü] L. Bröcker, H.W. Schülting: Valuation theory from the geometric point of view. *J. reine angew. Math.* **365**, 12-32 (1986)

[Br-St]   L. Bröcker, G. Stengle: On the Mostowski number. *Math. Z.* **203**, 629-633 (1990)

[Bg-To]   F. Broglia, A. Tognoli: Approximation of C^∞-functions without changing their zero-set. *Ann. Inst. Fourier* **39**, 611-632 (1989)

[Bw]   R. Brown: The reduced Witt ring of a formally real field. *Transactions AMS* **230**, 257-292 (1977)

[Bw-Mr]   R. Brown, M. Marshall: The reduced theory of quadratic forms. *Rocky Mountain J. Math.* **11**, 161-175 (1981)

[Bh-Ca1] F. Bruhat, H. Cartan: Sur la structure des sous-ensembles analytiques réels. *C.R. Acad. Sci. Paris* **244**, 988-990 (1957)

[Bh-Ca2] F. Bruhat, H. Cartan: Sur les composantes irréductibles d'un sous-ensemble analytique réel. *C.R. Acad. Sci. Paris* **244**, 1123-1126 (1957)

[Bf1]   G.W. Brumfiel: Partially ordered rings and semi-algebraic geometry. *Lecture Notes of the London Math. Soc. 37*, Cambridge Univ. Press, Cambridge 1979

[Bf2]   G.W. Brumfiel: Real valuation rings and ideals. *In* Géométrie algébrique réelle et formes quadratiques, Proceedings, Rennes 1981. *Springer LNM* **959**, 55-97 (1982)

[Bf3]   G.W. Brumfiel: Witt rings and K-theory. *Rocky Mountain J. Math.* **14**, no. 4, 733-765 (1984)

[Bf4]   G.W. Brumfiel: The real spectrum of an ideal and KO-theory exact sequences. *K-Theory* **1**, 211-235 (1987)

[Bs-Is-Vg]   R. Brüske, F. Ischebeck, F. Vogel: Kommutative Algebra. B.I. Wissen-
              schaftsverlag, Mannheim Wien Zürich 1989

[Ca]   H. Cartan: Variétés analytiques réelles et variétés analytiques com-
              plexes. *Bull. Soc. Math. France* **85**, 77-99 (1957)

[Cs1]   A. Castilla: Sums of $2n$-th powers of meromorphic functions with com-
              pact zero-set. *In* Real algebraic geometry, Proceedings, Rennes 1991.
              Springer LNM **1524**, 174-177 (1992)

[Cs2]   A. Castilla: Artin-Lang property for analytic manifolds of dimension
              two. *Math. Z.* **217**, 5-14 (1994)

[Cs-An]   A. Castilla, C. Andradas: Connected components of global semian-
              alytic subsets of 2-dimensional analytic manifolds. *J. reine angew.
              Math.*

[C-T]   J.-L. Colliot-Thélène: Variantes du Nullstellensatz réel et anneaux
              formellement réels. *In* Géométrie algébrique réelle et formes quadra-
              tiques, Proceedings, Rennes 1981. *Springer LNM* **959**, 98-108 (1982)

[Co]   M. Coste: Ensembles semi-algébriques et fonctions de Nash. Univ.
              Paris Nord 1981

[Co-Ry1]   M. Coste, M.F. Roy: Topologies for real algebraic geometry. *In* Topos
              theoretic methods in geometry, *Various Publ. Ser.* **30**, 37-100. Mat.
              Inst. Aarhus Univ. (1979)

[Co-Ry2]   M. Coste, M.F. Roy: La topologie du spectre réel. *In* Ordered fields
              and real algebraic geometry. *AMS Contemporary Math.* **8**, 27-59
              (1981)

[Co-Rz-Sh]   M. Coste, J.M. Ruiz, M. Shiota: Approximation in compact Nash
              manifols. *Amer. J. Math.* **117**(1995)

[Cr]   T. Craven: Witt rings and orderings of skew fields. *J. Algebra* **77**,
              74-96 (1982)

[Df]   H. Delfs: The homotopy axiom in semialgebraic cohomology. *J. reine
              angew. Math.* **355**, 108-128 (1985)

[Df-Kn1]   H. Delfs, M. Knebusch: Semialgebraic topology over a real closed field
              II: Paths and components in the set of rational points of an algebraic
              variety. *Math. Z.* **177**, 107-129 (1981)

[Df-Kn2]   H. Delfs, M. Knebusch: Semialgebraic topology over a real closed field
              II: Basic theory of semialgebraic spaces. *Math. Z.* **178**, 175-213 (1981)

[Df-Kn3]   H. Delfs, M. Knebusch: On the homology of algebraic varieties over
              real closed fields. *J. reine angew. Math.* **335**, 122-163 (1982)

[Dz1]   C.N. Delzell: A continuous, constructive solution to Hilbert's 17th
              problem. *Invent. math.* **76**, 365-384 (1984)

[Dz2]   C.N. Delzell: Nonexistence of analytically varying solutions to Hilbert's
              17th problem. *In* Recent advances in real algebraic geometry and
              quadratic forms, Proceedings of the RAGSQUAD Year, Berkeley 1990-
              1991. *AMS Contemporary Math.* **155**, 107-117 (1994)

[D-C]   A. Díaz-Cano: Estabilidad de gérmenes semianalíticos básicos cerrados
              en dimensión 2: caso liso. Univ. Complutense de Madrid 1995

[Dk]   M.A. Dikmann: Sur les ouverts semialgébriques d'une clôture réelle.
              Univ. Paris VII 1983

[Ds]   A. Dress: On orderings and valuations of fields. *Geom. Dedicata* **6**, 259-266 (1977)

[Db]   D.W. Dubois: A Nullstellensatz for ordered fields. *Ark. Mat.* **8**, 111-114 (1969)

[Dg]   J. Dugundji: Topology. Allyn and Bacon, Boston London Sidney Toronto 1966

[Ef]   G. Efroymson: The extension theorem for Nash functions. *In* Géométrie algébrique réelle et formes quadratiques, Proceedings, Rennes 1981. *Springer LNM* **959**, 343-357 (1982)

[El-Lm]   R. Elman, T.Y. Lam: Quadratic forms under algebraic extensions. *Math. Ann.* **219**, 21-42 (1976)

[El-Lm-Pr]   R. Elman, T.Y. Lam, A. Prestel: On some Hasse principles over formally real fields, *Math. Z.* **134**, 291-301 (1973)

[El-Lm-Wd]   R. Elman, T.Y. Lam and A. Wadsworth: Orderings under field extensions. *J. reine angew. Math.* **306**, 7-27 (1979)

[En]   O. Endler: Valuation Theory. *Universitext*, Springer-Verlag, Berlin Heidelberg New York 1970

[Fz-Rc-Rz]   F. Fernández, T. Recio, J.M. Ruiz: The generalized Thom lemma in semianalytic geometry. *Bull. Polish Ac. Sc.* **35**, no. 5-6, 297-301 (1987)

[Fr]   J. Frisch: Points de platitude d'un morphisme d'espaces analytiques complexes. *Invent. math.* **4**, 118-138 (1967)

[Ft-Gl]   E. Fortuna, M. Galbiati: Séparation de semi-algébriques. *Geom. Dedicata* **32**, 211-227 (1989)

[Gm]   J.M. Gamboa: Un example d'ensemble constructible à adhérence non constructible. *C.R. Acad. Sci. Paris* **306**, 617-619 (1988)

[Gi-Je]   L. Gillman, M. Jerison: Rings of continuous functions. Van Nostrand, Princeton 1960

[Gr]   H. Grauert: On Levi's problem and the imbedding of real-analytic manifolds. *Annals of Math.* **68**, no. 2, 460-472 (1958)

[Gb]   M.J. Greenberg: Lectures on forms in many variables. *Math. Lecture Note Series 31*, Benjamin, London Amsterdam Tokyo 1969

[EGA]   A. Grothendieck, J. Dieudonné: Élements de géométrie algébrique IV: étude locale des schémas et des morphismes de schémas. *Publ. Math. I.H.E.S.* **20,24,28,32** (1964,65,66,67)

[Gu-Ro]   R. Gunning, H. Rossi: Analytic functions of several complex variables. Prentice Hall, Englewood Cliffs 1965

[Ht1]   R.M. Hardt: Stratifications of real analytic mappings and images. *Invent. Math.* **28**, 193-208 (1975)

[Ht2]   R.M. Hardt: Topological properties of subanalytic sets. *Transactions AMS* **211**, 57-70 (1975)

[Hr]   D. Harrison: Finite and infinite primes for rings and fields. *Memoirs AMS* **68**, 1968

[H]   D. Hilbert: Mathematische Probleme. *Arch. Math. Physik* 1, 44-64, 213-277 (1901). Also *Ges. Abh.* **3**, 290-323. Chelsea Publishing Company, New York 1965

[Hk1]   H. Hironaka: Resolution of singularities of an algebraic variety over a field of characteristic zero. *Annals of Math.* **79**, I:109-123, II:205-326 (1964)

[Hk2] H. Hironaka: Introduction to real-analytic sets and real-analytic maps. Istituto Mat. L. Tonelli, Pisa 1973

[Hö] L. Hörmander: On the division of distributions by polynomials. *Ark. Mat.* **3**, 555-568 (1958)

[Ja] B. Jacob: Fans, valuations, and hereditarily pythagorean fields. *Pacific J. Math.* **93**, 95-105 (1981)

[Jw1] P. Jaworski: Positive definite analytic functions and vector bundles. *Bull. Polish Ac. Sc.* **30**, no. 11-12, 501-506 (1982)

[Jw2] P. Jaworski: Extension of orderings on fields of quotients of rings of real analytic functions. *Math. Nachr.* **125**, 239-339 (1986)

[Jw3] P. Jaworski: The 17th Hilbert problem for noncompact real analytic manifolds. *In* Real algebraic geometry, Proceedings, Rennes 1991. *Springer LNM* **1524**, 289-295 (1992)

[Kl] F. Kalhoff: Anordnungsräume und Wittringe projektiver Ebenen. *Habilitationsschrift*, Univ. Dortmund 1988/89

[Kh] A. Khovanski: On a class of systems of transcendental equations. *Soviet. Math. Dokl.* **22**, no. 3, 762-765 (1980)

[Kn1] M. Knebusch: On the local theory of signatures and reduced quadratic forms. *Abh. Math. Sem. Univ. Hamburg* **51**, 149-195 (1981)

[Kn2] M. Knebusch: Generalization of a theorem of Artin-Pfister to arbitrary semilocal rings and related topics. *J. Algebra* **36**, 46-67 (1975)

[Kn3] M. Knebusch: On algebraic curves over real closed fields I. *Math. Z.* **150**, 189-205 (1976)

[Kn4] M. Knebusch: An invitation to real spectra. *In* Quadratic and hermitian forms, Proceedings, Hamilton 1983, *Canad. Math. Conf. Proc.* **4**, 51-105 (1984)

[Kn-Sch] M. Knebusch, C. Scheiderer: Einführung in die reelle Algebra. *Aufbaukurs Math.*, Vieweg-Verlag, Braunschweig Wiesbaden 1989

[Kv] J.-L. Krivine: Anneaux préordonnés. *J. Analyse Math.* **12**, 307-326 (1964)

[Kr] W. Krull: Allgemeine Bewertungstheorie. *J. reine angew. Math.* **167**, 160-196 (1931)

[Ku] W. Kucharz: Sums of $2n$-th powers of real meromorphic functions. *Monatsh. Math.* **107**, 131-136 (1989)

[Lm1] T.Y. Lam: The algebraic theory of quadratic forms (2nd edition). *Math. Lecture Note Series 54*, Benjamin, London Amsterdam Tokyo 1980

[Lm2] T.Y. Lam: Orderings, valuations and quadratic forms. AMS *Reg. Conf. Math.* **52** (1983)

[Lm3] T.Y. Lam: An introduction to real algebra. *Rocky Mountain J. Math.* **14**, 767-814 (1984)

[Ld] E. Landau: Über die Darstellung definiter Funktionen durch Quadrate. *Math. Ann.* **62**, 272-285 (1906)

[Lg1] S. Lang: The theory of real places. *Annals of Math.* **57**, 378-391 (1953)

[Lg2]   S. Lang: Diophantine geometry. *Tracts in Math.* *11*, Interscience, New York 1962

[Ls]   G. Lassalle: Sur le théorème des zéros différentiable. *In* Singularités d'applications différentiables, Plans-sur-Bex 1975. *Springer LNM* **535**, 70-97 (1975)

[Le-Lo]   J. Leicht and F. Lorenz: Die Primideale des Wittschen Ringes. *Invent. math.* **10**, 82-88 (1970)

[L1]   S. Lojasiewicz: Ensembles semi-analytiques. I.H.E.S. Bures-sur-Yvette 1964

[L2]   S. Lojasiewicz: Sur le problème de la division. *Studia Math.* **18**, 87-136 (1959)

[Mh]   L. Mahé: Une démonstration élémentaire du théorème de Bröcker-Scheiderer. *C. R. Acad. Sci. Paris* **309**, 613-616 (1989)

[Mr1]   M. Marshall: Classification of finite spaces of orderings. *Canad. J. Math.* **31**, 320-330 (1979)

[Mr2]   M. Marshall: Quotients and inverse limits of spaces of orderings. *Canad. J. Math.* **31**, 604-616 (1979)

[Mr3]   M. Marshall: The Witt ring of a space of orderings. *Transactions AMS* **298**, 505-521 (1980)

[Mr4]   M. Marshall: Spaces of orderings IV. *Canad. J. Math.* **32**, 603-627 (1980)

[Mr5]   M. Marshall: Spaces of orderings: systems of quadratic forms, local structure and saturation. *Comm. in Algebra* **12**, 723-743 (1984)

[Mr6]   M. Marshall: Minimal generation of basic sets in the real spectrum of a commutative ring. *AMS Contemporary Math.* **155**, 207-219 (1994)

[Mr7]   M. Marshall: Separating families for semi-algebraic sets. *Manuscripta math.* **80**, 73-79 (1993)

[Mr8]   M. Marshall: An axiomatic description of the real spectrum of a ring. Dept. of Math. and Statistics, Univ. of Saskatchewan 1994

[Mr9]   M. Marshall: Minimal generation of constructible sets in the real spectrum of a ring. *To appear*

[Mr-Wa]   M. Marshall, L. Walter: Minimal generation of basic semialgebraic sets over an arbitrary ordered field. *In* Real algebraic geometry, Proceedings, Rennes 1991. *Springer LNM* **1524**, 346-353 (1992)

[Mt]   H. Matsumura: Commutative algebra (2nd edition). *Math. Lecture Note Series 56*, Benjamin, London Amsterdam Tokyo 1980

[Me]   J.-P. Merrien: Idéaux de l'anneau des séries formelles à coefficients réels et variétés associées. *J. Math. pures et appl.* **50**, 169-187 (1971)

[Mz]   J. Merzel: Quadratic forms over fields with finitely many orderings. *In* Ordered fields and real algebraic geometry, Proceedings, San Francisco 1981. *AMS Contemporary Math.* **8**, 185-229 (1982)

[Mi]   J. Milnor: On the Betti numbers of real varieties. *Proceedings AMS* **15**, 275-280 (1964)

[Mo]   S.A. Morris: Pontryagin duality and the structure of locally compact abelian groups. *London Math. Soc. Lecture Note Series 29.* Cambridge Univ. Press, Cambridge London New York Melbourne 1977

[Mw]   T. Mostowski: Some properties of the ring of Nash functions. *Ann. Scuola. Norm. Sup. Pisa* **3**, 245-266 (1976)

[Ng]   M. Nagata: Local rings. *Intersc. Tracts Math.* *13*, John Wiley & Sons, New York London 1962

[Nh]   R. Narasimhan: Analysis on real and complex manifolds. *Advanced Studies Pure Math.* *1*, Masson North-Holland, Paris Amsterdam 1968

[Na]   J. Nash: Real algebraic manifolds. *Annals of Math.* **56**, 405-421 (1952)

[Po1]   D. Popescu: General Néron desingularization. *Nagoya Math. J.* **100**, 97-126 (1985)

[Po2]   D. Popescu: General Néron desingularization and approximation. *Nagoya Math. J.* **104**, 85-115 (1986)

[Po3]   D. Popescu: Letter to the editor. *Nagoya Math. J.* **118**, 45-53 (1990)

[Pf]   A. Pfister: Quadratische Formen in beliebigen Körpern. *Invent. math.* **1**, 116-132 (1966)

[Pr1]   A. Prestel: Quadratische Semi-Ordnungen und quadratische Formen. *Math. Z.* **133**, 319-342 (1973)

[Pr2]   A. Prestel: Lectures on formally real fields. IMPA, Rio de Janeiro 1975. Also *Springer LNM* **1093**, Berlin Heidelberg New York 1984

[Qz]   R. Quarez: The idempotency of the real spectrum implies the extension theorem for Nash functions. *Math. Z.* to appear.

[Rd]   M. Raynaud: Anneaux locaux henséliens. *Springer LNM* **169**, Berlin Heidelberg New York 1970

[Rc]   T. Recio: Una descomposición de un conjunto semi-algebraico. *Actas del V congreso de la Agrupación de Matemáticos de Expresión Latina*, Madrid 1978

[Ri]   P. Ribenboim: Théorie des valuations. *Presses Univ.*, Montreal 1964

[Rs]   J.-J. Risler: Le théoréme des zéros en géométries algébrique et analytique réelle. *Bull. Soc. Math. France* **104**, 113-127 (1976)

[Rn]   J. W. Robbin: Evaluation fields for power series. *J. Algebra* **57**, I:196-211, II:212-222 (1979)

[Ro]   A. Robinson: Germs. *In* Applications of model theory to algebra, analysis and probability, Proceedings, Pasadena 1967, 138-149. Holt Rinehart Winston, New York 1969

[Rb]   R. Robson: Nash wings and real prime divisors. *Math. Ann.* **273**, 177-190 (1986)

[Rt]   Ch. Rotthaus: On the approximation properties of excellent rings. *Invent. math.* **88**, 39-63 (1987)

[Ry]   M.F. Roy: Faisceau structural sur le spectre réel et fonctions de Nash. *In* Géométrie algébrique réelle et formes quadratiques, Proceedings, Rennes 1981, *Springer LNM* **959**, 406-432 (1982)

[Rz1]   J.M. Ruiz: Central orderings in fields of real meromorphic function germs. *Manuscripta math.* **46**, 193-214 (1984)

[Rz2]   J.M. Ruiz: A note on a separation problem. *Arch. Math.* **43**, 422-426 (1984)

[Rz3]   J.M. Ruiz: On Hilbert's 17th problem and real Nullstellensatz for global analytic functions. *Math. Z.* **190**, 447-454 (1985)

[Rz4]   J.M. Ruiz: Basic properties of real analytic and semianalytic germs. *Publ. Inst. Recherche Math. Rennes* **4**, 29-51 (1986)

[Rz5]   J.M. Ruiz: On the real spectrum of a ring of global analytic functions. *Publ. Inst. Rech. Math. Rennes* **4**, 84-95 (1986)

[Rz6]   J.M. Ruiz: Cônes locaux et complétions. *C. R. Acad. Sci. Paris* **302**, 67-69 (1986)

[Rz7]   J.M. Ruiz: On the connected components of a global semianalytic set. *J. reine angew. Math.* **392**, 137-144 (1988)

[Rz8]   J.M. Ruiz: A dimension theorem for real spectra. *J. Algebra* **124**, no. 2, 271-277 (1989)

[Rz9]   J.M. Ruiz: A going-down theorem for real spectra. *J. Algebra* **124**, no. 2, 278-283 (1989)

[Rz10]   J.M. Ruiz: On the topology of global semianalytic sets. *In* Real analytic and algebraic geometry, Proceedings, Trento 1988. *Springer LNM* **1420**, 237-246 (1990)

[Rz11]   J.M. Ruiz: A characterization of sums of $2n$-th powers of global meromorphic functions. *Proceedings AMS* **109**, 915-923 (1990)

[Rz12]   J.M. Ruiz: The basic theory of power series. *Advanced Lectures in Math.* Vieweg-Verlag 1993

[Rz-Sh]   J.M. Ruiz, M. Shiota: On global Nash functions. *Ann. Sc. Ec. Norm. Sup. Paris* **27**, 103-124 (1994)

[Schl1]   W. Scharlau: Quadratic forms. *Queen's Papers in Pure and Appl. Math.* **22**, Kingston, Ontario (1969)

[Schl2]   W. Scharlau: Quadratic and hermitian forms. *Grund. Math.* **270**. Springer-Verlag, Berlin Heidelberg New York 1985

[Sch1]   C. Scheiderer: Stability index of real varieties. *Invent. math.* **97**, 467-483 (1989)

[Sch2]   C. Scheiderer: Real algebra and its applications to geometry in the last ten years: some major developments and results. *In* Real algebraic geometry, Proceedings, Rennes 1991, *Springer LNM* **1524**, 75-96 (1992)

[Sch3]   C. Scheiderer: Purity theorems for real spectra and applications. *In* Real analytic and algebraic geometry, Proceedings, Trento 1992, 229-250. Walter de Gruyter, Berlin New York 1995

[Schü]   H.W. Schülting: Prime divisors on real varieties and valuation theory. *J. Algebra* **98**, no. 2, 499-514 (1986)

[Schw1]   N. Schwartz: Local stability and saturation in spaces of orderings. *Canad. J.* **35**, 454-477 (1983)

[Schw2]   N. Schwartz: The basic theory of real closed spaces. *Memoirs AMS* **397** (1989)

[Sy]   H. Seydi: Sur la théorie des anneaux de Weierstrass.I. *Bull. Sc. math.* **95**, 227-235 (1971)

[Se]   J.-P. Serre: Extensions de corps ordonnés. *C.R. Acad. Sci. Paris* **229**, 576-577 (1949)

[Sk]   M. Spivakovsky: Smoothing of ring homomorphisms, approximation theorems and the Bass-Quillen conjecture. Toronto Univ. 1994

[Sp]   T.A. Springer: Quadratic forms over fields with a discrete valuation. *Indag. Math.* **17**, 352-362 (1955)

[St1]   G. Stengle: A Nullstellensatz and a Positivstellensatz in semialgebraic geometry. *Math. Ann.* **207**, 87-97 (1974)

[St2]   G. Stengle: A measure for semialgebraic sets related to boolean complexity. *Math. Ann.* **283**, 203-209 (1989) .

[Sw]   R. Swan: Topological examples of projective modules. *Transactions AMS* **230**, 201-234 (1977)

[Tk]   A. Tarski: A decision method for elementary algebra and geometry. Prepared for publication by J.C.C. Mac Kinsey, Berkeley 1951

[Te]   B. Teissier: Résultats récents sur l'approximation des morphismes en algèbre commutative. Séminaire Bourbaki, 46ème année, 1993-94, no. 748.*Astérisque* **227**, 259-282 (1995)

[Th]   R. Thom: Sur l'homologie des variétés algébriques réelles. *In* Differential and combinatorial topology, Proceedings, Princeton 1964, 255-265. Princeton University Press 1965

[Tg]   J.C. Tougeron: Idéaux de fonctions différentiables. *Ergeb. Math. 71*, Springer-Verlag, Berlin Heidelberg New York 1972

[Ts]   A. Tschimmel: Über Anordnungsräume von Schiefkörpern. *Dissertation*, Univ. Münster 1981

[Wh]   H. Whitney: Local properties of analytic varieties. *In* Differential and combinatorial topology, Proceedings, Princeton 1964,205-244. Princeton University Press 1965

[Wh-Bh]   H. Whitney, F. Bruhat: Quelques propriétés fondamentales des ensembles analytiques réels. *Comment. Math. Helv.* **33**, 132-160 (1959)

[Wt1]   E. Witt: Zerlegung reeller algebraischer Funktionen in Quadrate, Schiefkörper über reellen Funktionenkörpern. *J. reine angew. Math.* **171**, 4-11 (1934)

[Wt2]   E. Witt: Theorie der quadratischen Formen in beliebigen Körpern. *J. reine angew. Math.* **176**, 31-44 (1937)

[Yo]   J. Yomdin: Metric properties of semi-algebraic sets and mappings and their applications in smooth analysis. Ben Gourion Univ. 1984

Glossary

Index